Sustainable Management of Electronic Waste

Scrivener Publishing
100 Cummings Center, Suite 541J
Beverly, MA 01915-6106

Publishers at Scrivener
Martin Scrivener (martin@scrivenerpublishing.com)
Phillip Carmical (pcarmical@scrivenerpublishing.com)

Sustainable Management of Electronic Waste

Edited by
**Abhishek Kumar
Pramod Singh Rathore
Ashutosh Kumar Dubey
Arun Lal Srivastav
T. Ananth Kumar**
and
Vishal Dutt

Scrivener Publishing

WILEY

This edition first published 2024 by John Wiley & Sons, Inc., 111 River Street, Hoboken, NJ 07030, USA and Scrivener Publishing LLC, 100 Cummings Center, Suite 541J, Beverly, MA 01915, USA
© 2024 Scrivener Publishing LLC
For more information about Scrivener publications please visit www.scrivenerpublishing.com.

All rights reserved. No part of this publication may be reproduced, stored in a retrieval system, or transmitted, in any form or by any means, electronic, mechanical, photocopying, recording, or otherwise, except as permitted by law. Advice on how to obtain permission to reuse material from this title is available at http://www.wiley.com/go/permissions.

Wiley Global Headquarters
111 River Street, Hoboken, NJ 07030, USA

For details of our global editorial offices, customer services, and more information about Wiley products visit us at www.wiley.com.

Limit of Liability/Disclaimer of Warranty
While the publisher and authors have used their best efforts in preparing this work, they make no representations or warranties with respect to the accuracy or completeness of the contents of this work and specifically disclaim all warranties, including without limitation any implied warranties of merchantability or fitness for a particular purpose. No warranty may be created or extended by sales representatives, written sales materials, or promotional statements for this work. The fact that an organization, website, or product is referred to in this work as a citation and/or potential source of further information does not mean that the publisher and authors endorse the information or services the organization, website, or product may provide or recommendations it may make. This work is sold with the understanding that the publisher is not engaged in rendering professional services. The advice and strategies contained herein may not be suitable for your situation. You should consult with a specialist where appropriate. Neither the publisher nor authors shall be liable for any loss of profit or any other commercial damages, including but not limited to special, incidental, consequential, or other damages. Further, readers should be aware that websites listed in this work may have changed or disappeared between when this work was written and when it is read.

Library of Congress Cataloging-in-Publication Data

ISBN 9781394166176

Front cover images supplied by Pixabay.com
Cover design by Russell Richardson

Set in size of 11pt and Minion Pro by Manila Typesetting Company, Makati, Philippines

Printed in the USA

10 9 8 7 6 5 4 3 2 1

Contents

Foreword		xvii
Preface		xix
1	**Integration of Artificial Intelligence Techniques for Energy Management**	**1**
	Bhanu Chander and Kumaravelan Gopalakrishnan	
	1.1 Introduction	2
	1.2 Summary of Artificial Intelligence Techniques	4
	1.2.1 Machine Learning	5
	1.2.2 Deep Learning (DL) Techniques	8
	1.3 Reasons for Applying AI in EMS	17
	1.4 ML in Renewable Energy	18
	1.4.1 ML for Renewable Energy Applications	19
	1.4.2 Countries Focusing on ML	20
	1.4.3 Notable ML Projects	20
	1.4.4 How ML is Renovating the Energy Industry	21
	1.4.5 Machine Learning in Renewable Energy	22
	1.5 Integration of AI in Smart Grids	24
	1.5.1 Load Balancing	25
	1.5.2 Power Grid Stability Assessment	26
	1.5.3 Smart Grid Challenges	27
	1.5.4 Future of AI in Smart Grids	27
	1.5.5 Challenges of AI in Smart Grids	28
	1.6 Parameter Selection and Optimization	29
	1.6.1 Strategies for Tuning Hyperparameter Values in a Machine Learning Model	30
	1.7 Biological-Based Models for EMS	32
	1.8 Future of ML in Energy	33
	1.9 Opportunities, Limitations, and Challenges	36
	1.9.1 Opportunities and Limitations	36
	1.9.2 Challenges	38

		1.10	Conclusion	39
			References	40

2 Artificial Neural Network Process Optimization for Predicting the Thermal Properties of Biomass: Recent Advances and Future Challenges 47
S. Dayana Priyadharshini and M. Arvindhan

    Abbreviations 47
2.1 Introduction 48
2.2 AI Technology and Its Application on Renewable Energy 51
2.3 Bioenergy and ANN 52
2.4 ANN Model Development 52
    2.4.1 Methodologies Used for Target Model 55
    2.4.2 Various Stages Involved in ANN Modelling 57
    2.4.3 Important Terms Used in ANN Modeling 58
2.5 Future Scope of ANN-Bioenergy 60
2.6 Conclusion 61
    References 61

3 E-Waste Management and Bioethanol Production 67
Anshu Sibbal Chatli

3.1 Introduction 68
3.2 Review 69
3.3 Degradation of Lignocellulose 70
3.4 Bioprocessing of Lignocellulosic Materials 72
    3.4.1 Fermentation 72
    3.4.2 Microorganisms for Ethanol Production 74
    3.4.3 Pretreatment of Lignocellulosic Material and Preparation of Extracts 74
        3.4.3.1 Analytical Method 74
        3.4.3.2 Estimation of Total Cellulose Contents 74
        3.4.3.3 Estimation of Hemicelluloses Content 74
    3.4.4 Preparation of Inoculum 75
    3.4.5 Estimation of Ethanol 75
3.5 AI Technologies 75
3.6 Conclusions 76
    References 76

Contents

4 A Novel-Based Smart Home Energy Management System for Energy Consumption Prediction Using a Machine Learning Algorithm — 79
N. Deepa, Devi T., S. Rakesh Kumar and N. Gayathri
    4.1 Introduction — 80
        4.1.1 Bayesian Linear Regression — 82
    4.2 Literature Review — 83
    4.3 Proposed Work — 87
        4.3.1 Enhanced Bayesian Linear Regression Machine Learning Algorithm (EBLRML) — 87
    4.4 Results and Discussion — 89
    4.5 Conclusion — 92
    References — 93

5 AI-Based Weather Forecasting System for Smart Agriculture System Using a Recurrent Neural Networks (RNN) Algorithm — 97
Devi T., N. Deepa, N. Gayathri and S. Rakesh Kumar
    5.1 Introduction — 98
        5.1.1 Deep Learning — 100
    5.2 Literature Review — 102
    5.3 Proposed Work — 105
        5.3.1 Dataset Description — 105
        5.3.2 Recurrent Neural Network (RNN) — 107
    5.4 Results and Discussion — 107
    5.5 Conclusion — 110
    References — 110

6 Comprehensive Review of IoT-Based Green Energy Monitoring Systems — 113
Aishwarya V.
    6.1 Introduction — 113
    6.2 IoT-Based Green Energy Monitoring Systems — 114
    6.3 Comparative Analysis — 130
    6.4 Conclusion — 136
    References — 137

7 The Contribution of Renewable Energy with Artificial Intelligence to Accomplish Organizational Development Goals and Its Impacts — 145
K. M. Baalamurugan and Aanchal Phutela
    7.1 Introduction — 146
        7.1.1 AI will Balance Millions of Assets on the Grid — 146

7.2	AI Contributions in Conventional and Renewable Energy	148
7.3	AI-Based Technology in Renewable Energy	148
	7.3.1 Challenges of the Renewable Energy Sector	150
	7.3.2 Artificial Intelligence in Wind Farming	150
7.4	A Look at Some Challenges Faced by the Renewable Energy Industry	151
7.5	Higher Computational Power & Intelligent Robotics	155
7.6	The Impact of Digital Technology on Energy Company Results	157
	7.6.1 Smart Match of Supply Through Demand	157
	7.6.1.1 Thermal Energy	159
7.7	Conclusion	160
7.8	Future Work	161
	References	162

8 Current Trends in E-Waste Management 167
Anjali Sharma, Devkant Sharma, Deepshi Arora and Ajmer Singh Grewal

8.1	Introduction	168
8.2	What is E-Waste?	168
8.3	How is E-Waste Recycled and Why Do Problems Exist?	169
8.4	Global E-Waste Management Market Restraints	169
8.5	Global E-Waste Management Market Opportunities	170
8.6	Management of E-Waste in India	170
8.7	New Solutions for E-Waste Excess	171
8.8	Recent Trends in E-Waste Management	172
	8.8.1 Public Health, Environment, and E-Waste	172
	8.8.2 Disposal and Management of E-Waste	172
8.9	E-Waste Regulation in India	174
	8.9.1 Treatment of E-Waste	176
	8.9.2 Amendments in E-Waste Management Rules 2018	176
8.10	Recovery of Resources from E-Waste	177
	8.10.1 Pyrometallurgy	177
	8.10.2 Hydrometallurgy	178
8.11	Generation and Management of Mobile Phone Waste	179
	8.11.1 Generation of Mobile Phone Waste	179
	8.11.2 Management of Waste from Mobile Devices	180
8.12	Players of the Market	181
8.13	Recent Developments	182

	8.14 Conclusion	182
	References	182
9	**Current E-Waste Management: An Exploratory Study on Managing E-Waste for Environmental Sustainability**	**187**
	Shweta Solanki and Pramod Singh Rathore	
	9.1 Introduction	188
	9.2 E-Waste Production	189
	9.3 The Present Predicament	190
	9.3.1 Prospective Developments	191
	9.3.2 Environmental Impacts	192
	9.3.2.1 Possible Adverse Effects on the Environment Resulting from Disposal of Electronic Waste	193
	9.3.2.2 Electronic Trash Disposal and Recycling Both Contribute to Contamination of the Environment	195
	9.4 Conclusion	196
	References	197
10	**Challenges in E-Waste Management**	**201**
	Himani Bajaj, Anjali Sharma, Deepshi Arora, Mayank Yadav, Devkant Sharma and Prabhjot Singh Bajwa	
	10.1 Introduction	201
	10.2 E-Waste: Meaning and Definition	202
	10.3 Environmental Sustainability in E-Waste Management	204
	10.4 Sustainable Management of E-Waste	205
	10.5 Life Cycle of E-Waste	206
	10.6 Terminology of E-Waste	206
	10.7 Key Stakeholders in the E-Waste Management System	207
	10.8 Status of E-Waste Management in India	207
	10.9 Challenges in E-Waste Management	209
	10.9.1 Discrepancies in Estimate of E-Waste	209
	10.9.2 Lack of Awareness	210
	10.9.3 The Dominance of the Informal Sector	211
	10.9.4 Inadequate Formal Recycling Sector	211
	10.9.5 Lack of Technological Advancement	212
	10.9.6 Involvement of Child Labour	212
	10.9.7 Ineffective Legislation	212
	10.9.8 Health Hazards	213
	10.9.9 Lack of Incentive Schemes	213

10.9.10	Reluctance of Authorities Involved	213
10.9.11	Security Implications	213
10.9.12	Lack of Research	213
10.10	E-Waste Policy and Regulation	213
10.11	E-Waste Recycling	214
10.12	Life Cycle Assessment (LCA) Analysis of E-Waste	214
10.13	Existing Laws Relating to E-Waste	215
10.14	Management Options	215
10.14.1	Responsibilities of the Government	215
10.14.2	Responsibility and Role of Industries	215
10.14.3	Responsibilities of the Citizen	216
10.15	Conclusion	216
	References	217

11 Recycling of Electronic Wastes: Practices, Recycling Methodologies, and Statistics — 221
Suresh Chinnathampy M., Ancy Marzla A., Aruna T., Dhivya Priya E. L., Rindhiya S. and Varshini P.

11.1	Introduction	222
11.2	Recycling E-Waste	223
11.2.1	Process of E-Waste Recycling	224
11.2.2	Collection and Transportation	224
11.2.3	Shredding, Classify, and Uncoupling	225
11.2.4	Most Effective Method to Remove Metals	225
11.3	Smart Phones at the End of Their Life	226
11.4	Recycling of Printed Circuit Boards (PCB)	227
11.5	Solar Panel Recycling	228
11.6	How Has E-Waste Management in India Evolved Through the Years?	230
11.6.1	Promoting Formal E-Waste Recycling	230
11.6.2	Training and Upskilling Informal Sector Players	230
11.6.3	International Statistics of E-Waste	231
11.6.4	National Statistics of E-Waste	232
11.7	Conclusion	233
	References	234

12 Sustainable Development Through the Life Cycle of Electronic Waste Management — 237
D. Magdalin Mary, S. Jaisiva, C. Kumar and P. Praveen Kumar

12.1	Introduction	238

		12.1.1	Electronics Production, Waste, and Impacts	238
		12.1.2	E-Production: Recycle	239
		12.1.3	Uses of Recycling	239
	12.2	Impact on the Environment		240
	12.3	Environmental Impact of Electronics Manufacturing		242
		12.3.1	Built-In Obsolescence is Also to Blame	242
		12.3.2	Usage of ISO 14001 Helps to Lessen How Much of an Impact Producing Electronics has on the Environment	243
	12.4	E-Waste Management Initiative		243
		12.4.1	Present Barriers in Recycling of E-Waste	245
		12.4.2	Global E-Waste Problem	245
	12.5	Issues with E-Waste in India		247
	12.6	Impact of E-Waste Recycling in Developing Nations		248
	12.7	Opportunities and Challenges in E-Waste Management in India		249
	12.8	Recent Investigations on Electronic Waste Management		250
	12.9	Conclusion		251
		References		251
13	**E-Waste Challenges & Solutions**			**255**
	K. Dhivya and G. Premalatha			
	13.1	Introduction		256
	13.2	Related Works		257
	13.3	E-Waste: A Preamble		259
	13.4	Six Categories of E-Waste		259
	13.5	Composition of Materials Found in Equipment		260
	13.6	Recycling of WEEE		261
	13.7	Procedures in the E-Waste Management		263
		13.7.1	Disposal Systems	264
		13.7.2	E-Waste Management Challenges	265
		13.7.3	Market Mismanagement for End-of-Life Products	265
	13.8	E-Waste (Management) Rules		266
		13.8.1	2016 E-Waste (Management) Rules	266
		13.8.2	Central Issues for Rules 2016 of E-Waste Management	269
	13.9	Report from the Central Pollution Control Board		269
	13.10	An Integrated Waste Management Systems Web Application		270
		13.10.1	Intent Behind the Application	270

13.11	E-Waste Management Rules 2016 Amendments	270
13.12	Management of Battery Waste Rules, 2022	271
13.13	Conclusion	273
	References	273

14 Global Challenges of E-Waste: Its Management and Future Scenarios — 277
Pranay Das and Swati Singh

14.1	Introduction	278
14.2	Worldwide Production of E-Waste	279
14.3	Global Availability of E-Waste: Additional Information	279
14.4	Environmental Impact of E-Waste	280
	14.4.1 E-Waste Impacts on Soil Ecosytem	281
	14.4.2 Impacts of E-Waste in the Water	281
	14.4.3 E-Waste Impacts on Air Pollution	281
	14.4.4 Final Considerations on the Effect of E-Waste on the Environment	282
14.5	Management of E-Waste	282
14.6	Concerns and Challenges	284
	14.6.1 Absence of Infrastructure	285
	14.6.2 Hazardous Impact on Human Health	285
	14.6.3 Lack of Enticement Schemes	285
	14.6.4 Poor Awareness and Sensitization	285
	14.6.5 High-Cost of Setting Up Recycling Facility	285
14.7	Future Scenarios of E-Waste	286
14.8	The Need for Scientific Acknowledgment and Research	286
14.9	Conclusion	286
	References	287

15 Impact of E-Waste on Reproduction — 293
Adrija Roy, Sayantika Mukherjee, Dipanwita Das and Amrita Saha

15.1	Introduction	293
15.2	Literature Review	295
15.3	Discussion	296
15.4	Conclusion	298
	References	299

16 Challenges in Scale-Up of Bio-Hydrometallurgical Treatment of Electronic Waste: From Laboratory-Based Research to Practical Industrial Applications   301
Ana Cecilia Chaine Escobar, Andrew S. Hursthouse and Eric D. van Hullebusch

    16.1   Introduction   302
    16.2   Methodology   302
    16.3   Results   302
        16.3.1   Pre-Treatment   302
                16.3.1.1   Physical   304
                16.3.1.2   Chemical   304
                16.3.1.3   Biological: Bioflotation   305
        16.3.2   Treatment: Metal Extraction   306
                16.3.2.1   Biohydrometallurgy: Bioleaching   307
        16.3.3   E-Waste Bioleaching Process   308
                16.3.3.1   Preparation   308
                16.3.3.2   Microorganism Adaptation to E-Waste Phase   308
                16.3.3.3   Bioleaching Methods   308
                16.3.3.4   Technical Considerations in the Bioleaching Process   309
        16.3.4   E-Waste Bioleaching Up-Scaling   309
                16.3.4.1   Bioleaching Process Scale-Up: Challenges   309
                16.3.4.2   Pre-Treatment   310
                16.3.4.3   Culture Conditions   324
        16.3.5   Omics and Bioinformatics   327
        16.3.6   Metal/Metalloids Recovery Techniques   328
                16.3.6.1   Biological Recovery Techniques: Immobilization Mechanisms   328
        16.3.7   Recovered Materials   331
                16.3.7.1   Rare Earth Elements (REEs)   331
                16.3.7.2   Nanomaterials   332
    16.4   Economic Feasibility   332
    16.5   Conclusions   333
        References   334

17 Current Advances in Recycling of Electronic Wastes — 341
Kumar Sagar Maiti, Irin Khatun, Serma Rimil Hansda and Dipankar Ghosh

17.1	Introduction	342
	17.1.1 Informal Recycling	345
	17.1.2 Formal Recycling	346
17.2	E-Waste: A General Description and Classification and Issues on the Environment and Health	347
	17.2.1 E-Waste Classification	348
	17.2.2 Issues on Environment and Health	349
17.3	Conventional Approaches to E-Waste Recycling, Advantages and Disadvantages	351
	17.3.1 Physical Methods	351
	17.3.1.1 Incineration	352
	17.3.1.2 Sieving	353
	17.3.1.3 Gravity Separation	353
	17.3.1.4 Magnetic Separation	353
	17.3.1.5 Electrochemical Separation	353
	17.3.1.6 Metallurgical Methods	354
	17.3.1.7 Pyrometallurgy	354
	17.3.1.8 Disadvantages of Pyrometallurgy	355
	17.3.1.9 Hydrometallurgy	356
	17.3.1.10 Biometallurgy	357
17.4	Advances in Approaches for Improving E-Waste Recycling for Value-Added Materials and Biomaterials Generation	358
17.5	Conclusion	362
	Acknowledgement	364
	References	364

18 E-Waste: The Problem and the Solutions — 375
Krati Taksali and Pramod Singh Rathore

18.1	Introduction	375
18.2	India's Electronic Waste Crisis	378
18.3	Inadequate Infrastructure for Refurbishing E-Waste	383
	18.3.1 Financial Incentives and Lack of Awareness	384
	18.3.2 Limited Available Statistics on Rates of Creation of Electronic Trash	385
	18.3.3 Market Mismanagement for End-of-Life Products	385
	18.3.4 Environmentally Irresponsible Practices in the Informal Sector	385
	18.3.5 Poorly Designed and Enforced Regulations	386

18.4		Enhancing India's E-Waste Management	386
	18.4.1	Providing Price Information on the Market for E-Waste	387
	18.4.2	Promoting the Recycling of Formal E-Waste	387
	18.4.3	Players in the Informal Sector are Trained and Upgraded	388
	18.4.4	Utilizing Technology for Recycling that is Both Finished and Cutting-Edge	388
	18.4.5	Creating New Technology & Methods for Processing New Types of E-Waste	389
18.5		Effects of E-Waste Recycling in Developing India Like Nations	389
18.6		Opportunity for Managing E-Waste in India	390
	18.6.1	What Assistance Can Governments, City Management, and Citizens Provide?	391
18.7		Management of Electronic Waste	392
		References	393

19 Contribution of E-Waste Management in Green Computing   397
Shweta Sharma and Vishal Dutt

19.1		Introduction	398
19.2		Concept of Green Computing	398
	19.2.1	E-Waste Management and Recycling	399
19.3		A History of Green Computing	401
19.4		Benefits of Recycling in Green Computing	401
19.5		E-Waste Management Steps	402
19.6		E-Waste Recycling: An Approach Towards Green Computing	404
	19.6.1	How is Green Computing Achieved?	404
	19.6.2	How Does E-Waste Recycling Affect Green Computing?	404
	19.6.3	How Does Veracity World Differ?	405
19.7		Harmful Effects of E-Waste	406
	19.7.1	Effects on Air Quality	406
	19.7.2	Effects on Humans	406
	19.7.3	Concerning Global Data on E-Waste	406
19.8		E-Waste and the Sustainable Development Goals of the 2030 Agenda	407
19.9		Significant International E-Waste Agreements	408
	19.9.1	The International Convention to Prevent Pollution from Ships (MARPOL)	408

	19.9.2	The Ozone Depletion Protocol of the Montreal Protocol (1989)	409
	19.9.3	The Durban Declaration of 2008	409
	19.9.4	India's E-Waste Production	409
	19.9.5	2016 E-Waste Management Regulations	409
19.10	Conclusion		410
	References		411

Index   **413**

Foreword

Background

This book gives a better understanding of how the management of e-waste systems functions. This book provides comprehensive, state-of-the-art coverage of all the essential aspects of modeling and simulating physical and conceptual systems. Various real-life examples show how simulation is crucial in understanding real-world systems. Practical uses of AI and data mining techniques are presented to show successful applications of the modeling and simulation techniques given.

Objectives

The number of researchers studying e-waste and trends for e-Waste management is significant due to their attractiveness in the challenges they pose for researchers and their applications for the benefit of mankind. However, due to the overlapping of the two domains, which maintain their distinct culture, providing tools for this overlap to be appreciated by both communities has become more than necessary. This book provides one such opportunity, providing a platform of appreciation for both communities.

The book collects important research articles from experts in the field after a thorough screening process to provide a collective effort to enhance the contributions of the combined communities. It is based on a wealth of years of research from dedicated researchers.

About the Book

The book has a collection of research articles from different authors from all over the world. It truly represents the uniqueness of the research as a common platform from diverse backgrounds and cultures.

The book starts with a chapter on the contribution of renewable energy and artificial intelligence, a critical topic. The second chapter applies to current trends in e-waste management, a standard tool employed in AI. As seen previously, the traditional method used is machine learning. The third chapter deals with current e-waste management through an exploratory study on managing e-waste for environmental sustainability. The sixth, eighth, tenth, and eighteenth chapters deal with e-waste management issues and challenges. Chapter 5 deals with the paradigm of machine learning that includes deep learning and provides some useful case studies. Chapter 9 extends this paradigm and discusses big data, a common buzzword associated with health data. AI associated with diagnostics has provided smart tools leading to smart informatics; Chapter 10 deals with this as applied to health. Chapters 14 and 15 give a useful resource on the impact of e-waste on reproduction. Chapter 16 comprehensively challenges the scale-up of the bio-hydrometallurgical treatment of electronic waste from laboratory-based research to practical industrial application. Chapter 17 provides one of the essential benefits of current electronic waste recycling advances. Chapter 18 discusses e-waste's addition to our ever-growing hazardous solid waste. Finally Chapter 19 discusses e-waste management as crucial to the development of green computing.

Preface

This book is organized into nineteen chapters. In Chapter 1, a study explores the published works of AI with EMS, RES, and SG, their effects, and their reasons. AI technology has evolved quickly in the last several decades, and its applications have rapidly increased in modern industrial systems. Similarly, nature-inspired and biological systems denote an unlimited inspiration source for developing technical systems that lead human civilization's progress and shape our thinking style.

In Chapter 2, the author introduces the concept of the progress of artificial intelligence and blockchain technology unlocking prospects toward thermal marketing accuracy. The Artificial Neural Network (ANN) is crucial for improving biomass energy prediction research. This study emphasizes the steps in modeling and using ANN in forecasting biomass thermal values. Several research gaps in the present state of the investigation on ANN, in terms of biomass and guidance for additional research, are identified.

Chapter 3 will also go over growing environmental concerns over the use and depletion of non-renewable fuel sources, together with the increasing price of oil and instabilities in the oil markets, which have recently stimulated interest in producing sustainable energy sources in the form of biofuels derived from plants. Ethanol developed from lignocellulosic biomass has characteristic benefits of safety, economics, and being more environmentally friendly than fossil fuels. The lignocellulosic materials comprise 50% cellulose, 25% hemicellulose, and 25% lignin.

In Chapter 4, a novel approach for a Smart Home Energy Management System for Energy Consumption Prediction is proposed using an Enhanced Bayesian Linear Regression Machine Learning algorithm (EBLRML). The residual sum of squares shows the significant linear model difference from the existing application of ML techniques where the coefficient based on correlation is lower, and the energy consumption is calculated to achieve the best-fit model.

In Chapter 5, the author explains that using artificial intelligence today improves all the limitations outlined in the problem statements above. Machine learning algorithms have been used to predict crop cultivation, optimization metrics, irrigation levels, and so on, but very little has been done to predict weather and forecast agriculture. By using advanced sensors connected to a network, it is possible to identify bad weather that may threaten agriculture itself at an earlier stage. A live farm's features can improve crop exploitation earlier in the growing cycle.

In Chapter 6, the author explains that energy consumption has risen exponentially, putting pressure on the utilities to increase energy production. One of the significant concerns of the current day is that energy conservation and monitoring systems have been developed to optimize the increasing demand for energy and its consumption. Energy management systems help to decrease current consumption, prevent energy wastage, and enable the optimized utilization of available resources.

In Chapter 7, the author explains a significant impediment to the widespread adoption of these energy sources in their high integration costs. However, artificial intelligence (AI) explanations and data-intensive expertise are currently being deployed in various sectors of the electrical significance chain and, given the future smart grid's increasing complexity and data generation capacity, have the probability of adding substantial value to the system.

Chapter 8 provides detailed explanations of critical issues influencing India's entire e-waste value chain, including a lack of data inventorization, unlawful disposal, and treatment choices. As a result, this study focuses on strategic interventions that comply with existing legislation and are necessary for a long-term e-waste value chain, secure resources, social well-being, reduced environmental consequences, and overall sustainable development. Apart from these, other methods for recycling e-waste are also incorporated to maintain the environment's integrity.

Chapter 9 discusses the environmental concerns linked with discarded electronic equipment, sometimes known as "e-waste," in detail. In addition, the development of e-waste both now and in the future, as well as any environmental difficulties that may come from its management and disposal approaches, are investigated, and the existing programs for the management of e-waste are discussed.

Chapter 10 discusses electronic devices, such as televisions, smartphones, and refrigerators, that have limited useful lives and must therefore be replaced frequently, creating e-waste. E-waste has the fastest growth rate among municipal solid garbage, producing 20 to 50 million tons annually globally. As a result, several nations are currently managing enormous amounts of e-waste. An essential issue in handling e-waste is environmental health. Worldwide, governments struggle to increase public awareness and make significant steps to protect the environment from rapid degradation. Because of the reasons above, effective e-waste management is required constantly.

Chapter 11 discusses the goals pertaining to the disposal of devices nearing the end of their "useful life." Increasing transboundary secondary resource movement and Asia's rapid economic growth will require both 3R initiatives (reduce, reuse, recycle) in every nation and effective management of the global material cycle. As a result, electrical and electronic garbage management, or "e-waste," has gained significant attention. For material processes from the national and international environmental and resource preservation perspectives, digital goods are at an all-time high in the developing digital world, and it is difficult to envision our everyday lives without them. Although they are most significant throughout their lifetime, they could endanger the environment if burned or disposed of in landfills.

Chapter 12 will include information on how the overall amount of trash managed locally and worldwide may be decreased by improving the lifecycle management of electronics by reducing the source of materials consumed, enhancing reuse, restoration, extending product life, and recycling electronics. The EPA's waste management hierarchy aligns with the life cycle approach. The hierarchy underlines that reducing, reusing, and recycling is essential components of sustainable materials management and grades the different management options from greenest to least green.

Chapter 13 discusses that e-waste is a significant problem in the technological world. To put it simply, e-waste refers to any unwanted or obsolete electronic equipment. Obsolete technology is inevitably phased out, reducing amounts of WEEE garbage. Large appliances and electronics such as refrigerators, air conditioners, computers, and mobile phones are all broken. This trash has the potential to kill humans. Humans and ecosystems alike are negatively affected by improper waste management. Polluting the environment by incinerating, burying, or dumping electronic waste is unacceptable. Rare earth elements used in reusable electronics are reusable.

Chapter 14 highlights modern developments in e-waste production and streams, current recycling technologies, human health, and environmental impacts of recycled materials and processes. The background, challenges, and problems of e-waste disposal and its proper management are also discussed, along with the implications of the analysis.

Chapter 15 shows associations between exposure to e-waste and physical health outcomes, including thyroid function, reproductive health, lung function, growth, and changes to cell functioning. Several researchers have investigated the consequences of pregnancies in communities exposed to e-waste. In most investigations, there have been consistent effects of exposure with increases in spontaneous abortions, stillbirths, premature deliveries, lower birthweights, and birth durations, despite diverse exposure settings and toxins being examined.

Chapter 16 will assess peer-reviewed data gathered to establish the technology readiness level of biohydrometallurgy for material recovery from e-waste at a pilot scale, concluding that bioleaching at a commercial scale currently faces diverse operational challenges that hamper its scale-up and industrial implementation.

Chapter 17 presents diverse e-waste recycling technologies, including conventional (physical and chemical treatments) and modern (biological or microbial treatments) approaches to combat environmental pollution and community health hazards. However, conventional and modern techniques suffer from multiple lacunae related to the efficacies of e-waste recycling techniques. To this end, the current literature review deals with a general outline of e-waste generation and categorization with special emphasis on e-waste recycling processes in great detail.

Chapter 18 discusses e-waste's addition to our ever-growing hazardous solid waste. Electronics and electrical equipment are part of e-waste. Many countries, especially developing countries like India, are facing infinite challenges in the management of e-waste which imported illegally or generated internally. India is also one of the countries fighting for e-waste management.

Chapter 19 discusses e-waste management as crucial to the development of green computing. We can lessen technology's adverse environmental effects and save resources by properly disposing of and recycling electronic equipment. However, for the rising issue of e-waste to be solved sustainably, governments and businesses must collaborate to develop effective e-waste management regulations and initiatives.

Dr. Abhishek Kumar
Associate Professor
Department of Computer Science
SMIEEE, Chandigarh University, India

Pramod Singh Rathore
Assistant Professor
Department of Computer and Communication Engineering,
Manipal University Jaipur, India

Dr. Ashutosh Kumar Dubey
Associate Professor
Department of Computer Science and Engineering
Chitkara University School of Engineering and Technology,
Chitkara University, Himachal Pradesh, India
SMIEEE, SMACM

Dr. Arun Lal Srivastav
Associate Professor
Chitkara University School of Engineering and Technology,
Chitkara University, Himachal Pradesh, India

Dr. T. Ananth Kumar
Associate Professor
Computer Science and Engineering
IFET College of Engineering,
Tamilnadu, India

Vishal Dutt
Department of Computer Science and Engineering
Chandigarh University, Mohali, Punjab, India

1
Integration of Artificial Intelligence Techniques for Energy Management

Bhanu Chander* and Kumaravelan Gopalakrishnan

Department of Computer Science and Engineering, Pondicherry University, Karaikal Campus, Puducherry, India

Abstract

Artificial intelligence (AI) is a scientific application of knowledge used to build intellectual devices, specifically intelligent computer programs. Innovations in AI-based techniques are extensively applied in Energy Management Systems (EMS), Renewable Energy Systems (RES), and Smart Grids (SG) and are a cutting-edge frontier in power electronics and power engineering with powerful tools for design, control, fault diagnosis, and simulation. In particular, this study explores the published works of AI with EMS, RES, and SG, as well as their effects and reasons. AI technology has quickly evolved in the last several decades and its applications have rapidly increased in modern industrial systems. Similarly, nature-inspired and biological systems denote an unlimited inspiration source for developing technical systems that lead human civilization's progress and shape our thinking style.

Nowadays, using renewable energy to condense climate revolution and global warming has become a growing trend and various AI-based prediction techniques have been developed to improve the prediction ability of renewable energy. An intelligent grid must predict the amount of power, integrate renewable sources, and manage an intelligent grid with optimal sizes, all of which are challenging tasks. Optimization techniques in ML also increase the maximum power output of a particular source and minimize computational costs. Parameter selection is another huge task in ML since it influences the performances of ML models. Hence, appropriate optimization and parameter selection techniques in EMS are discussed. This chapter will summarize AI techniques in EMS and RES with real-time applications. Integration of AI with SG, optimization, and parameter

---

*Corresponding author: gujurothubhanu@gmail.com

Abhishek Kumar, Pramod Singh Rathore, Ashutosh Kumar Dubey, Arun Lal Srivastav, T. Ananth Kumar and Vishal Dutt (eds.) Sustainable Management of Electronic Waste, (1–46) © 2024 Scrivener Publishing LLC

selection in AI-based techniques to improve the energy system are elaborated. Finally, the chapter concludes with novel research limitations and future outlooks.

Keywords: Artificial Intelligence (AI.), energy management system (EMS), smart grid, renewable energy

1.1 Introduction

At present, the abstraction of energy from renewable sources has increased. Technological advances made in sun, geothermal, water, wind, and many other natural renewable sources have gained notoriety as reasonable energy sources. Conservative energy sources such as coal, gas, and a mixture of crude oils, with their negative impression on the environment and strong connection with the respective country's economy, have become political and financial weapons to create pressure [1-3]. So, there is a need for a replaceable and higher ratio of renewable energy sources. Harnessing energy from renewable sources has gained significant attention from the research community because of its advantages over conservative energy sources and its range for single-homes to large-scale power plants. However, renewable power plants are not fully controlled or planned due to the inherent features and power generation dependent on the ecological boundaries. For example, consider a power grid where the generation of power is created by renewable sources where capacity management alterations could impact grid physical health and quality of life [2-3]. Moreover, renewable energy sources continue to expand depending on the available sources, so it is necessary to determine grid localization, features, outlines, and optimal sizes. Management of an intelligent grid covering renewable energy plants is hard to integrate with other sources.

Energy sustainability and security are considered one of the most important challenges faced by the world. Specifically, it will be considered a leader for economic volatility. The limited fossil-based fuel sources and their adverse burning effects have boosted interest in developing sustainable energy sources. Solar thermal, wind, geothermal, biomass, tidal, and solar photovoltaic power are some of the renewable energy markets that are rapidly gaining attention. Around the world, major countries are hunting to expand their future works and share of renewables [2-5]. In conservative energy sources, the generation process and manufacturing rely on the energy demand from users; power constancy depends on demand and supply. If the demand and supply increase power-grid work quality, overall results degrade and the chances of power failure increase in some regions.

If the demand and supply are less than needed, energy will be lost and the chances for futile costs will reduce. So, more research is needed to produce sufficient power at the right time and smooth, safe grid running for higher economic aid. Appropriate research suggests that sufficient capacity per the requirement minimizes energy wastage [3-6]. Most renewable power sources are incredibly flexible on ecological fluctuations.

Over the past few decades, Artificial Intelligence (AI) approaches such as statistics, mathematics, data mining, machine learning, deep learning, artificial neural networks, and optimization methods have been useful in various domains to solve data-driven problems. AI is a scientific application of knowledge to build intelligent devices, specifically intelligent computer programs. AI is generally explained as extraordinary intelligence demonstrated through learning, reasoning, and problem-solving [4-8]. With the development of various modern technologies, the world is likely to establish systems incorporated with knowledgeable human features such as the capacity to think, reason, simplify, differentiate, find meaning, and learn from previous knowledge to correct errors. AI techniques like decision-making systems, fuzzy logic, neural networks, and nature-inspired model deployment advance the progression limits in power electronics and engineering. The approaches mentioned above provide impressive designs, controls, simulation tools, error detection, and optimization in various energy-based systems. Nowadays, AI is one of the hot research topics and its applications have rapidly increased in modern manufacturing systems.

Machine Learning (ML) generally attempts to learn the relationship between the provided data (input) and the output data by employing mathematical complications. Once the ML models are trained to fit the training dataset, end-users get satisfied values and feed them to an expert system for forecasting results. Here, data pre-processing plays a crucial role in improving ML performance resourcefully. ML has three fundamental schemes: supervised, unsupervised, and reinforcement. In supervised learning, a model gains an advantage from labeled data during training, then produces the best possible results. In unsupervised learning, models inevitably classify input data points into clusters through certain principles and standards for training data that have not been categorized in advance [5-8]. For example, take clustering as an example, the amount of clusters mostly relies on the clustering principles employed. Reinforcement learning (RL) models learn from interactions with the outer atmosphere to attain feedback to capitalize on the predictable benefits. There are countless learning principles, rules, and theoretical mechanisms defined by researchers based on their application area. With rapid progression in hardware, software, and mathematical advances, a new learning method is booming: DL, which is considered a sub-field of ML. Apart from the learning ML models mentioned, DL can grasp characteristic nonlinear features and high-level invariant data formations. Thus, DL has been helpful in numerous research arenas to find adequate performances.

It is a known fact that renewable energy resources such as wind, heat, and solar light are extremely inconstant. The subsequent variations in generation capacity can cause variability in the power grid since the production of the grid plant is distinct with issues such as intensity of solar radioactivity, wind speed, and other related aspects such as solar power only being offered in the daytime. Therefore, we need to focus on power generation when the resources exist and a solution to store renewable sources for later use [6-9]. For example, sources like wind, water, and solar are incredibly hazardous and expensive to store. In addition, if the dimensions of creating natural resources are inadequate to meet demand, gas and fuel-based power plants are used to overcome the shortage issues. Many of the difficulties mentioned above push the inclusion of AI-based approaches for energy management and optimized consumption. The research community suggested numerous AI approaches depending on the requirements and characteristics of renewable energy sources.

1.2 Summary of Artificial Intelligence Techniques

The term Artificial Intelligence (AI) is well-defined as a set of high-tech structures which accurately execute numerous tasks generally associated with human beings. Like a human brain, AI approaches adapt their behaviour without direct reprogramming. In some cases, AI will reach or surpass human intelligence depending on how it accomplishes human-like levels of cognitive functions, learning, perception, and interaction. Here, intelligence indicates the capability to absorb, study, then learn from another environment and put on that data to conditions that have, until that specific point in time, not happened. Similarly, a machine can sense information from its surrounding atmosphere and automatically decide on probable future data built on past data [5-9]. Hence, AI can be fixed as the procedure for representing absorbed intellect. However, the story behind AI started in medieval times. Between 360-300 BC, a famous researcher, Aristotle, stated the fundamental governing principles of the human brain with conceptual logic. After that, from the 15th century, numerous researchers expanded brain logic concepts.

A breakthrough was reached in 1943 when two famous research scientists, Warren McCulloch and Walter Pitts, projected an artificial neuron (neural) model. Authors who studied their research concluded that appropriately constructed neurons could learn and then make predictions. After that, in 1949, Donald Hebb came up with new rules on inter-neuron modifications and connectivity which inspired Alan Turing to design Turing Test to measure machine intelligence. Finally, in 1956, the term artificial intelligence was coined by John McCarthy. From the 1990's, the world saw the growth and importance of intelligent neural networks, which mimics the concepts of human work. With the rapid progression in various hardware, software, and miniature sensing devices, building AI-based real-life applications became possible [3-5, 7-10]. At the end of the 19th-century, AI-based statistical learning models were employed for decision making. Here, fuzzy logic and rule-based models were highly used. From the 19th century until now, AI-based systems successfully adopted numerous real-time applications, from washing machines to controlling high-speed bullet trains. At present, AI itself is incorporated into many streams of human and industrial infrastructure.

1.2.1 Machine Learning

Machine Learning (ML) is a sub-part of AI that produces computational and flexible methods which allow the evolution of superior machine performance (See Figure 1.1). This means ML designs boundaries that allow humans to analyse the behaviour of AI schemes for practical actions. Deep learning is another concept of AI, which is described as a sub-part of ML. ML generally produces the best results with processed data, however, getting pre-processed data from real-time appliances is impossible. Here, DL with deep neural networks with many hidden layers extracts the valuable data layer by layer, then produces efficient data for decision making.

Decision Tree (DT): The decision tree is one of the prominent classification techniques in ML. The classification results in decision trees that are separated into groups of choices based on the input topographies. It builds a tree-like structure; the procedure starts from the base feature and then develops like the tree. Here, the tree's structure is built from a base feature by distributing the source root node of the tree into sub-branch nodes. The tree partition was decided based on principles determined by the properties of the set and the target classification. The decision tree model is employed in several areas of intelligent energy management, demand, and supply of smart grids for the prediction of anomalies detection, planning and energy management, risk detection, and optimization.

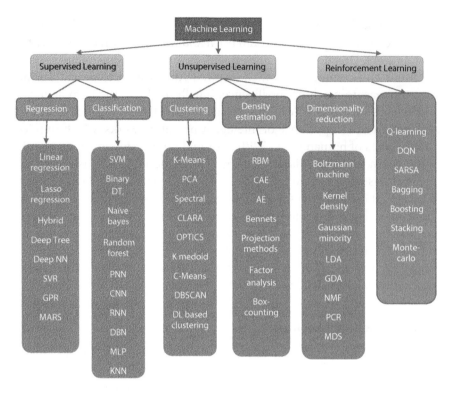

Figure 1.1 Pictorial representation of machine learning approaches.

Random Forest (RS): Random Forest models are identified as an extension of decision trees that are mainly applied for categorization and regression and to progress the prediction precision of decision trees. Bootstrap collection is one of RS's available variants, which decreases the variance of a DT by estimating a quantity from a data sample. RS algorithms work by forming a bootstrap sample of the training set, then use it to train a decision tree. Topographies are arbitrarily chosen at each node of the DT. The probability of incorrect classification is applied to select the optimal choice and the procedure continues until an efficient tree is formed. RS variant models are highly employed in structuring energy schemes to forecast hourly energy ingesting.

Wavelet Neural Network (WNN) combines wavelet analysis and neural structures. In this model, wavelet concepts considered from the generalization of windowed Fourier transform are effectively employed for predicting time series data in renewable energy sources to optimize the cost and battery.

Naïve Bayes: Bayesian theorem-based naïve Bayes calculates the likelihood of a guess (presumption) when fed with a piece of prior knowledge. The variants of the naïve model are broadly applied to solve building energy problems like analysis of building energy efficiency, weather predictions, prediction of photolytic and photovoltaic energy, and forecasts of energy usage on an hourly basis.

Artificial Neural Networks (ANNs): ANN was developed based on the human neuron structure and is broadly applied for nonlinear modeling processes. ANN is broadly applied to decision making, regression problems, natural language processing, dimension reduction, prediction, computer vision, intrusion, and anomaly detection. In the energy sector, ANN is applied to predict electricity consumption, battery optimization, cooling, and load sharing, which assists interior energy consumption by analyzing indoor climate and real-time energy monitoring.

Multiple Linear Regression (MLR): MLR is a statistical linear regression model widely applied for predicting smart buildings' heating and cooling processes and non-linear energy demand functions. The principal objective of MLP-based models is to detect functions from the analysis of training data such as weather conditions, the impact of solar radiation, and humidity.

Logistic Regression (LR): Logistic Regression was designed with the statistical logistic function acknowledged as the sigmoid function. LR is mainly employed for classification events requiring likelihood occurrence forecasts. LR in the energy sector estimates weather predictions, fault detection, energy consumption, etc.

Genetic Algorithm (GA): Genetic algorithms work extensively for a heuristic search approach employed extensively in complex models because of their capacity to deal with non-linear topographies. In addition, GA models efficiently solve constrained and unconstrained optimization issues. Hence, GA variants are extensively useful in the scheduling of housing power loads to diminish the total energy cost in a dynamic pricing system.

Fuzzy Logic (FL): FL is a formula where the truth values of variables might be any actual number between 0 and 1 or comprehensively. Its primary function is to handle the perception of partial truth, where the truth value may choose between values that are totally true or false. Because of its simple design and complexity, FL is adopted in many more appliances than other models. FL in the energy sector is used in prediction, optimization, fault detection, data anomaly, and intrusion detection.

Particle Swarm Optimization (PSO): PSO is a nature-inspired meta-heuristic optimization approached design built on the cooperation of birds and fish. PSO employs particles of swarms crisscrossing in a multi-dimensional exploration to find the optimum position. Every particle has a probable solution influenced by the involvement of its neighbour particles. PSO is employed for the optimal solution in hybrid ML approaches for energy estimation, optimization, supply, and demand.

K-Nearest Neighbour (KNN): The KNN model is trained with available datasets that then estimate the label of a new example based on the tags of its closet neighbours in the training set. KNN helps build energy management, demand, and supply, as well as monitors energy consumption in a particular time region.

Principal Component Analysis (PCA): PCA is an unsupervised ML model used to reduce dimensionality by plotting a particular function from higher-dimensional into lower-dimensional space. PCA methods in the energy sector are used to predict carbon dioxide emission, progress the housing load disaggregation of energy, and analyse power ingesting.

Hybrid Models (HM): HM models are a collection or combination of multiple ML algorithms. HM is mostly applied for data pre-processing and solving optimization issues. Day-based power predictions and generation, energy consumption, reduced energy model complexities, and power data cleaning are some areas where HM is applied in the energy sector.

1.2.2 Deep Learning (DL) Techniques

DL is considered as a sub-section of ML and is defined as a representation learning model that learns through numerous hidden layers and levels of representation that obtain results when each transforms the representation at one level. Shallow models consist of 1-to-3 levels of non-linear operations, whereas DL consists of more than 4 levels. DL-based techniques are broadly applied for classification issues and their usage in the past few years in various research fields has grown. Some reasons force us to employ DL in various research fields, for example DL with numerous hidden layers can handle large datasets, improve model performance, and have feature extraction capabilities. DL models with numerous hidden layers easily cope with large datasets and extract high-rated features to make accurate predictions. In addition, unlike orthodox ANNs, DL models can hold and store more information within the neurons.

Autoencoder (AE): An auto-encoder (AE) is a representation-based neural network with many unseen hidden layers. AE involves two functions: encoding and decoding. AE aims to absorb a representation of input data through training and then reconstruct the output. For example, if A is the input dataset, the encoder designs the input data to a representation or illustration of hidden layer h = f(x), and the decoder gives a hidden representation or illustration to construct the output g(h). Naturally, AE copies are helpful for dimensionality reduction/feature finding for training features in big datasets.

Recurrent Neural Networks (RNN): RNN is a classification-based neural network specifically applied in classification data. It uses a response/feedback loop which is associated to their previous calculations. RNN stores the feedback/response information in its memory database and uses it for previous outputs over time. In time series-based RNN models, the value calculated at a particular time (t) is affected by the standards impressed in the previous steps (t-1). RNN can study the model and time-based behaviours are revealed inside the time-series data. Then, the response values are used to recall the preceding steps. RNN and its variants like gated recurrent unit (GRU) and Long-short-term-memory (LSTM) are successfully applied to predict building energy, weather predictions, and power generation.

Convolutional Neural Networks (CNN): CNN models are experts in processing incoming data features with grid-shaped structures. These kinds of models are mainly applied for analyzing large-quantity complex datasets. A single CNN model contains four major parts: (a) a convolutional layer builds the feature maps of the input data; (b) next, a pooling layer is employed to decrease the dimensionality of complicated features; (c) a flattering-based technique is applied to adjust the data into a column vector; and (d) finally, a completely allied hidden layer estimates the loss function.

Deep Belief Networks (DBN): DBM consists of algorithms based on probability and unsupervised learning algorithms to produce the outputs. Table 1.1 described the classification and comparison of AI, ML, and DL

Table 1.1 Classification and comparison of AI, ML, and DL applications in energy sector.

Model	General application domain	Application domain in energy sector	Advantages	Disadvantages
Decision Trees	Classification	Building energy management, energy storage planning, refining operative efficiency	Practically accurate, reasonable speed, scalable	Complex, low user-friendliness
Linear Discriminative Analysis (LDA)	Classification	Energy prediction and power-data analysis, demand response management	Good speed and precision	Low accessibility
ANN	Modelling, prediction, and curve-fitting of non-linear progressions	HVAC energy consumption modelling, demand response management, failure probability modelling, smart grid management, sector coupling	High precision, intelligent speed, decent for noisy data	Extremely complex, low availability
SVM	Data arrangement, high-accuracy predictions	Building energy consumption and prediction, dynamic energy management, security and theft detection	High precision, scalable	Highly complex, low accessibility, low speed

(*Continued*)

Table 1.1 Classification and comparison of AI, ML, and DL applications in energy sector. (*Continued*)

Model	General application domain	Application domain in energy sector	Advantages	Disadvantages
Random Forest	Event estimating, data organization	Energy consumption forecasting, power-data analysis	Practically accurate, diminishes over-fitting	Reasonably compound, low accessibility, little speed
Deep Learning	Data prediction, pattern modelling	Energy efficient system design and modelling, dynamic energy management, security and theft detection	Reasonably user-friendly, high precision, moderate speed, best classifier interactions	Highly complex
Mixture Discriminative Analysis (MDA)	Dependability analysis, classification	Demand response management, preventive equipment analysis	Highly user-friendly, simple structure, high speed, feature dependencies	Low accessibility
WNN	Time series event calculation	HRES operating cost optimization, wind and solar power prediction	High accuracy, scalable	Low speed, low user-friendliness, reasonably complex

(*Continued*)

Table 1.1 Classification and comparison of AI, ML, and DL applications in energy sector. (*Continued*)

Model	General application domain	Application domain in energy sector	Advantages	Disadvantages
Fuzzy Logic	Control claims	Power point tracking, control and monitoring, outage prediction, preventive equipment analysis	Almost user-friendly, nearly accurate, high speed	Reasonably complex
Least Absolute Shrinkage and Selection Operator (LASSO)	Dependability analysis, classification	Power-data analysis, dynamic energy management, security and theft detection	Highly user-friendly, simple structure, high speed, feature dependencies	Low speed, low user-friendliness, reasonably complex
Multi-Dimensional Scaling (MDS)	High-accuracy predictions, dependability analysis, classification	Demand response management, preventive equipment analysis	Extremely user-friendly, modest structure, high speed, feature dependencies	Low accuracy
Hybrids	High-accuracy forecasts	Load prediction, energy generation foretelling, refining operational efficiency	High speed, generality, accuracy	Practically complex

(*Continued*)

Table 1.1 Classification and comparison of AI, ML, and DL applications in energy sector. (*Continued*)

Model	General application domain	Application domain in energy sector	Advantages	Disadvantages
Regression	Forecast of the possibility of occurrence	Energy ingesting and predicting, power-data analysis	Highly user-friendly, simple structure, high speed, feature dependencies	Little precision
Auto-encoder	Data representation, error calculations	Outage prediction, preventive equipment analysis	High speed, high accuracy, good generality	Reasonably composite
Convolution Neural Network (CNN)	Classification, high-accuracy predictions	Outage prediction, preventive equipment analysis, demand response management	High speed, high accuracy, high speed, good generality	Highly complex, low availability
Long-Short-Term-Memory (LSTM)	Classification, time-series analysis	Demand response management, outage prediction, preventive equipment analysis, failure probability modelling, smart grid management, sector coupling	Highly user-friendly, simple structure, high speed, feature dependencies	Highly complex, low availability

(*Continued*)

Table 1.1 Classification and comparison of AI, ML, and DL applications in energy sector. (*Continued*)

Model	General application domain	Application domain in energy sector	Advantages	Disadvantages
Recurrent Neural Network (RNN)	High-accuracy predictions, time-series analysis	Outage prediction, preventive equipment analysis, failure probability modelling, smart grid management, sector coupling	Highly user-friendly, simple structure, high speed, feature dependencies	Highly complex, low availability
Deep Belief Network (DBN)	High-accuracy predictions	Demand response management, dynamic energy management, security and theft detection	High speed, high accuracy, high speed, good generality	Highly complex, low availability
Stacked Generalization (Blending)	High-accuracy predictions, optimization	Power-data analysis, demand response management	Extremely user-friendly, modest structure, high speed, feature dependencies	Low accessibility
Genetic Algorithm	Problematic optimization	Optimum load scheduling, smart grid management, sector coupling	High precision, used in hybrid mode	Low-slung speed

(*Continued*)

Table 1.1 Classification and comparison of AI, ML, and DL applications in energy sector. (*Continued*)

Model	General application domain	Application domain in energy sector	Advantages	Disadvantages
Multivariate Regression Analysis	Classification, clustering, data reduction	Power-data analysis, dynamic energy management, security and theft detection	Highly user-friendly, simple structure, feature dependencies	Low availability and low speed
PSO	Problem optimization	Operational cost optimization and energy forecast	High accuracy, used in hybrid mode	Low speed, low convergence rate, fall in local optimum
KNN	Classification, forecast	Building energy ingesting study	High speed, user-friendly	Low accuracy
Bagging	Problem optimization, prediction	Demand response management	High speed, accuracy, good generality	Low accuracy
Naïve Bayes	Calculating the probability of occurrence	Building energy efficiency study, energy generation forecast	High speed, user-friendly	Low accuracy

(*Continued*)

Table 1.1 Classification and comparison of AI, ML, and DL applications in energy sector. (*Continued*)

Model	General application domain	Application domain in energy sector	Advantages	Disadvantages
K-Medoids	Classification, clustering	Preventive equipment analysis, sector coupling, demand response management	Practically accurate, reasonable speed, scalable	Low convergence rate, fall in local optimum
Principal Component Analysis	Data reduction	Failure probability monitoring, power-data analysis	Highly user-friendly, Simple structure, High speed, feature dependencies	Low convergence rate, fall in local optimum
Self-Organising Map (SOM)	Classification, clustering, data reduction	Operating cost optimization and energy scheduling, smart grid management	High speed, high accuracy, high speed, good generality	Low speed
Least Angle Regression (LARS)	Data reduction, classification	Outage prediction, demand response management, smart grid management	Highly user-friendly, Modest structure, High speed, feature dependencies	Low speed
Hierarchical Clustering	Classification, clustering, data reduction	Dynamic energy management, improving operational efficiency, power consumption analysis	Practically accurate, Reasonable speed, Scalable	Low accuracy

applications in energy sector. The restricted Boltzmann machines (RBM) are a fundamental characteristic of the DBN. The RBM is a light two-layer neural net used to study likelihood circulations over its input data space so that its formation can display needed resources. The 1st layer of the RBM is called the visible/input layer and the 2nd is the hidden layer. DBN is broadly applied for data size decrease, feature learning, regression, classification, and collaborative filtering.

1.3 Reasons for Applying AI in EMS

Here, we mentioned the articles that inspired the integration of AI and ML approaches in the energy sector. Mondal *et al.* [9] designed a game theory-based energy management approach for smart grids and simulation results show models that maximize advantages in cost and Negeri supply. Elseid *et al.* [10] proposed energy management in a smart grid, which automatically optimizes energy demand. Leonori *et al.* [11] fabricated an adaptive neural fuzzy interference scheme by employing Echo-state networks as a series analyst. The authors tried to minimize energy exchange with the smart grid and experimental results show that the model achieved over 30 percent prediction for a particular time. De Santis *et al.* [12] presented an interconnected fuzzy logic-based Mamdani scheme for energy supervision in a smart grid. The authors focused on decision-making based on energy management and storage tasks. Venaygamoorthy *et al.* [13] reduced carbon emissions using evolutionary adaptive dynamic programming and a neural learning scheme. Model performance analyzed battery life and renewable energy load. Ma *et al.* [14] fabricated game-theory-based leaders and followers for accurate energy management, maximizing the benefits of active consumers and ensuring the optimal distribution in the smartgrid. Arcos-Aviles *et al.* [15] designed a low-complexity-based fuzzy logic controller for intelligent buildings and residential grids. Aldaouab *et al.* [16] proposed a genetic scheme-based optimization for commercial smart grids and primarily used micro-turbines and diesel generators. Liu *et al.* [17] presented the Stackelberg model for energy management. Nnamdi and Xiaohua [18] designed an incentive-based demand-supply model for energy management in a smart grid for analyzing connectional grid operations. The experimental results demonstrate that the proposed model maximizes the grid operations and minimizes the furl transaction costs.

1.4 ML in Renewable Energy

Due to the excessive usage of fossil fuels to increase economic growth, it has been documented that extreme consumption of fossil fuels will not only fast-track the decrease of power sources but also negatively impact the environment. Most of the effects on human health threaten global climate change. Hence, renewable energy, which can quickly be recovered or reproduced with its characteristics and low environmental pollution, has attracted attention from research communities. With the development of renewable energy systems, most of the current energy problems like consistency of energy source and solving regional energy supply need to be addressed. High-rate accuracy of energy monitoring can progress the efficacy of the energy organization. Hence, power prediction plays an energetic role in energy organization approaches. Numerous studies prove that AI-based approaches have been applied for renewable-energy forecasts. In addition, hybrid-AI-ML technologies increase the prediction accuracy of renewable energy with periodical intervals like months, weeks, days, hours, and minutes. Prediction accuracy and proficiency is characteristically operated to estimate the performance of AI-ML-DL models in renewable-energy predictions.

Nowadays, technologies like AI and ML provide a valuable contribution, opportunities for energy management and cooperation, and a chance for investors to implement effective, creative methods for more significant assets and compassionate energy conversion. ML is a kind of technology that can categorize through a set of rules and data in such a way that it studies and expands its methods through improved knowledge/practice. Industrial experts and data specialists state that AI will emerge in the future of energy management. Numerous AI-based approaches are already theoretically considerable for potential changes in the energy sector. Investments in the AI-based energy sector will surpass over 800 billion US dollars by the end of 2025. It is difficult to overlook the opportunities for development ML provides. Many industrial companies are now integrating AI and ML into their company tactics to catch the promise it offers.

The reports mentioned that the energy sector denotes around 2% of capital for AI in Europe and it specifies that there is still much room for development. The AI and ML industries are household to many start-ups with plenty to offer, and companies and industries are primed to take primary steps towards future AI technology. China invested nearly 400 billion US dollars in the AI landscape and 200 million in emerging technologies like energy and smart grid software. The UK government invested 130 million in 2020 and the US invested 50 million in funding for the energy department for AI and ML-based energy techniques. From the above information, there are plenty of investment opportunities which are expected to increase soon. The number of appliances for ML and AI in the energy management sector is practically limitless. Now, it is more a matter of discovering which requests will be the most beneficial, maintainable, and money-making. This will be predominantly accurate for solar and wind power, historically hampered by meteorological conditions and patterns that are hard to forecast and have numerous variable quantities to study.

1.4.1   ML for Renewable Energy Applications

Solar Energy: AI-based techniques have benefited most renewable energy-related applications. In recent years, solar power production has benefited more from AI and ML approaches, particularly in weather predictions. With accurate weather forecasts, it is easy to make the grid supply more accurate than before. The US Energy Department signed MoU with IBM to develop the AI-based Watt-Sun technique which analyses ecological data reports, then tries to reduce excess energy and decrease energy production. Primary simulation results show that this technology can raise the precision of prediction by up to 30%. However, the inconsistency of sunlight is inevitable, but its unpredictability can be reduced. Micro-grid challenges like variable electricity supply and load balancing are handled with ML. By analysing existing power flows, operators can work to moderate blocks in the grid network and help choose whether energy should be stored, distributed, or sold in an electricity exchange at a specific location, which results in a more excellent energy consumption proficiency.

Wind Energy: In weather forecasting, wind energy faces more challenges than solar power. In 2004, Google considered its importance and acquired AI-based DeepMind for 500 million US dollars. The ML techniques designed by DeepMind can expect an output 36 hours ahead of time. It helps Google to reach its goal of enhancing the value of its energy up to 30%. Soon the UK government also implies DeepMind technologies in their smart-grid operations. Condition-based weather monitoring is another critical ML application that dramatically reduces energy production costs, especially for wind turbines.

Appropriate AI and ML models can accurately monitor blade faults or generator temperature, permitting quick maintenances before it degrades. It ensures operations run smoothly at offshore wind farms where crucial monitoring and regulations are the two key aspects. There are numerous traditional power curve models for wind predictors, however, they incorporate drawbacks for reliable data analysis.

1.4.2 Countries Focusing on ML

Due to the rapid development of various AI and ML-based technologies in countless fields, countries worldwide are realizing ML's potential and trying to incorporate it into their companies.

Countries incorporating AI and ML are as follows:

Australia: Investing $25 million into AI and targeting to achieve more excellent results in the energy sector than others.

China: China stands first in the AI race; they have already adopted numerous widespread intelligent grids across the country by employing its current innovative proficiencies in AI.

Italy: Started installment of 41 million smart meters as part of a scheme in early 2001. Italy today has the most innovative smart grid compared to other competitive countries.

UK: The UK invested nearly $130 million into an AI-based robotics program for robotics research for risky atmospheres such as seaward oil and gas sites.

USA: After China, the US shows more interest in AI technology in the energy sector and the US sector of energy has a publicized subsidy of $30 million focused toward ML and AI.

1.4.3 Notable ML Projects

Exxon Mobil: Manchester Institute of Technology (MIT) is working on AI-based integrated self-learning androids that will photograph ocean flows to discover oil, natural gases, and undisclosed treasures.

US Department of Energy: Financed over 4.5 billion US dollars for modern intelligent grid infrastructure and management projects. It contains over 15 million smart meters for monitoring energy demand and supply.

Google DeepMind: Google with DeepMind has been analysing 700MW of wind energy to understand how it might be improved. The ML network used data from notable climate predictions and turbine data, permitting them to forecast wind energy production before 36 hours.

1.4.4 How ML is Renovating the Energy Industry

AI and ML are the two buzzwords that have impacted business since industries and business-based companies are eagerly looking for ways to implement them to mechanize their core processes. Here, the energy sector is getting more and more interested in AI applications. Renewable energy sources have greatly benefited from the power of ML approaches over the years. ML models manage the company's energy sector to lower costs and increase performance. We have mentioned some popular energy sector applications with ML for your consideration.

Grid Management: Grid management is one of the significant uses of AI in energy management. Electric power is continuously delivered to customers through a complex network called a power grid. However, there are significant issues that both the power generation and power demand must match over time, otherwise there are chances of system failures. On one hand, there are plenty of chances to store energy, however, we still have not found efficient methods for pumped hydroelectric storage. When dealing with renewable energy, it is hard to expect the grid's electricity production dimensions. In some cases, it depends on numerous factors like sunlight and wind.

Demand Response: In some cases, suddenly, large swings in demand occur, which becomes very expensive for organizing countries that produce most of their energy through the renewable energy sector. In recent days, Germany has started a project to cover all of its energy companies' renewable energy sources by 2050. It is a perfect example of countries changing to green energy for effective prediction and response. However, swings in demand and ecological volatility are the two major issues in this process.

Solving Demand Response Issues: Some solutions can tackle the issues mentioned earlier: weather forecasting and electricity demand. Many countries are collaborating with AI-based companies to analyze and forecast how much solar and wind energy can be expected at a particular time. It will help countries make up for additional electricity demand by using non-renewable energy in essential time. In this process, companies use many historical datasets for wind turbines and solar planes to train their ML.

Predictive Maintenance: As discussed in the above sections, AI improves energy production and consumption. It is also helpful in the reliability and robustness of power grids. A massive blackout occurred in Ohio City in 2003 because of a low-hanging high-voltage power line grazing against an overrun tree. Additionally, there was no proper alarm system when the incident occurred. The electric company did not identify similar accidents, which resulted in the entire grid going down. Here, ML models can be employed to implement predictive maintenance with data collected from deployed sensor nodes. With the acquired data, ML approaches can also forecast cable failure time or estimate when a subsequent failure may occur and recommend the remaining useful life of a particular cable or machinery. From the above, it is evident that the main aim of these ML approaches is to forecast machine efficiency and cable failures, avoid blockages and downtimes, and optimize activities, thus decreasing maintenance costs.

Energy Source Exploration: Exploration and detecting for drilling of fossil fuel energy sources is another usable application of AI and ML in the energy sector. The Massachusetts Institute of Technology (MIT) recently developed auto-learning submarine robotics to discover the deepest ocean locations containing oil and natural gases. These robots will be equipped with advanced ML approaches that not only help them learn from their faults while conducting studies but also carry out similar work to what a scientist would without any hazards.

Energy Consumption: Shifting to renewable energy sources is not just a need for companies and governments to overcome financial crisis; the main reason is to save the ecological system and prevent negative impacts on human life. Companies like Google and Microsoft have recently tried to impact the atmosphere by lowering their total energy consumption. In particular, Google maintains massive data centers worldwide, producing a high volume of heat that requires a vast amount of electricity to cool down. Companies like DeepMind designed AI-based techniques to diminish 40 percent of energy to cooling data centers and also reduce the carbon tax by 20 percent.

1.4.5 Machine Learning in Renewable Energy

Mellit *et al.* [19] showed a systematic literature review on photovoltaic-based power forecasting using AI, ML, and DL approaches. Authors noticed that using feature extraction-based DL approaches on numerical weather forecasts enhances long-term photovoltaic power-generation and then guesstimates the time-dependence data in photovoltaic power forecasts. Zendehboudi *et al.* [20] analyzed SVM-based appliances to forecast wind and solar energy and pointed out that SVM outperformed other models regarding commutation and prediction accuracy. Das *et al.* [21] studied the forecasting techniques in solar-power generations based on SVM and ANN. Wang *et al.* [22] revised various renewable energy forecasting approaches based on DL algorithms. Bermejo *et al.* [23] explored ANN variants in forecasting energy and consistency. The authors focused on energy sources like solar, hydraulic, and wind. Mosavi *et al.* [24] reported hybrid ML approaches in energy systems and Ahmed *et al.* [25] revised the prediction of renewable energy sources based on power transmitting schemes, energy storing, policy, and marketplaces along with optimal reserve sizes. Perez-Ortiz *et al.* [26] studied classification models for renewable energy difficulties and suggested open research issues. Khare *et al.* [27] conducted a comprehensive study on renewable-energy schemes based on modelling, sizing, control and consistency, requests of the evolutionary system, and game theory.

For hydro-power predictions, a hybrid model based on gray wolf optimization and ANN is used to forecast power management. In [28, 29], it was stated that ML procedures, linear regression (LR) and K nearest neighbours regression (KNN), support-vector-machine-regression (SVMR), and decision-tree-regression (DTR) could be used to model classification without additional modifications. In [30], the authors use 4 ML models to estimate the CO, CO_2, CH_4, and H_2 during biomass gasification. [31] built an improved ensemble-based hybrid model to predict sea wave heights and in [32], the authors employed Bayesian optimization with genetic and an extreme ML algorithm to predict the wave height and energy fluctuations. For prediction, the authors use SVR based on wavelet transformations. [33] designed a cluster-based power flow forecast model to obtain harmonic power fluctuations. The authors use a combination of wavelet, ANN, and Fourier series combinations. [34] uses a PPN-based ML model to predict power flow in the hydrologic system. The authors of [35] employed an LSTM encoder-decoder structure to guesstimate geothermal energy. An LSTM encoder and decoder were useful to contract with past geothermal energy creation and forecasts of the upcoming geothermal-energy generations. In [36], the authors studied the heat exchanges of coal fields by employing time-dependent neural networks (TDNN). [37] planned a forecasting scheme (LMS-BSDP) using inflow forecasts with numerous lead time intermissions to increase hydropower stations' performance in hydropower calculations. The authors of [38] projected that an innovative hybrid model based on SVM and PSO (SVM-PSO) was accessible to estimate the higher heating value (HHV) in biomass-energy generation. The authors of [39] employed the gaussian process (GP) for forecasting short-term waves with the help of ANN and auto-regressive (AR) schemes. Experimental outcomes show that the planned model through a numerical dataset outperformed other existing models. The authors of [40] designed multiple regression and ANN-based models to guess hyper-parameters of penetration rates (ROP) while generating geothermal energy. In [41], a novel RF algorithm was applied to data composed from geographic info schemes to forecast very shallow geothermic chances. The study showed that the RF is a reasonable way to estimate geothermal energy when datasets such as landscape, climate, and soil are offered.

1.5 Integration of AI in Smart Grids

According to the reports of the USA Department of Energy Management and Smart Grids, the perception of the smart grid is transitioning the traditional electric power grid from an electromechanically measured system to an electronically controlled system. In detail, the smart grid allows the collection of high-dimensional data for electric grid operations through integrating control and communication technologies and field devices that synchronize several electric procedures. As we mentioned above, outdated technologies face many challenges and restrictions in processing and storing data, hence the appliances of AI-based approaches in the smart grid turn out to be more perceived than before. For the last 5-6 years, the integration of AI into smart-grid applications like grid planning and operation has increased rapidly. AI is effectively applied in five major areas of the smart grid: (a) to observe and share data among devices and power-based connected systems; (b) for monitoring communication procedures, measuring data and sending information back to the operation centres, and trying to respond mechanically to regulate the ongoing operations; (c) there are chances for power-grid failure, fault detection, and smart-grid security which can be made easily answerable with AI-based approaches; (d) to analyse the data coming from the distinct network and help operators and digital technologies; and (e) to increase the optimization in load balancing, grid stability assessment ,and overall performance of the power grid. Some of the traditional models have many limitations in solving the activities mentioned earlier, hence, applications of AI-based approaches have been increased in the smart grid.

1.5.1 Load Balancing

Load balancing in innovative grit is challenging with the integration of unreliable renewable sources like solar, tide, and wind powers. Load balancing plays a crucial role in stabilizing the power system, which is vital for forecasting and operating modern power grids. In addition, accurate ecological predictions are also crucial for reducing costs and saving electric power [41]. Based on time levels, load-balancing is categorized into three types, as follows:

Short-Level Load Balance is employed to predict the power load from minutes to hours. Numerous AI and ML-based approaches are employed for short-term load-balancing. Lie *et al.* [42] designed a new ensemble algorithm that integrates AI concepts and simulation results show that the proposed model accurately predicts the short-level load balance. However, the authors have not made many comparisons with existing models and need further validation in the ensemble's choice of base model selection. Shi *et al.* [43] projected a DL-based RNN-pooling to solve the over-fitting issues. Here, the authors use deep-hidden layers to address the time-consuming problem. Moon *et al.* [44] fabricated ensemble-based DNN models with hidden layers and eliminated the poorly performed models to enhance overall performance. He *et al.* [45] modeled a novel DBN embedded with parametric techniques for predicting the hourly load of a power grid. Experimental outcomes show the efficiency of the technique by linking it with DT, ANN, PCA, and SVR. Hafeez *et al.* [46] designed a hybrid model using Factored Conditional RBM (FCRBM) model in association with Genetic Wind Driven Optimization (GWDO). The proposal is authenticated by outstripping the existing algorithm. Aly [47] manufactured a novel hybrid clustering approach built on a wavelet neural network (WNN) and the results exposed higher performance, which was then compared with other clustering procedures.

Mid-Term Level Balancing predicts power loads from a few hours to weeks and long-term level balancing predicts power loads for years. Mid-term load-balancing is effectively applied for load balance, generation, dispatch, and maintenance, but it affects historical load, demographic, and weather data. Jiang *et al.* [48] designed a DBN-based approach to estimate the ultimate power load for the upcoming year. In Askari and Keynia [49], the authors arranged a DNN scheme with an optimized training procedure containing dual search systems for MTLF and offered the model's efficiency. Rai and De [50] proposed an improved SVR model for MTLF with an average tiniest absolute mean error. Gul *et al.* [51] projected a solution combining CNN and LSTM methods. Dudek *et al.* [52] propose a hybrid DL technique for MTLF that correlates with progressive LSTM along with ensemble methods. It is a competitive method that also practices the ensemble tactic.

Long-Term Load Forecasting predicts upcoming years or subsequent decades of power consumption, scheduling, planning, and expansion of power systems. Researchers proposed numerous AI and ML-based approaches for long-term load-scheduling since it requires a huge initial investment. Ali *et al.* [53] proposed a new hybrid fuzzy-neuro model for accurate prediction. In addition, the authors employed LSTM for the final decision-making process. Zheng *et al.* [54] designed LSTM-RNN based on the long-term feature dependencies for the electric load time series in which the system had brilliant performance accuracy. Dong *et al.* [55] presented a hybrid technique built on LSTM, then employed gated recurrent unit (GRU) long-term load-scheduling. Bouktif *et al.* [56] came up with an improved LSTM-RNN model and Sangrody *et al.* [57] employed six regularly used ML skills: ANN, SVM, RNN, KNN, and GPR, then a generalized-regression-neural-network (GRNN) for performance and parameter-based analysis. Simulation results show that ANN performs better than the other five methods in long-term load-balancing.

In general, mid-term and long-term load-balancing models function with historical power consumption data, weather reports, customer types, and demographic data. Short-term loads are primarily applied in dissimilar appliances like real-time control, demand, and energy transfer scheduling. Based on the data provided by smart meters, numerous methods are used to predict the functional operation of mid-term and long-term load balancing.

1.5.2 Power Grid Stability Assessment

The power grid stability is considered fundamental for ensuring the reliability and security of the power system since it compromises transient, signal, frequency, and voltage stability. In the literature, there are numerous traditional models designed for stability assessments. However, most of them are complex in calculation and need significant computing resources since they profoundly depend on exact real-time energetic power scheme models. Several data-driven AI, ML, and DL constancy analysis methods have been practical for power grid stability analysis. Voltage Stability Assessment (VSA) tracks voltage fluctuations that could tremendously modify the stability of the power grid. Ashraf et al. [58] used an ANN approach to guess the loading boundary of power schemes and got relevant results. Amroune et al. [59] employed a hybrid scheme with SVR with an optimization technique. Mohammadi et al. [60] proposed an improved SVM technique for online VSA. Experimental results show that the misclassification charges of the SVMs are lower than 2% for power grids. Liu et al. [61] fabricated a feature selection-based random forest model using partial mutual information (PMI). Small-signal stability assessment (SSSA) ensures the system's ability to maintain synchronization continuously when it is in trouble. The authors of [62] designed a CNN approach for OSA with experimental results showing that the proposed approach is robust and system performance will not be condensed as the scheme raises dimensionality. Xiao et al. [63] utilized the multivariate random forest regression (MRFR) approach and Kamari et al. [64] proposed a particle swarm optimization SO scheme to accelerate the determination of OSA.

1.5.3 Smart Grid Challenges

1. Security and data privacy are two important challenges; data leakage will expose the operations of any network. Hence, a secure, reliable, and resilient communication scheme is needed in the smart grid, which improves the M2M connections to the last elements integrated into the smart grid.
2. Finding low complex security rules and protocols for avoiding possible security risks like intrusion and anomaly detection in smart grids is also a big challenge.
3. Efficient bi-directional communication and control facilities among numerous smart grid elements is a huge task.
4. Development of appropriate strategies for device-oriented security platforms and their integration with the current status of the power network is a challenge.
5. There is a need for strict control of propagation delay on the functional network to maintain the grid system's real-time capabilities.
6. Incentive systems are compulsory to build up for optimization of energy proficiency by influencing the prosumers.

1.5.4 Future of AI in Smart Grids

The main aim of AI-based smart-grid is to make an adaptive, self-motivated, fully automotive, and cost-effective system. We listed some future directions and opportunities below that are available to create an advanced smart grid system.

Transfer Learning: The absence of appropriate power-based data is still one of the significant contests faced for smart grid technical analysis. Here, the advanced transfer learning techniques efficiently reduce training data requirements, which solves the insufficient data and attracts research communities.

Integration with Cloud Computing: Integration of grid management with cloud computing technologies also enhances the self-learning of the smart-grid system, improves security and robustness, and minimizes power outages.

Fog Computing: The smart grid contains numerous connected devices that collect massive amounts of raw data. Here, fog computing makes efforts to pre-process the raw data locally before transferring it to cloud servers with on-demand resources fog-computing effectively in terms of scalability and flexibility.

Consumer Behaviour Prediction: The learning factors of power consumption and consumer behaviour will significantly contribute to consumer demand and supply tasks. Innovations like 5G networks and human-machine interaction create easier management of participation users in the smart grid.

1.5.5 Challenges of AI in Smart Grids

As mentioned in above sections, older power-grid systems are complex in numerical calculations, analysis, and control depending on their physical structure. Designing environment-friendly renewable intelligent grids and transforming old-styled power-grids into integrated innovative systems is challenging. Since most communication networks use old techniques where a huge quantity of data has high variabilities, additional uncertainties are added to modern smart grid systems. In addition, researchers are highly focused on robustness, adaptiveness, and automotive AI and ML algorithms. However, intelligent grids are facing severe challenges, as mentioned below.

Integration of Renewable Energy: Resource-friendly, highly combined renewable energy is a significant feature of the smart-grid, but it has numerous challenges in variability, unpredictability, and output that can vary in short periods and recurrently.

Preserving Data Security and Privacy: The smart grid connects with numerous devices where communications happens in a two-way path. Wireless communications are easily prone to numerous attacks since it is directly exposed to several malicious users. The authors designed numerous cyber security schemes to quickly identify system risks, false data injection, power theft, systems data thefts, and much more. Although, operating systems, different devices, network protocols, physical structure, and ecological behaviour, of the existing smart grid still reveal the system to an extensive variety of outbreaks. The current AI, ML, and DL-based smart grid cybersecurity solutions also have security and performance trade-offs.

Big Data Storage and Analysis: Smart grids with numerous networks and devices continuously collect vast raw data. Here, storing and retrieving a massive amount of smart grid data for analysis is another significant challenge in the AI community.

Limitations and Explainability of AI Algorithms: Recent advances in AI, ML, and DL technologies tremendously affect the deployment of AI to innovative grid systems and renewable energy management. However, AI, ML, and DL approaches suffer from black box issues since most of them are not explainable and interpretable. Moreover, every approach has limitations and barriers that must be considered before applying them to the smart grid.

1.6 Parameter Selection and Optimization

AI, ML, and DL approaches involve predicting and classifying data and based on the datasets, numerous models are employed. These models are parameterized, so the behavior of selected problem model parameters can be tuned. Search-based problems are the best way to choose the appropriate combination of parameters. In ML language, the parameter is a configuration variable closely related to the model and the value of the parameter is estimated from the given dataset. If we use a particular model on specific datasets , we can find parameters to predict upcoming new data. Weights in ANN, the coefficient in linear regression, and support vectors in SVM are some examples of model parameters.

On the other hand, model hyperparameters are the parameters external to the model and their values cannot be estimated from data. Data specified by the applicant can assist in tuning given predictive modeling issues and estimating model parameters. It is hard to know the best model hyper-parameters for a given problem; to know them, we have to apply search or trial-and-error. When applying any AI or ML-based model for a specific problem, it is easy to tune the model's hyperparameters to discover the model parameters that must result in the optimal forecasts. In some cases, model hyperparameters are considered as model parameters that make it confusing and inaccurate. Many researchers stated that every model has its important parameters which are not directly estimated from the available data.

1.6.1 Strategies for Tuning Hyperparameter Values in a Machine Learning Model

When we employ any ML approach, we are presented with numerous design choices to define our model structure. In that process, selecting the best parameters is like a model setting that can optimize the overall performance, like tuning the channel on the right frequency for clear picturization. In this context, selecting the right set of values is pronounced as "Hyperparameter optimization" or "Hyperparameter tuning."

Grid Searching of Hyperparameters: Grid search is considered a basic approach for hyper-parameter-tuning, which systematically shapes and estimates a model for each mixture of model parameters specified in the grid structure. This approach is straightforward and does an exhaustive search for hyperparameters as the applicant sets.

Random Searching of Hyperparameters: Random searching for hyperparameters was proposed by James Bergstra and Yoshua Bengio. It differs from grid searching because it uses a discrete, randomized set of values to explore each hyperparameter. The search process is continuous until the desired accuracy is reached. Experimental results show that random search is more effective than grid search.

Bayesian Optimisation: Recently, the Bayesian model emerged as a well-organized tool for hyper-parameter tuning and selecting model parameters in AI and ML algorithms. It chooses a unique optimization model for many deep learning models since it was efficient for highly complex black-box type functions with no idea about it. In addition, Bayesian models are broadly applied to learn optimal robot mechanics, synthetic gene design, and sequential experimental design.

Evolutionary Algorithms: Evolution models work on modifications in a set of candidate solutions based on specific rules or operators. These models can use any conditions since they are straightforward and independent from underlying issues. Evolutionary algorithms produce better accuracy speeds than the grid and random-based models.

Gradient-Based Optimization: Gradient-based optimization is a procedure that optimizes numerous hyperparameters based on the computation of the gradient of an ML model selection criterion of hyperparameters. It is mainly applied when we face differentiability and continuity circumstances of the training criterion being fulfilled.

Keras' Tuner is a library that allows for finding appropriate optimal hyper-parameters for ML and DL models. It effectively searches for acceptable kernel sizes and finds the learning rate for optimization.

Population-Based Optimisation: These models are a series of random search methods based on genetic algorithms, such as particle swarm optimization, ant colony, and evolutionary algorithms, among others.

Majid Dehghani *et al.* [65] proposed a naturally inspired gray wolf optimization model for parameter selections to predict hydropower generation. Experimental results are accurate and improve forecasting performance. Fan *et al.* [66] employed whale optimization and a Bat procedure based SVM-BAT model for model parameter detection in SVM for solar radiation predictions. Experimental results show excellent prediction accuracy. Demicran *et al.* [67] proposed artificial bee colony variants for global solar radiation prediction. The authors considered the duration and angles of sunlight to make predictions. Li *et al.* [68] showed short-term wind-power forecasts with an enhanced approach to choose the limitations of the SVM model. García Nieto *et al.* [69] designed a novel mixed model of SVM and simulated annealing (SA) to predict the HHV of the biomass. Here, the authors used SA to choose SVM parameters, and the simulation results of the proposed approach produced promising forecasting results. For accurate wind predictions, Wu *et al.* [70] fabricated extreme-learning machines with natural inspired-based multi-objective grey wolf optimization (MOGWO). Lin *et al.* [71] came up with an improved month optimization algorithm (IMFO) to improve the constraints of the SVM model for forecasting photovoltaic-power groups. Lin *et al.* [72] designed a hybrid model with a DBN and genetic algorithms (DBNGA) to guess wind speeds. Zhou *et al.* [73] offered a novel hybrid model by combining variational mode decomposition (VMD), backtracking search procedures (BSA), and regularized extreme learning machine (RELM) procedures for wind-speed estimation. Cornejo-Bueno *et al.* [74] employed Bayesian optimization to get ELM limitations in the ocean-wave forecasting scheme. Papari *et al.* [75, 76] designed an adapted harmony search (AHS) to control parameters in SVR for power-flow calculations.

In general, optimization techniques are helpful to capitalize maximum power out from each source, minimize electricity costs, and maximize storage system efficiency. Across the world, numerous researchers have proposed countless AI, ML, and DL optimization techniques for different applications. Energy management and optimization controls are crucial objectives in power supply. This can be evaluated by minimizing maintenance, operation, degradation, fuel costs, storage devices, batteries, and their capacitor functions. In addition, minimization and maximization of load balancing are also considered to improve optimization. Because of multi-constraints and multi-dimensional features, various authors proposed metaheuristic techniques to solve the optimization issues. Some other authors designed stochastic dynamic programming and game-theory-based approaches for optimization in energy management [77, 78].

1.7 Biological-Based Models for EMS

The word "biological" denotes the technical sources for developing human evolution. So, the biological world demonstrates the various metaphors that shape our thinking style by revealing the patterns of natural techniques. Every living being in this world has a structure which could be the inspiration for the next era of technology. Structures of biosystems are imitated in technical systems and the complex functionalities along with respective metaphors argue the necessity to solve the complex issues and find solutions for modern systems. Biological methods to achieve smart grid requirements use features like self-healing, learning models, different types of control, and energetic strategies. These characteristics, as mentioned earlier, exist in living beings and are vital for smart grid energy management. Biological schemes explain countless metaphors utilized for shaping our thinking style to discover patterns of nature in modern technologies. The significant benefit of biosystems is that they mock complicated functionalities and interventions of technical schemes, making easy solutions for modern systems. In recent years, the smart grid has improved its energy-management performances by integrating biological models using learning models, self-healing, distinct kinds of control, and energy strategies.

As we mentioned above, a smart power grid is an innovative deviation through an enlarged combination of electric power generation. Biological patterns inspire the possible control approaches for future smart grids based on biological living systems. The smart grid's operational security can be achieved through biological systems, for example, preventing and limiting damages, fast recovery, and frequently checking loss energy balances. Central pattern generators are activated sets of subsequent actions at a specific time. For example, consider spinal reflex arcs controlled by the limbic system to produce specific patterns. These kinds of designs are accountable for actions like walking, swimming, breathing, chewing, and digestion. Integrating these actions into power grid network structures like elements, lines, topologies, and switching fundamentals will progress energy management. In [16], the authors designed a complete theoretic energetic framework for identifying essential sources. The authors of [17] studied the capelin population for the result of a specific nutrition chain and used it for energy management. [18] presented a prediction-based property based on Kooijman's Theory and experimental results showing great accuracy in a growth phase and energy starvation [79].

1.8 Future of ML in Energy

Even with the integration of AI and ML, there is a long way to go in the energy sector. Most developed countries are already focusing on an entire green economy with efficient maintenance on a reliable smart grid. In the process, the intelligent grid plays a significant role, combining AI and ML to create a digital power grid that makes two-way communication accessible among consumers and utility companies. The modern smart-grid employs intelligent sensors and alarming devices that continuously gather and display customer data to improve energy consumption behavior and prevent system failures. As the techniques for AI and ML improve, the problem-solving techniques will solve the problems better, increasing accuracy and speed. Energy consumption frequently appears in AI approaches and here, it must be noted that less computing with fewer datasets means less energy disbursed. This tactic drives specific architectural ideas to lower energy usage. Some future energy appliances with AI and ML are described as follows:

Smaller Models: Simple and smaller AI-based models contain fewer layers and filters and require less computing and memory than complex models. However, choosing a smaller model for colossal data is not an acceptable solution. Researchers show great interest in smaller models and if they are efficient, there is no need for bigger models. Highly sophisticated decision-making with different training samples is required for bigger models, however, there is a lack of ideas and suggestions that make the model smaller. Researchers suggested a CNN-based MobileNet and ResNet for better accuracy [80].

Moving Less Data: Reducing the less data is one of the perceptible energy mitigations, which touches the workload on memory management. Memory management moves the data across the area with shorter distances, hence architecture style can be a determining feature for a given quantity of data. In theory, if computing is funnelled over a single core that takes x energy, the energy spent for y calculations will be x times y. For every calculation, data must be raised, stored, and worked for later usage. In addition, commands are also made and then decrypted for a given set of operations. Building large chips to contain entire model weights and maintaining weights stationary at one position are some of the affordable solutions. Minimizing central computations also reduces the data loss since less data will be fed from the far reaches of the station.

Less Computing: Less computation will increase the optimization of the proposed model and there are numerous ways to calculate a result with fewer actions. Here, a selection of activity functions is also an important aspect. The same selection of execution model is also essential since it consumes less energy than a complex model straight to the virtual hardware.

Batching Helps: In some cases, the collection of frames and operations will increase the high efficiency of the batch-size model. This means the given processing resources like weights and activation kernels are repaid with lower data resulting and instruction fetches. However, batching applies to data centers where large amounts of data is sampled, but not to edge-based applications where they receive a single data packet at a time. More importantly, batching does not affect accuracy.

Data Formats Matter: Model the highest accuracy attained using the most precise data patterns in the computational process. Here, floating-point numbers are a key part of the training model. However, floating-point circuits consume more energy than integer-based circuits. Researchers focus on smaller bits like 4 bits, 2 bits, and 1 bit (binary neural networks) since they use less data storage and movement. In addition, accuracy can also be determined by distinct layers with dissimilar precisions. IBM has tried and made numerous efforts to find approaches to reduce data sizes without harming precision.

Sparsity Can Help: Neuron computations are done with vector and matrix calculations. The concepts of neural network calculations are defined through matrix sizes. From mathematical principles, there is the chance for one raw model logically to be sparser than some added replicas. However, rather than matrices or vectors being truly sparse, they may have entries that are minimal numbers with negligible impact on computations. Hence, it can be employed to real-time applications, leading to sparser stimulation vectors.

Use Compression: For a given compression level, it can diminish the quantity of data transferred to appropriate regions. Generally, weight can be compressed while designing a model based on the decompression hardware units. Like the same, at run time, selected inputs or initiations can be compressed at the expenditure of the compression and decompression computer hardware. This compression, arrogant and lossless, should not influence model precision.

Focusing on Events: A videotape generally consists of images in sequence order and every frame is different. However, a single object may move in the video frame. Nevertheless, there is no need to process every frame, which consumes more time and energy and focuses on what is altering in the stream, which typically contains limited pixels and allows for less computation and memory usage. Neural models are effectively applied in these kinds of problems; the difference can be taken as standard among the first and last frame, then directed to find the portions that changed. These changes can be identified as events and the approach is called an event-based approach.

Using Analog Circuitry: Recent simulation results show that implementing analogue signals consumes less energy than other digital versions. Lots of energy is consumed for converting analogue signals to digital. So, if the implementation is run with analogue signals, it's possible to save lots of energy. Nowadays, modern systems are equipped with resource-constrained sensors to capture stream data. This can be predominantly effective if the received sensor stream has insufficient relevant data. Digital data must be transformed into analogue for computation and then, the outcome results are precisely rehabilitated to digital, reducing overall energy savings.

Using Photons Instead of Electrons: In recent times, electronic computation has been done on silicon photonics because it consumes less energy. In photonic computations, the laser approach is utilized and some of this energy is split into distinct phases. However, there will be limits to how mysterious the stage efficiencies can be leveraged for lower costs.

Optimize for Your Hardware: Modifying computer hardware also assists computation capabilities. In a few cases, we train the model first to get something to work as fast as possible. In the process, once the hardware adopts the training process, it will result in a more resourceful execution.

1.9 Opportunities, Limitations, and Challenges

1.9.1 Opportunities and Limitations

Smart energy is effectively applied to smart buildings and then interconnected with AI to allied groups which intermingle in real-time via the power and water grids. Moreover, by empowering software technologies that study massive collected datasets by detecting noise and anomalies and identifying decision-based features, AI can improve the opportunities for intelligent buildings.

1. **AI for Renewable Energy Prediction:** There is a constant contest in weather-dependent energy management sources, which swing in their strength. Weather is volatile, which causes disturbances in the power supply generated from renewable sources. AI and ML-based algorithms analyse the data extracted from the local satellite and meteorological stations to estimate wind, humidity, density, temperature, and lightning conditions. Its benefits harmonize the energy source and demand innovative structures prepared with hybrid micro-grids.
2. **AI for Energy Proficiency:** Most traditional mechanization and energy managing approaches focus on building monitoring and alarm signals. Recent advances in smart building construction technologies present a chance for developing

a central analytical stage to produce more intuitions from collected data. Advanced AI-based monitors continuously absorb building information, then control, estimate, and accomplish energy consumption in buildings. This data analysis can help control energy practices, decrease energy during peak hours, classify and recognize signal glitches, and notice equipment failures before they arise. In some arid areas, connected communities resourcefully share the power and water grids with distributed solar power to the connected communities and plants that deliver water. The inclusion of AI technology progressed the integrative designs of the community with dispersed power and water, allowing future designers and builders to develop effective strategies that have been attained and tested at a specific scale.

3. **AI for Energy User-Friendliness**: With AI's increasing usage and acceptable results, decarbonization, independent power producers, and decentralization utilities motivated AI into the energy sector to manage the imbalances in demand and supply. Moreover, advanced technologies like cloud, big data, blockchain, fog computing, and IoT support the dynamic organization of power grids by enlightening the user-friendliness of renewable energy sources in houses.

4. Researchers, academics, and scientists mention that AI is a revolution in the 20th century based on its social and ethical impacts on society. The ongoing advanced AI technologies resourcefully progress fully-automated driving vehicles, IoT, blockchain, business strategies, security and privacy of big data, cloud management, and robotics. Hence, AI is a significant innovation in humankind's current 20th-century evolution. However, depending on human living comforts, AI requires development policies; such policies must confirm that all upcoming developments in AI are trustworthy human-centered schemes.

5. Policymaking in the area of AI is recognizable and reflects the realization that there are undeniable aids from AI appliances in today's society in addition to the possible profits of future prospects.

6. The robustness of AI raises the consistency and trustworthiness of the smart schemes employed in AI to process and execute everyday jobs appropriately that uphold the benefits of civilization. It contains topographies like impartiality and

promises that the secrecy of persons, relations, and societies is protected. One of the tasks in the methodological basics of AI schemes is founding human–operator potentials. For a society to trust it, the system must prove beyond reality, contributing to society's benefit.
7. Recent studies focused on a single ML to predict renewable energy sources. However, it is challenging to design a single ML-based renewable model due to different datasets, periods, measuring indicators, and prediction rages. To progress the prediction performance, some researchers employed hybrid ML models in renewable-energy predictions.
8. In order to increase the faith/trust of humankind, AI-based models are essential to solving the complicated issues humans face. The triumph of AI would significantly strengthen human society's confidence in AI schemes in the imminence of intelligent structures to enhance energy supplies rendering the data collected from distinct houses and other appliances. At the same time, intelligent buildings characterize a revolution in society with advanced functional values for handling several device capabilities of handling info. In addition, to make all the comfortable users of smart buildings not lose their data due to unsuitable action being taken, it is essential to ease security anxieties with additional events.
9. With billions of devices operating simultaneously in smart buildings worldwide, the most significant encounters are in the storage, defense, and scrutiny of the collected power data.

1.9.2 Challenges

1. Most academic researchers are non-registered or non-published, which makes it interesting to replicate outcomes for evaluation and case studies for further study.
2. Numerous books and journals are published, however, there is no proper normal or model evidence like forecast horizons and hyper-parameters for accurate forecasting. In addition, numerous performance metrics and error calculations are employed. In some papers, it was detected that most papers altered their hyperparameters over trial and error.
3. We noticed no proper guidelines for the DL model's design, training, testing, and development. This lack of guidelines

is significantly interesting to further develop and compare with additional models. In some situations, an automated procedure or set of rules will assist in reproducing results for various models.
4. Some DL models can enhance the prediction, but there is a lack of practical implementations.
5. The enhancement of DL methods across various building types and appliances of DL representations to case studies is not focused much on lightning, sub-meter/components, etc.
6. Inspection is needed for understanding of the ambiguity of the DL representations, along with the formation of strategies for DL model expansions like computerization and hyperparameter range.
7. The formation of scalable DL-based representations is established and adjusted in a sensible method for practical executions across dissimilar schemes.
8. The expansion of strong models can deliver precise predictions in the event of variations in the act, sensor failure, etc.
9. Execution of the different DL-based methods is difficult for absolute appliances and control schemes like model analytical controllers, management forecast, optimization etc.

1.10 Conclusion

The main contribution of this chapter is to provide a systematic introduction to artificial intelligence techniques in the energy sector. Nowadays, renewable energy sources are combined with non-renewable sources to allow use in electric grids, but it causes various challenges because of its interference and instability. AI, ML, and DL-based approaches for energy forecasting play crucial roles in resolving these causes. While forecasting energy consumption and energy management, it is essential to determine an appropriate prediction model based on expected forecasting results. We provided a complete introduction to energy management and reasons for applying AI-based approaches in the energy sector. Then we discussed, ML and its role in the renewable sector and ongoing projects and application areas. We provided precise ML-based approaches with a categorization diagram and table for their advantages and application domain in the energy sector. Next, the integration of AI in smart grid management with load balancing challenges and applications is described. Finally, parameter selection for various ML models and future opportunities of ML in the energy sector with challenges and limitations are demonstrated.

References

1. Olufemi A. Omitaomu, and Haoran Niu, (2021). Artificial Intelligence Techniques in Smart Grid: A Survey, Smart Cities, 4, pp. 548–568.
2. Weslei Gomes de Sousaa, Elis Regina Pereira de Meloa, Paulo Henrique De Souza Bermejo, Rafael Araújo Sousa Fariasa, Adalmir Oliveira Gomesa, (2019), How and where is artificial intelligence in the public sector going? A literature review and research agenda, Government Information Quarterly, 36 (2019)101392.
3. Bim K. Bose, (2017). Artificial Intelligence Techniques in Smart Grid and Renewable Energy Systems— Some Example Applications, 2262 Proceedings of the IEEE Vol. 105, No. 11, November 2017.
4. Dasheng Lee, Chin-Chi Cheng, (2016). Energy savings by energy management systems: A review, Renewable and Sustainable Energy Reviews, 56 (2016), pp. 760-777.
5. Lanre Olatomiwa, Saad Mekhilef, M.S. Ismail, M. Moghavvemi. (2016). Energy management strategies in hybrid renewable energy systems: A review, Renewable and Sustainable Energy Reviews, 62, (2016), pp. 821-835.
6. Prince Waqas Khan, Yung-Cheol Byun, Sang-Joon Lee, Dong-Ho Kang, Jin-Young Kang and Hae-Su Park. (2020). Machine Learning-Based Approach to Predict Energy Consumption of Renewable and Nonrenewable Power Sources, Energies 2020, 13.
7. Kumar, A., Dubey, A.K., Ramírez, I.S., Muñoz del Río, A., Márquez, F.P.G. (2022). A Review and Analysis of Forecasting of Photovoltaic Power Generation Using Machine Learning. In: Xu, J., Altiparmak, F., Hassan, M.H.A., García Márquez, F.P., Hajiyev, A. (eds) Proceedings of the Sixteenth International Conference on Management Science and Engineering Management – Volume 1. ICMSEM 2022. Lecture Notes on Data Engineering and Communications Technologies, vol 144. Springer, Cham. https://doi.org/10.1007/978-3-031-10388-9_36
8. Dubey, A.K., Kumar, A., Ramirez, I.S., Marquez, F.P.G. (2022). A Review of Intelligent Systems for the Prediction of Wind Energy Using Machine Learning. In: Xu, J., Altiparmak, F., Hassan, M.H.A., García Márquez, F.P., Hajiyev, A. (eds) Proceedings of the Sixteenth International Conference on Management Science and Engineering Management – Volume 1. ICMSEM 2022. Lecture Notes on Data Engineering and Communications Technologies, vol 144. Springer, Cham. https://doi.org/10.1007/978-3-031-10388-9_35

9. Syed Saqib Ali and Bong Jun Choi. (2020). State-of-the-Art Artificial Intelligence Techniques for Distributed Smart Grids: A Review, Electronics 2020, 9, 1030.
10. Jung-Pin Lai, Yu-Ming Chang, Chieh-Huang Chen and Ping-Feng Pai. (2020). A Survey of Machine Learning Models in Renewable Energy Predictions, Appl. Sci. 2020, 10, 5975; doi:10.3390/app10175975.
11. Rathore, P.S., Chatterjee, J.M., Kumar, A. et al. Energy-efficient cluster head selection through relay approach for WSN. J Supercomput 77, 7649–7675 (2021). https://doi.org/10.1007/s11227-020-03593-4
12. Mondal, Misra, S. Patel, L.S. Pal, S.K. Obaidat, M.S. (2018). DEMANDS: Distributed energy management using noncooperative scheduling in smart grid. IEEE Syst. J. 2018.
13. Elsied, M. Oukaour, A. Gualous, H. Hassan, R. (2015). Energy management and optimization in microgrid system based ongreen energy. Energy 2015.
14. A. Dubey, S. Narang, A. Srivastav, A. Kumar, V. Díaz, Woodhead Publishing, Science Direct, Artificial Intelligence for Renewable Energy Systems. Paperback ISBN: 9780323903967
15. A. Dubey, S. Narang, A. Srivastav, A. Kumar, V. Díaz, Woodhead Publishing, Science Direct, a Visualization Techniques for Climate Change with Machine Learning and Artificial Intelligence. ISBN: 9780323997140
16. Leonori, S. Rizzi, A. Paschero, M. Mascioli. (2018). Microgrid Energy Management by ANFIS Supported by an ESN Based Prediction Algorithm. In Proceedings of the International Joint Conference on Neural Networks (IJCNN), Rio de Janeiro, Brazil, 8–13 July 2018.
17. De Santis, E. Rizzi, A. Sadeghian. (2017). Hierarchical genetic optimization of a fuzzy logic system for energy flows management in microgrids. Appl. Soft Comput. J. 2017.
18. Venayagamoorthy, G.K. Sharma, R.K. Gautam, P.K. Ahmadi, A. (2016). Dynamic Energy Management System for a Smart Microgrid. IEEE Trans. Neural Netw. Learn. Syst. 2016.
19. Ma, L. Liu, N. Zhang, J. Tushar, W. Yuen, C. (2016). Energy Management for Joint Operation of CHP and PV Prosumers Inside a Grid-Connected Microgrid: A Game Theoretic Approach. IEEE Trans. Ind. Inform. 2016.
20. Arcos-Aviles, D. Pascual, J. Guinjoan, F. Marroyo, L. Sanchis, P. Marietta, M.P. (2017). Low complexity energy management strategy for grid profile smoothing of a residential grid-connected microgrid using generation and demand forecasting. Appl. Energy 2017.
21. Aldaouab, I. Daniels, M. Hallinan, K. (2017). Microgrid cost optimization for a mixed-use building. In Proceedings of the 2017 IEEE Texas Power and Energy Conference (TPEC), College Station, TX, USA, 9–10 February 2017.
22. Liu, N. Yu, X. Wang, C. Wang, J. (2017). Energy Sharing Management for Microgrids with PV Prosumers: A Stackelberg Game Approach. IEEE Trans. Ind. Inform. 2017.

23. Nwulu, N.I. Xia, X. (2017). Optimal dispatch for a microgrid incorporating renewables and demand response. Renew. Energy 2017.
24. Mellit, A. Massi Pavan, A. Ogliari, E. Leva, S. Lughi, V. (2020). Advanced methods for photovoltaic output power forecasting: A review. Appl. Sci. 10, 487.
25. Zendehboudi, A. Baseer, MA Saidur, R. (2018) Application of support vector machine models for forecasting solar and wind energy resources: A review. J. Clean. Prod. 2018, 199, pp. 272–285.
26. Das, U.K. Tey, K.S. Seyedmahmoudian, M. Mekhilef, S. Idris, M.Y.I. Deventer, W.V. Horan, B. Stojcevski, A. (2018). Forecasting of photovoltaic power generation and model optimization: A review. Renew. Sustain. Energy Rev. 2018, 81, pp. 912–928.
27. Wang, H.Z. Lei, Z.X. Zhang, X. (2019). A review of deep learning for renewable energy forecasting. Energy Convers. Manag. 198, 111799.
28. Bermejo, J.F. Fernández, J.F.G. Polo, F.O. Márquez, A.C. (2019). A review of the use of artificial neural network models for energy and reliability prediction. a study of the solar PV, hydraulic and wind energy sources. Appl. Sci. 2019, 9, 1844.
29. Mosavi, A. Salimi, M. Ardabili, S.F. Rabczuk, T. Shamshirband, S. Varkonyi-Koczy, A.R. (2019). State of the art of machine learning models in energy systems, a systematic review. Energies 2019, 12, 1301.
30. Ahmed, A. Khalid, M. (2019). A review on the selected applications of forecasting models in renewable power systems. Renew. Sustain. Energy Rev. 2019, 100, pp. 9–21.
31. Pérez-Ortiz, M. Jiménez-Fernández, S. Gutiérrez, P.A. Alexandre, E. Hervás-Martínez, C. Salcedo-Sanz, S. (2016). A review of classification problems and algorithms in renewable energy applications. Energies, 9, 607.
32. Khare, V. Nema, S. Baredar, P. (2016). Solar–wind hybrid renewable energy system: A review. Renew. Sustain. Energy Rev 2016, 58, pp. 23–33.
33. Shayan, E.; Zare, V.; Mirzaee, I. (2018). Hydrogen production from biomass gasification; a theoretical comparison of using different gasification agents. Energy Convers Manag. 2018, 159, pp. 30–41.
34. Ozbas, E.E.; Aksu, D.; Ongen, A.; Aydin, M.A.; Ozcan, H.K. (2019). Hydrogen production via biomass gasification, and modeling by supervised machine learning algorithms. Int. J. Hydrogen Energy 2019, 44, pp. 17260–17268.
35. Elmaz, F.; Yücel, O.; Mutlu, A.Y. (2019). Predictive modeling of biomass gasification with machine learning-based regression methods. Energy 2019, 191, 116541.
36. Ali, M.; Prasad, R. (2019). Significant wave height forecasting via an extreme learning machine model integrated with improved complete ensemble empirical mode decomposition. Renew. Sustain. Energy Rev. 2019, 104, pp. 281–295.

37. Cornejo-Bueno, L.; Garrido-Merchán, E.C.; Hernández-Lobato, D.; Salcedo-Sanz, S. (2018). Bayesian optimization of a hybrid system for robust ocean wave features prediction. Neurocomputing 2018, 275, pp. 818–828.
38. Hamed, H.H.A. (2019). A novel approach for harmonic tidal currents constitutions forecasting using hybrid intelligent models-based on clustering methodologies. Renew. Energy 2019, 147, pp. 1554–1564.
39. Michael, D.; Thomas, A.; Adcock, A. (2018). Prediction of tidal currents using Bayesian machine learning. J. Ocean Eng. 2018, 158, pp. 221–231.
40. Gangwani, P.; Soni, J.; Upadhyay, H.; Joshi, S. (2020). A deep learning approach for modeling of geothermal energy prediction. Comput. Sci. Inf. Secur. pp 62–65.
41. Baruque, B.; Porras, S.; Jove, E.; Calvo-Rolle, J.L. (2019). Geothermal heat exchanger energy prediction based on time series and monitoring sensors optimization. Energy 2019, 171, pp. 49–60.
42. Zhang, X.; Peng, Y.; Xu, W.; Wang, B. (2017). An Optimal Operation Model for Hydropower Stations Considering Inflow Forecasts with Different Lead-Times. Water Resour. Manag. 2017, 33, pp. 173–188.
43. Nieto, P.G.; Garcia-Gonzalo, E.; Paredes-Sánchez, J.P.; Sánchez, A.B.; Fernández, M.M. (2019). Predictive modelling of the higher heating value in biomass torrefaction for the energy treatment process using machine-learning techniques. Neural Comput. Appl. 2019, 31, pp. 8823–8836.
44. Shi, S.; Patton, R.J.; Liu, Y. (2018). Short-term wave forecasting using gaussian process for optimal control of wave energy converters. IFAC PapersOnLine 2018, 51, pp. 44–49.
45. Diaz, M.B.; Kim, K.Y.; Kang, T.-H.; Shin, H.-S. (2018). Drilling data from an enhanced geothermal project and its pre-processing for ROP forecasting improvement. Geothermics 2018. 72, pp. 348–357.
46. Assouline, D.; Mohajeri, N.; Gudmundsson, A.; Scartezzini, J.-L. (2019) A machine learning approach for mapping the very shallow theoretical geothermal potential. Geotherm Energy 2019, 7, 19.
47. Li, T.; Qian, Z.; He, T. (2020). Short-term load forecasting with improved CEEMDAN and GWO-based multiple kernel ELM Complexity, 1209547.
48. Shi, H.; Xu, M.; Li, R. (2017). Deep learning for household load forecasting—A novel pooling deep RNN IEEE Trans. Smart Grid 2017, 9, pp. 5271–5280.
49. Moon, J.; Jung, S.; Rew, J.; Rho, S.; Hwang, E. (2020). Combination of short-term load forecasting models based on a stacking ensemble approach. Energy Build. 2020, 216, 109921.
50. He, Y.; Deng, J.; Li, H. (2017). Short-term power load forecasting with deep belief network and copula models. In Proceedings of the 2017 9th International Conference on Intelligent Human-Machine Systems and Cybernetics (IHMSC), Hangzhou, China, 26–27 August, Volume 1, pp. 191–194.

51. Hafeez, G.; Alimgeer, K.S.; Khan, I. (2020). Electric load forecasting based on deep learning and optimized by heuristic algorithm in smart grid. Appl. Energy 2020, 269, 114915.
52. Aly, H.H. (2020). A proposed intelligent short-term load forecasting hybrid models of ANN, WNN and KF based on clustering techniques for smart grid. Electr. Power Syst. Res.182, 106191.
53. Jiang, W.; Tang, H.; Wu, L.; Huang, H.; Qi, H. (2019). Parallel processing of probabilistic models-based power supply unit mid-term load forecasting with apache spark. IEEE Access 2019, 7, pp. 7588–7598.
54. Askari, M.; Keynia, F. (2019). Mid-term electricity load forecasting by a new composite method based on optimal learning MLP algorithm. IET Gener. Transm. Distrib. 2019, 14, pp. 845–852.
55. Rai, S.; De, M. (2021). Analysis of classical and machine learning based short-term and mid-term load forecasting for smart grid. Int. J. Sustain. Energy 2021, pp. 1–19.
56. Gul, M.J.; Urfa, GM; Paul, A.; Moon, J.; Rho, S.; Hwang, E. (2021). Mid-term electricity load prediction using CNN and Bi-LSTM. J. Supercomput. 2021, pp. 1–17.
57. Dudek, G.; Pełka, P.; Smyl, S. (2021). A Hybrid Residual Dilated LSTM and Exponential Smoothing Model for Midterm Electric Load Forecasting. IEEE Trans. Neural Networks Learn. Syst. 2021.
58. Ali, D.; Yohanna, M.; Ijasini, P.M.; Garkida, M.B. (2018). Application of fuzzy–Neuro to model weather parameter variability impacts on electrical load based on long-term forecasting. Alex. Eng. J. 2018, 57, pp. 223–233.
59. Zheng, J.; Xu, C.; Zhang, Z.; Li, X. (2017). Electric load forecasting in smart grids using long-short-term-memory based recurrent neural network. In Proceedings of the 2017 51st Annual Conference on Information Sciences and Systems (CISS), Baltimore, MD, USA, 22–24 March 2017; pp. 1–6.
60. Dong, M.; Grumbach, L. (2019). A hybrid distribution feeder long-term load forecasting method based on sequence prediction. IEEE Trans. Smart Grid 2019, 11, pp. 470–482.
61. Bouktif, S.; Fiaz, A.; Ouni, A.; Serhani, M.A. (2020). Multi-sequence LSTM-RNN deep learning and metaheuristics for electric load forecasting. Energies 2020, 13, 391.
62. Sangrody, H.; Zhou, N.; Tutun, S.; Khorramdel, B.; Motalleb, M.; Sarailoo, M. (2018). Long term forecasting using machine learning methods. In Proceedings of the 2018 IEEE Power and Energy Conference at Illinois (PECI), Champaign, IL, USA, 22–23 February 2018; pp. 1–5.
63. Ashraf, S.M.; Gupta, A.; Choudhary, D.K.; Chakrabarti, S. (2017). Voltage stability monitoring of power systems using reduced network and artificial neural network. Int. J. Electr. Power Energy Syst, 87, pp. 43–51.
64. Amroune, M.; Bouktir, T.; Musirin, I. (2018). Power system voltage stability assessment using a hybrid approach combining dragonfly optimization algorithm and support vector regression. Arab. J. Sci. Eng, 43, pp. 3023–3036.

65. Mohammadi, H.; Khademi, G.; Dehghani, M.; Simon, D. (2018). Voltage stability assessment using multi-objective biogeography-based subset selection. Int. J. Electr. Power Energy Syst. 2018, 103, pp. 525–536.
66. Liu, S.; Shi, R.; Huang, Y.; Li, X.; Li, Z.; Wang, L.; Mao, D.; Liu, L.; Liao, S.; Zhang, M. (2021). A data-driven and data-based framework for online voltage stability assessment using partial mutual information and iterated random forest. Energies 2021, 14, 715.
67. Shi, Z.; Yao, W.; Zeng, L.; Wen, J.; Fang, J.; Ai, X.; Wen, J. (2020). Convolutional neural network-based power system transient stability assessment and instability mode prediction. Appl. Energy 2020, 263, 114586.
68. Xiao, H.; Fabus, S.; Su, Y.; You, S.; Zhao, Y.; Li, H.; Zhang, C.; Liu, Y.; Yuan, H.; Zhang, Y. (2020). Data-Driven Security Assessment of Power Grids Based on Machine Learning Approach; Technical Report; National Renewable Energy Lab.(NREL): Golden, CO, USA.
69. Kamari, N.; Musirin, I.; Ibrahim, A.; Halim, S. (2019). Intelligent swarm-based optimization technique for oscillatory stability assessment in power system. IAES Int. J. Artif. Intell. 2019, 8, 342.
70. Dehghani, M.; Riahi-Madvar, H.; Hooshyaripor, F.; Mosavi, A.; Shamshirband, S.; Zavadskas, E.K.; Chau, K.-W. (2019). Prediction of hydropower generation using grey wolf optimization adaptive neuro-fuzzy inference system. Energies 2019, 12, 289.
71. Fan, J.; Wu, L.; Ma, X.; Zhou, H.; Zhang, F. (2020). Hybrid support vector machines with heuristic algorithms for prediction of daily diffuse solar radiation in air-polluted regions. Renew. Energy 2020, 145, pp. 2034–2045.
72. Demircan, C.; Bayrakçı, H.C.; Keçeba¸s, A. (2020). Machine learning-based improvement of empiric models for an accurate estimating process of global solar radiation. Sustain. Energy Technol. Assess. 2020, 37, 100574.
73. Li, LL; Zhao, X.; Tseng, M.L.; Tan, R.R. (2020). Short-term wind power forecasting based on support vector machine with improved dragonfly algorithm. J. Clean. Prod. 2020, 242, 118447.
74. García Nieto, P.J.; García-Gonzalo, E.; Sánchez Lasheras, F.; Paredes-Sánchez, J.P.; Riesgo Fernández, P. (2019). Forecast of the higher heating value in biomass torrefaction by means of machine learning techniques. J. Comput. Appl. Math. 2019, 357, pp. 284–301.
75. Wu, C.; Wang, J.; Chen, X.; Du, P.; Yang, W. (2019). A novel hybrid system based on multi-objective optimization for wind speed forecasting. Renew. Energy 2019, 146, pp. 149–165.
76. Lin, G.-Q.; Li, L.-L.; Tseng, M.-L.; Liu, H.-M.; Yuan, D.-D.; Tan, R. (2020). An improved moth-flame optimization algorithm for support vector machine prediction of photovoltaic power generation. J. Clean. Prod. 2020, 253, 119966.
77. Lin, K.P.; Pai, P.F.; Ting, Y.J. (2019). Deep belief networks with genetic algorithms in forecasting wind speed. IEEE Access 2019, 7, pp. 99244–99253.

78. Yu, R.; Liu, Z.; Li, X.; Lu, W.; Ma, D.; Yu, M.; Wang, J.; Li, B. (2019). Scene learning: Deep convolutional networks for wind power prediction by embedding turbines into grid space. Appl. Energy. 2019, 238, pp. 249–257.
79. Cornejo-Bueno, L.; Garrido-Merchán, E.C.; Hernández-Lobato, D.; Salcedo-Sanz, S. (2018). Bayesian optimization of a hybrid system for robust ocean wave features prediction. Neurocomputing 2018, 275, pp. 818–828.
80. Papari, B.; Edrington, C.S.; Kavousi-Fard, F. (2017). An effective fuzzy feature selection and prediction method for modeling tidal current: A case of Persian Gulf. IEEE Trans. Geosci. Remote Sens. 2017, 55, pp. 4956–4961.

2

Artificial Neural Network Process Optimization for Predicting the Thermal Properties of Biomass: Recent Advances and Future Challenges

S. Dayana Priyadharshini[1] and M. Arvindhan[2]*

[1]*Independent Researcher, Greater Noida, India*
[2]*Department of Computer Science and Engineering, Galgotias University, Greater Noida, India*

Abstract

Biomass is one of the alternative energy sources that is most plentiful and dependable. Hence, comprehensive knowledge of its inherent potential is imperative to investigate the highest potential of biomass. However, the most scientific techniques involve extremely advanced and costly instrumentation. The progress of artificial intelligence and block chain technology unlocks future prospective towards thermal marketing accuracy. The Artificial Neural Network (ANN) is a crucial instrument for improving biomass energy prediction research. This study emphasizes the steps in the modeling of ANN and use of ANN in the forecasting of biomass thermal values. Several research gaps in the present state of the investigation on ANN in terms of biomass and guidance for additional research are identified.

Keywords: ANN, biomass, renewable energy, energy consumption, thermal properties

Abbreviations

ANN   Artificial Neural Network
HHV   High Heating Value

Corresponding author: saroarvindmster@gmail.com

Abhishek Kumar, Pramod Singh Rathore, Ashutosh Kumar Dubey, Arun Lal Srivastav, T. Ananth Kumar and Vishal Dutt (eds.) Sustainable Management of Electronic Waste, (47–66) © 2024 Scrivener Publishing LLC

MBPD   Million Barrels per Day
FC   Fixed Carbon
C   Carbon
H   Hydrogen
O   Oxygen
S   Sulphur
CO_2 Carbon Dioxide
CO   Carbon Monoxide
A   Ash
M   Moisture
VM   Volatile Matter
BR   Bayesian Regularization
LCB   Lignocellulosic Biomass
LHV   Low Heating Value

2.1 Introduction

In today's world, social systems seek to expand their reliance on renewable and sustainable energy sources in order to minimize their reliance on fossil fuels and reduce environmental pollution. Biomass is often regarded as the only carbon-neutral renewable alternative to fossil fuels. Nevertheless, biomass is characterized by a high degree of physicochemical diversity, making energy consumption difficult [1]. Due to the sensitivity of bioenergy transformation procedures to significant fluctuation in feed material characteristics (MC) and the requirement for continuous control, a non-destructive technique capable of measuring biomass in legitimate time is required [2], [3]. Thermochemical processes are among the most sophisticated methods for using biomass's energy density. Coal supplies are only expected to be in use for the next 150 years, however oil and gas supplies are only expected to last about 52 years, assuming the rate of electricity demand remains constant. Furthermore, the globe has been rapidly globalizing, leading to a lower real-life expectancy for non-renewable energy than previously predicted. As per the guidelines of United States Energy Information Administration, worldwide energy utilization will rise by 28% by 2040 [4].

Among the most efficient ways of using biomass' energy content is thermochemical procedures. The earliest method for pyrolysis of biomass is combustion. Pyrolysis, vapor reformation, and gasification are further methods. The synthesis (design and operation) of biomass energy generating or energy recovery systems relies largely on certain biomass properties.

Heating value is one of these essential biomass features. Heating value is normally shown as a lower (LHV) or higher heating value (HHV). Figure 2.1 explicitly depicts global energy production and utilization.

It is estimated that 96.2 MBPD of petroleum and liquid fuels were used globally in May 2021, up 11.9 MBPD from May 2020, but down 3.7 MBPD from May 2019. It is predicted that overall utilization of petroleum and liquid fuels would be approximately 97.7 MBPD on average throughout 2021, an increase of 5.4 MBPD over 2020. Global consumption of petroleum and liquid fuels is expected to grow by 3.6 MBPD to an estimate of 101.3 MBPD in 2022 (Figure 2.2a & 2.2b). OPEC crude oil output is expected to average 26.9 MBPD in 2021 and 28.7 MBPD in 2022. According to the projection, OPEC crude oil output would increase from 25.0 MBPD in April to an average of 28.0 MBPD in 3Q21. Our forecast of increased OPEC production is based on the premise that OPEC would boost output by about 1 MBPD in June and July in response to growing global oil demand and seasonal improvements in oil consumption for power generation by some OPEC members. Additionally, it is predicated on the expectation that Iran's crude oil output would keep growing this year. Although sanctions targeting Iran's crude oil exports remain in effect, crude oil shipments and production from Iran are higher from the majority of 2020, as per ClipperData, LLC ("Short-Term Energy Outlook - U.S. Energy Information Administration (EIA)," accessed on 26.6.2021).

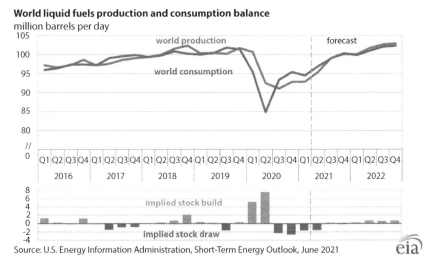

Figure 2.1 Global energy production, utilization, and forecast ("Short-Term Energy Outlook - U.S. Energy Information Administration (EIA)," accessed on 26.6.2021).

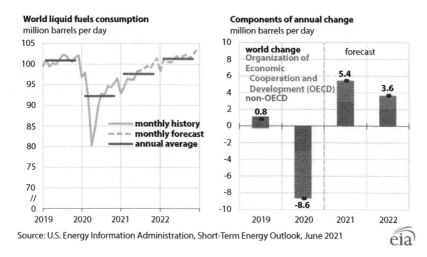

Figure 2.2a Utilization of liquid fuels globally and its forecast.

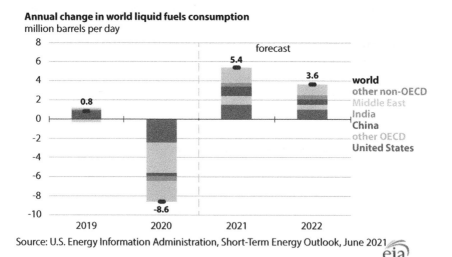

Figure 2.2b Utilization of liquid fuels globally and its forecast.

As a result of growing energy consumption and ecological concern, circulated and scattered power production is gaining importance. Conventional power plants are consolidated and frequently need the transmission of electric energy across great distances. On the contrary, distributed power production from renewable energy sources including solar and wind energy, biomass, and bio-gas are becoming more and more essential for the electric power distribution system [5], [6]. Gasification is a potential technique because it effectively converts solid biomass to combustible gases, such as H_2, CH_4, CO_2, hydrocarbons, and char. The resulting gaseous combinations, syngas, could be utilized in a straight line as a gaseous fuel or additionally managed to create power and heat. Additionally, gasification transforms low-value feedstocks into lucrative energy sources by enabling the generation of energy from non-traditional feedstocks including waste materials produced in various fields including agricultural, forest, poultry and municipal solid waste [7], [8].

A small number of mathematical models for kinetic analysis analyze potential similar and diverse reactions that have been created to assess the influence of operational parameters on the composition and calorific value of syngas produced in the gasification process. However, because such mathematical models involve transport viz., heat, mass, and momentum and kinetic equations, their formulation requires considerable work. As a result, solving these models is a lengthy and iterative procedure. CFD may be used to simulate 'different flow systems,' viz., combustion chamber and chemical reactors, as well as other procedures involving blending. As a result, Computational Fluid Dynamics and thermodynamic equilibrium models by means of a restricted type of process was created for gasification of biomass, although such models are more suited to represent a fluidized type of reactor than a fixed bed reactor [9].

The utilization of lignocellulosic biomass (LCB) as a source of efficient renewable energy source in energy generation needs a thorough understanding of its characteristics. The energy content of the lignocellulosic biomasses is determined directly by the level of heating value (HHV) of the biomass. Moreover, the experimental technique for these studies necessitates the use of extremely complicated and expensive instruments, as well as a constant electrical supply. AI technology has advanced tremendously and this has created a significant opportunity for the research of biomass characteristics with the goal of optimizing bioresource usage [10]. Artificial intelligence is being used for a wide range of disciplines of research, such as supply chain, risk evaluation, the fields of health and social care, business analytics, and energy modeling [11].

2.2 AI Technology and Its Application on Renewable Energy

Artificial intelligence has changed drastically how several sectors, including energy, produce and distribute their items and solutions. According to recent Roland Berger research, AI has the potential to increase utility business efficiency by fivefold in the next five years, yet more than three-quarters of the firms surveyed said that they had no urgent plans to set their sights on AI technology. Nonetheless, during the last several years, the money spent on digital power generation and software has grown at a rate of over 20% per year, hitting $47 billion in 2016. In 2016, the overall global capital in gas-fired power generation was $34 billion, but the money spent on that venture was nearly 40% greater. This aims to estimate how much of the energy industry has indeed been built using AI technology [12].

2.3 Bioenergy and ANN

Lignocellulosic biomass and ANNs are used to forecast a variety of processes. ANNs have been effectively utilized in a variety of disciplines, including mathematics, engineering, healthcare, marketing, and neuroscience. Additionally, AI has been utilized in the field of bioenergy, as described by [2]. Among its several advantages, it is capable of handling noisy and inadequate data, which is typical of the majority of renewable energy data. It will integrate hidden data in a huge pile, which may have a major impact on the model and can manage a large amount of data while being adaptable to parameter changes. Artificial neural networks may perform difficult tasks such as estimations, modeling, recognition, optimization, prediction, and monitoring once they have learned the pattern [6]. Numerous investigators have addressed the issue of overfitting and under fitting related to unsystematic selection of hidden nodes. However, such a random selection is not incorporated in the majority of biomass models. The majority of models created for biomass prediction are empirical in nature. The detailed survey on the application of ANN in the prediction of biomass characterization is tabulated in Table 2.1.

2.4 ANN Model Development

Information as data is the fuel required to build any type of model. Two types of data are necessary for the construction of a model:

- Predictor Data: (Alternately referred to as analyst variables) This data format is utilized to do predictive analytics. In this example, the predictor variables are proximal values like

Table 2.1 Various applications of ANN in biomass characterization.

S. no.	Types of material and analysis	Input variables	Output variables	Types of method used in ANN	Reference
1	Woody Biomass Gasification Process	Proximate analysis	CO, CO_2, CH_4, H_2, HHV, calorific value	Binary least squares support vector machine	[13]
2	Biomass	Light intensity, photoperiod, temperature, and initial pH	Biomass of *Scenedesmus* sp.	Feed forward network - error back propagation	[14]
3	Biomass	Proximate analysis	HHV	Bayesian Regulated back propagation SVR-Gaussian-kernel function	[7]
4	Bamboo Biomass	Proximate and ultimate analysis	HHV	Multilayer perceptron (MLP) ANN	[15]
5	Biomass	Fixed carbon (FC), volatile matters (VM), and ash content	HHV	Levenberg– Marquardt	[16]
6	Wheat Straw	Temperature, (flow rate of water) F_w, (flow rate of biomass) F_b	H_2, CH_4, CO	Feed forward backpropagation	[11]
7	Various Coal and Biomass Samples	FC, M, VM, A, C, H, N, O, and S	The elemental composition of coal and biomass	MLP-FF-BP	[3]

(*Continued*)

Table 2.1 Various applications of ANN in biomass characterization. (*Continued*)

S. no.	Types of material and analysis	Input variables	Output variables	Types of method used in ANN	Reference
8	Lignocellulosic Forest Residue and Olive Oil Residue	Temperature (°C) and Residue (%)	Activation energy	Transfer function: tansig, logsig Training algorithm: TRAINLM	[17]
9	Algal Nanomaterials	SBET, VPor, Dave, Vmic, Vmes, ID/IG, Rs, N (%), CL (%), rL/D)	Specific capacitance (Cp)	Levenberg–Marquart back propagation algorithm	[18]
10	Biomass	Proximate analysis data	HHV	MLP-ANN	[1], [19]
11	Biomass	C, H, O, S, N	HHV	MLP-ANN	[20]
12	Various Biomass	Proximate analysis	HHV	ANN and empirical correlation models	[21]
13	Performance Evaluation of Gasification System Efficiency	A, MC, VM, C, H, N, O, S, and LHV	CO CO_2, H_2, CH_4, gas yield, LHV of gas, CGE, and CCE.	Levenberg-Marquardt (LM) backpropagation, Bayesian regularization (BR)	[22]
14	Biomass	Extraction temperature, time, particle size range, and solid loading	Lignin extraction	Levenberg-Marquardt (LM) backpropagation,	[8]

carbon content, volatile matter, moisture levels, M, and ash content.
- Target Data: This is the function or behavior for which a prediction is being made. HHV is the goal factor in this case because it influences the enthalpy changes of LCB.

Correlations between predictive and target parameters are established using a suitable mathematical or statistical approach. Once a relation between the resultant models is established, it may be used to calculate the objective function of a new instance when the predictor data is available [21], [24].

2.4.1 Methodologies Used for Target Model

Predictive modeling is an analytical approach that uses historical and present data to predict a future event or behavior. Additionally, it may be described as a process that employs data mining and probability to forecast the future. Each model has a number of predictor factors that may have an effect on the future projection. The initial stage is data collection, followed by model design. When the data has been obtained, a model is constructed, which may be linear, multiple regression, or AI. There are several ways to develop a model, but their usefulness is dependent on the size and amount of the data. Below Table 2.2 describes the models in ANN. Figure 2.3 is described below which shows the diagramatic representation of ANN. Table 2.3 described the types of algorithm used in ANN. Generally, models may be categorized into six main categories, which are as follows:

Table 2.2 Types of models in ANN.

Types of model	References
Linear	[23]
Support Vector Machine (SVM)	[23]
Kernel Nearest Neighbor (K-NN)	[23]
Multiclass Support Vector Machine (MCSVM)	[25]
Expert Systems (ES)	[26]
Decision Tree (alternatively referred to as CART)	[27]
Cluster Model	[28]

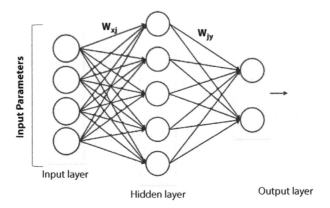

Figure 2.3 Diagrammatic representation of ANN.

Table 2.3 Types of algorithms used for training the data in ANN.

Training algorithm	Characteristics	Reference
Qasi-Newton Method	The loss function derivatives are used to construct an approximation to the inverted Hessian in all iterations.	[29]
Gradient Descent Method	Although this isn't path-driven, it is population-driven. The iteration time is excessive and is concerned with finding stuck in the local minima.	[30], [31]
Newton's Method	The computational cost is high. More precise guidance is needed to improve training results with a 2nd derivative of the loss function.	[32]
Gauss-Newton	Use the squared Jacobian to simplify the Hessian computation. It should not have to keep information from past iterations in order to determine the direction of the current iteration. The simpler version of the Hessian calculation is to use squared Jacobian. This doesn't require previously-calculated values to determine the update direction.	[31]

(Continued)

Table 2.3 Types of algorithms used for training the data in ANN. (*Continued*)

Training algorithm	Characteristics	Reference
	If you struggle with indefinite Hessian, the training will be stopped and a significant increase in training progress was made. It also acquires important training skills.	
Scaled Conjugate Gradient	It is more rapid than the steepest decline.	[31]
Levenberg–Marquardt	Mathematically, it is possible to determine the gradient and the Jacobian matrix, which leverages smaller- and medium-sized data efficiently at a high rate of convergence having to stay within the local minima	[33]
Modified Levenberg–Marquardt Method	With appropriate convergence there is less of an oscillation when it comes to learning. There is also a lowered training iteration count.	[33]
Approximate Greatest Descent	It uses 2-phase spherical optimization to get trajectory by using control theory. Relative step length is regulated by the built spherical areas and the step size is adjusted depending on that. The adaptive approach is also useful.	[31]
Genetic Algorithm	For suitable execution, it could be changed to include the mutation operation.	[34], [35]

2.4.2 Various Stages Involved in ANN Modelling

Collection of data and attribute selection, noises, inadequate data, insufficient sets of data, as well as how to separate the data source for learning, training, and verification are only a few of the challenges researchers might face while doing ANN modeling work [36].

1) Initial Development of the Model: Data gathering, standardization of data, and data partition into training and test datasets are all requirements

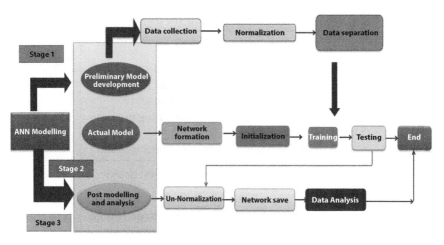

Figure 2.4 Stages involved in ANN model development.

of machine learning. But, in a situation in which the software is created to randomly split the data, the data partition might as well be considered part of the model creation process.

2) Actual Development of the Model: The model is constructed when training, testing is performed, and error functions are generated to evaluate the correctness of the models established.

3) Post Modeling: Converting the data back to its initial condition before normalization and storing the network, followed by a significant discussion, will allow the GUI to be built. Figure 2.4 shows the stages involved in ANN model development.

2.4.3 Important Terms Used in ANN Modeling

a) Algorithms: Neural Networks are provided with an algorithm that establishes a series of rules in order for the system to learn without human interference. An advanced algorithm that directs NN to do anything smartly may occur as a result of programming commands or phrases. To cluster, classify, suggest, or regret data, algorithms may be created.

b) Classification: This may be characterized as a method of data organization that places data based on how similar or close, they appear. Using this technique, the NN allows the class allocation to be based on a training model's results [37].

c) Clustering: Clustering is a classifying technique that emphasizes the importance of group similarity in multivariate data sets obtained from different disciplines. The primary key to a successful clustering method is to offer the maximum possible similarity between the items it contains. This possible link may be based on the data set's distance. Distance variables are commonly used in clustering algorithms to find similarities or differences between two items. Hierarchical clustering classifies data into hierarchical groups, while partitioning classifies data into region-based segmentation groups.

d) Training Data: Training data is data that is utilized to assess input variables or illustrate how well the data distribution is altered as fresh data sets are introduced to the model. Training typically seeks to conform to a steady value in the algorithm's perspective, such as precision or a cost function. Convergence is the indicator that additional data does not further increase performance [31], [33].

e) Testing: Sometimes testing data is also known as validation data. Such data may be used to evaluate the model's efficacy without using it. Model training should be stopped while this data is being delimited clearly.

f) Hidden Layer: This layer consists of artificial neurons that are programmed to add weight values to their input to generate an output via an activation function. In this case, the "hidden layer" is more accurately described as the "cockpit" of ANN because that includes the transference of weight, bias, and training functions.

g) Output Layer: An AI network has an output layer that consists of the last set of neurons in the system that provides a particular set of outputs for the program.

h) Overfitting: A primary issue seen in NN training is overfitting. In this situation, the training data and assumptions do not support generalization. To put it another way, the algorithms are successful in producing accurate results, but not efficient when a fresh dataset is provided to them. We believe that overfitting can be exacerbated or minimized by the incidence of noise, for example the noise that's found in municipal waste-based biomass feedstock [7].

i) Underfitting: This is a characteristic of a model that cannot use both the training data to form an accurate model and be of any use in classifying the newly given dataset. In order to fix underfitting and overfitting, utilize regularization algorithms to optimize the amount of hidden layers and training time [39].

j) The generalization objective is usually to apply a hypothesis derived from examples to new data. Leveraging previously acquired information isn't difficult. On the other hand, the algorithm must be capable of utilizing the new data. In order to accomplish this generalization, facts alone are not sufficient. It is nearly impossible to determine how well a learner will do based on simple assertions like the training data unless the learner works something beyond random predictions.

k) When it comes to building a system or setting criteria for a system, people frequently assume various things, whether these assumptions are brought to light or remain implicit. If the assumptions under which the analysis and verification were done are violated, the results of the analysis and verification might be invalidated. While attempting to prevent an unintentional violation, make sure that the assumption used throughout the model's construction is explicitly specified [22].

l) Normalization: Data science normalization is used to bring together disparate data components into a coherent whole. The key priority for application developers is reducing and removing of duplication. Data normalization is the exact reverse of un-normalization. For the analysis and interpretation, the data are restored to their original format [38], [40].

2.5 Future Scope of ANN-Bioenergy

It is eminent that the partition of a dataset will have a major effect on the ANN model's precision and enactment. Therefore, more research should be conducted to determine the appropriateness and influence of existing dataset division techniques on the effectiveness of the regression analysis for thermal efficiency and chemical compositions [20], [39]. A significant design variable is the training algorithm, which must be used throughout the training phase. Because Levenberg-Marquardt has been commonly used to forecast heating values through a small number of other tested algorithms, alternative training methods must be investigated in light of the requirement to deal with noise and outliers which are frequently associated with certain categories of biomass data. A comparison analysis of several training algorithms should be conducted to ascertain its generalizability [41], [42]. The number of nodes and hidden layers is mostly determined by the guideline for predicting biomass characteristics. To create or apply the standard technique for selecting the optimal nodes in the hidden layer, the standard method might be developed or implemented. The selection of hidden neurons and nodes is critical since it has an effect on the model's intricacy, predictive capabilities, and training time. To a more general point, the majority of models presented lacked comprehensive information that may aid with model replication. The investigation determination should be directed towards concentrating on the hidden layers, since it is critical to the model's success [43], [44]. Therefore, additional research is required that might potentially result in the creation of global correlations capable of reliably predicting the characteristics of biomass derived from a variety of sources. Model development elements including data division, network design, and model improvement for tiny datasets would advance the estimation of thermal properties for biomass.

2.6 Conclusion

The Artificial Neural Network has eased the modeling of biomass thermal properties and offers several benefits in terms of increased prediction accuracy compared to prior approaches. The purpose of this study was to examine contemporary prediction methods for the analysis of biomass thermal characteristics. The models were constructed using data from a variety of sources. The data utilized to build these models were mostly derived from ultimate, proximal, and thermogravimetric analyses. Extensive use of ANNs will ultimately result in the development of strong dedicated software capable of legitimate prediction of biofuel thermal properties.

References

[1] J. O. Ighalo, A. G. Adeniyi, and G. Marques, "Application of artificial neural networks in predicting biomass higher heating value: an early appraisal," *Energy Sources, Part A: Recovery, Utilization and Environmental Effects*. 2020.

[2] J. Xing, K. Luo, H. Wang, Z. Gao, and J. Fan, "A comprehensive study on estimating higher heating value of biomass from proximate and ultimate analysis with machine learning approaches," *Energy*, vol. 188, 2019.

[3] A. I. Lawal, A. E. Aladejare, M. Onifade, S. Bada, and M. A. Idris, "Predictions of elemental composition of coal and biomass from their proximate analyses using ANFIS, ANN and MLR," *Int. J. Coal Sci. Technol.*, vol. 8, no. 1, 2021.

[4] S. L. Y. Lo, B. S. How, W. D. Leong, S. Y. Teng, M. A. Rhamdhani, and J. Sunarso, "Techno-economic analysis for biomass supply chain: A state-of-the-art review," *Renewable and Sustainable Energy Reviews*, vol. 135. 2021.

[5] "Short-Term Energy Outlook - U.S. Energy Information Administration (EIA)." [Online]. Available: https://www.eia.gov/outlooks/steo/report/global_oil.php. [Accessed: 26-Jun-2021].

[6] F. M. Monticeli, R. M. Neves, and H. L. Ornaghi Júnior, "Using an artificial neural network (ANN) for prediction of thermal degradation from kinetics parameters of vegetable fibers," *Cellulose*, vol. 28, no. 4, pp. 1961–1971, Mar. 2021.

[7] M. U. Ahmed et al., "A machine learning approach for biomass characterization," in *Energy Procedia*, 2019, vol. 158.

[8] T. Rashid et al., "Enhanced lignin extraction and optimisation from oil palm biomass using neural network modelling," *Fuel*, vol. 293, 2021.

[9] M. Das Ghatak and A. Ghatak, "Artificial neural network model to predict behavior of biogas production curve from mixed lignocellulosic co-substrates," *Fuel*, vol. 232, 2018.

[10] B. A. Souto, V. L. C. Souza, M. T. Bitti Perazzini, and H. Perazzini, "Valorization of acai bio-residue as biomass for bioenergy: Determination of effective thermal conductivity by experimental approach, empirical correlations and artificial neural networks," *J. Clean. Prod.*, vol. 279, p. 123484, Jan. 2021.

[11] A. Mohammadidoust and M. R. Omidvar, "Simulation and modeling of hydrogen production and power from wheat straw biomass at supercritical condition through Aspen Plus and ANN approaches," *Biomass Convers. Biorefinery*, 2020.

[12] M. I. Jahirul et al., "Investigation of correlation between chemical composition and properties of biodiesel using principal component analysis (PCA) and artificial neural network (ANN)," *Renew. Energy*, vol. 168, pp. 632–646, May 2021.

[13] A. Y. Mutlu and O. Yucel, "An artificial intelligence based approach to predicting syngas composition for downdraft biomass gasification," *Energy*, vol. 165, 2018.

[14] M. Nayak, G. Dhanarajan, R. Dineshkumar, and R. Sen, "Artificial intelligence driven process optimization for cleaner production of biomass with co-valorization of wastewater and flue gas in an algal biorefinery," *J. Clean. Prod.*, vol. 201, 2018.

[15] S. Pattanayak, C. Loha, L. Hauchhum, and L. Sailo, "Application of MLP-ANN models for estimating the higher heating value of bamboo biomass," *Biomass Convers. Biorefinery*, 2020.

[16] S. Hosseinpour, M. Aghbashlo, M. Tabatabaei, and M. Mehrpooya, "Estimation of biomass higher heating value (HHV) based on the proximate analysis by using iterative neural network-adapted partial least squares (INNPLS)," *Energy*, vol. 138, pp. 473–479, Nov. 2017.

[17] Ö. Çepelioğullar, İ. Mutlu, S. Yaman, and H. Haykiri-Acma, "Activation energy prediction of biomass wastes based on different neural network topologies," *Fuel*, vol. 220, 2018.

[18] J. Wang, Z. Li, S. Yan, X. Yu, Y. Ma, and L. Ma, "Modifying the microstructure of algae-based active carbon and modelling supercapacitors using artificial neural networks," *RSC Adv.*, vol. 9, no. 26, 2019.

[19] A. Dashti, A. S. Noushabadi, M. Raji, A. Razmi, S. Ceylan, and A. H. Mohammadi, "Estimation of biomass higher heating value (HHV) based on the proximate analysis: Smart modeling and correlation," *Fuel*, vol. 257, 2019.

[20] A. Darvishan, H. Bakhshi, M. Madadkhani, M. Mir, and A. Bemani, "Application of MLP-ANN as a novel predictive method for prediction of the higher heating value of biomass in terms of ultimate analysis," *Energy Sources, Part A Recover. Util. Environ. Eff.*, vol. 40, no. 24, 2018.

[21] I. Estiati, F. B. Freire, J. T. Freire, R. Aguado, and M. Olazar, "Fitting performance of artificial neural networks and empirical correlations to estimate higher heating values of biomass," *Fuel*, vol. 180, 2016.

[22] M. Ozonoh, B. O. Oboirien, A. Higginson, and M. O. Daramola, "Performance evaluation of gasification system efficiency using artificial neural network," *Renew. Energy*, vol. 145, 2020.

[23] L. Fehrmann, A. Lehtonen, C. Kleinn, and E. Tomppo, "Comparison of linear and mixed-effect regression models and a k-nearest neighbour approach for estimation of single-tree biomass," *Can. J. For. Res.*, vol. 38, no. 1, pp. 1–9, Jan. 2008.

[24] J. Zhang, H. Tian, D. Wang, H. Li, and A. M. Mouazen, "A Novel Approach for Estimation of Above-Ground Biomass of Sugar Beet Based on Wavelength Selection and Optimized Support Vector Machine."

[25] C. Yin et al., "Auto-classification of biomass through characterization of their pyrolysis behaviors using thermogravimetric analysis with support vector machine algorithm: case study for tobacco," *Biotechnol. Biofuels*, vol. 14, no. 1, p. 106, Dec. 2021.

[26] J. Mohd Ali, M. A. Hussain, M. O. Tade, and J. Zhang, "Artificial Intelligence techniques applied as estimator in chemical process systems - A literature survey," *Expert Systems with Applications*, vol. 42, no. 14. Elsevier Ltd, pp. 5915–5931, 15-Aug-2015.

[27] L. Torre-Tojal, J. M. Lopez-Guede, and M. M. Graña Romay, "Estimation of forest biomass from light detection and ranging data by using machine learning," in *Expert Systems*, 2019, vol. 36, no. 4, p. e12399.

[28] V. Vijay, P. M. V. Subbarao, and R. Chandra, "An evaluation on energy self-sufficiency model of a rural cluster through utilization of biomass residue resources: A case study in India," *Energy Clim. Chang.*, vol. 2, p. 100036, Dec. 2021.

[29] B. Karaçalı, "An efficient algorithm for large-scale quasi-supervised learning," *Pattern Anal. Appl.*, vol. 19, no. 2, pp. 311–323, May 2016.

[30] F. Dkhichi and B. Oukarfi, "Neural Network Training By Gradient Descent Algorithms: Application on the Solar Cell," *Int. J. Innov. Res. Sci. Eng. Technol.*, vol. 03, no. 08, pp. 15696–15702, Aug. 2014.

[31] Hong Hui Tan and King Hann Lim, "Review of second-order optimization techniques in artificial neural networks backpropagation," *IOP Conf. Ser. Mater. Sci. Eng.*, vol. 495, p. 012003, 2019.

[32] D. Thomson and L. Corte, "Newton's Method Backpropagation for Complex-Valued Holomorphic Neural Networks: Algebraic and Analytic Properties Recommended Citation "Newton's Method Backpropagation for Complex-Valued Holomorphic Neural Networks: Algebraic and Analytic," 2014.

[33] N. Mahmudah, A. Priyadi, A. L. Setya Budi, and V. L. Budiharto Putri, "Photovoltaic Power Forecasting Using Cascade Forward Neural Network Based On Levenberg-Marquardt Algorithm," 2021, pp. 115–120.

[34] K. Sudha, N. Kumar, and P. Khetarpal, "GA-ANN hybrid approach for load forecasting," *J. Stat. Manag. Syst.*, vol. 23, no. 1, pp. 135–144, Jan. 2020.

[35] F. Ahmad, N. A. Mat-Isa, Z. Hussain, R. Boudville, and M. K. Osman, "Genetic Algorithm - Artificial Neural Network (GA-ANN) hybrid intelligence for cancer diagnosis," in *Proceedings - 2nd International Conference on Computational Intelligence, Communication Systems and Networks, CICSyN 2010*, 2010, pp. 78–83.

[36] O. Olatunji, S. Akinlabi, and N. Madushele, "Application of Artificial Intelligence in the Prediction of Thermal Properties of Biomass," in *Green Energy and Technology*, 2020.

[37] O. Obafemi, A. Stephen, O. Ajayi, and M. Nkosinathi, "A survey of artificial neural network-based prediction models for thermal properties of biomass," in *Procedia Manufacturing*, 2019, vol. 33.

[38] A. Karaci, A. Caglar, B. Aydinli, and S. Pekol, "The pyrolysis process verification of hydrogen rich gas (H-rG) production by artificial neural network (ANN)," *Int. J. Hydrogen Energy*, vol. 41, no. 8, 2016.

[39] J. Abdulsalam, A. I. Lawal, R. L. Setsepu, M. Onifade, and S. Bada, "Application of gene expression programming, artificial neural network and multilinear regression in predicting hydrochar physicochemical properties," Bioresour. Bioprocess., vol. 7, no. 1, 2020.

[40] Kumar, A., Dubey, A.K., Ramírez, I.S., Muñoz del Río, A., Márquez, F.P.G. (2022). A Review and Analysis of Forecasting of Photovoltaic Power Generation Using Machine Learning. In: Xu, J., Altiparmak, F., Hassan, M.H.A., García Márquez, F.P., Hajiyev, A. (eds) Proceedings of the Sixteenth International Conference on Management Science and Engineering

Management – Volume 1. ICMSEM 2022. Lecture Notes on Data Engineering and Communications Technologies, vol 144. Springer, Cham. https://doi.org/10.1007/978-3-031-10388-9_36

[41] Dubey, A.K., Kumar, A., Ramirez, I.S., Marquez, F.P.G. (2022). A Review of Intelligent Systems for the Prediction of Wind Energy Using Machine Learning. In: Xu, J., Altiparmak, F., Hassan, M.H.A., García Márquez, F.P., Hajiyev, A. (eds) Proceedings of the Sixteenth International Conference on Management Science and Engineering Management – Volume 1. ICMSEM 2022. Lecture Notes on Data Engineering and Communications Technologies, vol 144. Springer, Cham. https://doi.org/10.1007/978-3-031-10388-9_35

[42] Rathore, P.S., Chatterjee, J.M., Kumar, A. *et al.* Energy-efficient cluster head selection through relay approach for WSN. J Supercomput 77, 7649–7675 (2021). https://doi.org/10.1007/s11227-020-03593-4

[43] A. Dubey, S. Narang, A. Srivastav, A. Kumar, V. Díaz, Woodhead Publishing, Science Direct, Artificial Intelligence for Renewable Energy Systems. Paperback ISBN: 9780323903967

[44] A. Dubey, S. Narang, A. Srivastav, A. Kumar, V. Díaz, Woodhead Publishing, Science Direct, a Visualization Techniques for Climate Change with Machine Learning and Artificial Intelligence. ISBN: 9780323997140

3

E-Waste Management and Bioethanol Production

Anshu Sibbal Chatli

Farm Value Foods Pvt. Ltd., Ludhiana (Pb), India

Abstract

Electronic waste (e-waste) is either an electronic product or a product containing electronic components that is no longer usable and can be disposed of only through a certified e-waste hauler or recycler due to its potential effect on the earth's air, soil, and water and human health.

Growing environmental concerns over the use and depletion of non-renewable fuel sources, together with the increasing price of oil and instabilities in the oil markets, have recently stimulated interest in producing sustainable energy sources in the form of biofuels derived from plants. Ethanol developed from lignocellulosic biomass has characteristic benefits of being safer, more economic, and more environmentally friendly than fossil fuels. Lignocellulosic materials are comprised of 50% cellulose, 25% hemicellulose, and 25% lignin. Ethanol made from renewable resources such as rice, wheat, corn, and wood industry byproducts can be blended with gasohol. The production of bioethanol requires four steps: pretreatment, hydrolysis of cellulose, fermentation of C5 (Xylose) and C6 (Glucose) sugars, and distillation. Both bacteria and yeasts have been used for ethanol production. Among the bacteria, the most widely used organism is *Zymomonas mobilis*, while *Saccharomyces cerevisiae* is the most commonly used yeast. The commercial application of ethanol includes the production of vinegar, extracts (food grade), and pharmaceutical products. The technology for the production of bioethanol is selected on the basis of economics, environmental safety, and energy efficiency.

Bioethanol production is an intricate system and Artificial Intelligence (AI) based technologies can facilitate prediction and diagnostics of bioethanol production from lignin and cellulose biomass. The application of AI technologies in the production phase help in the optimization of quality, process conditions,

---

Email: manansh@hotmail.com

Abhishek Kumar, Pramod Singh Rathore, Ashutosh Kumar Dubey, Arun Lal Srivastav, T. Ananth Kumar and Vishal Dutt (eds.) *Sustainable Management of Electronic Waste*, (67–78) © 2024 Scrivener Publishing LLC

and quantity, whereas the consumption phase helps in the regulation of emissions composition, motor temperature, and performance.

***Keywords*:** Lignocellulosic wastes, ethanol, sugars, AI

3.1 Introduction

E-waste contains toxic materials such as lead, zinc, nickel, flame retardants, barium, and chromium and these toxic materials can seep into the soil and groundwater, affecting not only our health, but also land and sea animals. E-waste is growing at a very fast rate throughout the world and forms 2-5% of total municipal waste in the USA and Europe per reports from the EPA.

The increasing prices of fossil oil in the national and international market have forced the industry to explore strategies for the development of substitute fuels. The active strategy is the utilization of fermentable sugars extracted from waste materials such as lignocellulose for the development of biofuel, organic acid, and animal feed (Haq *et al*, 2016).

Ethanol is a high-octane alcohol manufactured from rice/wheat/corn/wood industry byproducts and various additional biomaterials. Ethanol is blended with gasoline at 10% level to develop gasohol which is utilized as a fuel. Gasohol acts as a natural antifreeze and burns with greater efficiency in combustion engines. In addition, ethanol is also utilized as a food grade vinegar and extract and in various pharmaceutical products (Lamichhane *et al*, 2021).

Retrospective views collected from the literature clearly showed that the majority of the research is focused on the development of an economical and ecofriendly process protocol for ethanol production. Major emphasis is being given to renewable resources, especially agricultural and forestry residues and other forms of lignocellulosic biomass (Arunachalam, 2007).

After bagasse, rice husk is probably the largest mill- generated source of biomass available for energy use. As large quantities of rice husks are normally available at the rice mills, there are no additional efforts or costs involved in the collection of this biomass for use as an energy source. Generally, the lignocellulosics contain 50% cellulose, 25% hemicelluloses, and 25% lignin.

The production of bioethanol, however, requires pretreatment of the feedstock to enable fermentation of the sugars contained in the biomass. Pretreatment is an important tool for the cellulose conversion process. The fermentation itself needs to be adapted and, furthermore, new kinds of enzymes are required to convert the C5 sugars into ethanol. The process consists of four steps: pretreatment, hydrolysis of cellulose, fermentation of C5 (xylose) and C6 sugars (glucose), and distillation.

Both yeasts and bacteria have been used for ethanol production. Among the bacteria, the most widely used organism is *Zymomonas mobilis* (Amutha *et al.*, 2001). *Saccharomyces cerevisiae* is the most commonly used yeast.

The selection of technology for bioethanol production is based on the factors of sustainable environment, energy, and economics (Chandel *et al*, 2007, Kang *et al*, 2014).

AI refers to ability of machines to perform activities that mimic human intelligence with the use of different techniques of computer science, such as machine learning, heuristic algorithms, and fuzzy logic (FL) (Xing *et al*, 2021). It can be used to predict biomass properties and process efficiency using different conversion pathways and technologies. AI tools are also used to assess biofuel properties and its performance in vehicles and motors.

Therefore, the present chapter is written with the following objectives to produce ethanol from agricultural waste, i.e., rice husks, and its utilization for the production of biofuel (ethanol):

1. To evaluate the different aspects of lignocellulosic waste with the help of different pre-treatment methods such as acid, alkali, and steam pre-treatments
2. To find out the percentage of constituents present in the lignocellulosic waste
3. To find out the possibility of fermentable sugar from 2 different extracts
4. To find out the possibility of ethanol from different extracts collected from different pre-treatments that are applied
5. To elucidate the application of AI technologies to augment the process and performance efficiency of bioethanol production from biomass

3.2 Review

Lignocellulose, a renewable plant biomass, is a structural module of both woody and non-woody plants (including grass) and it is generated as a byproduct of agro-industries viz. paper pulp, timber, etc. and through various forestry and agricultural practices (Table 3.1). Thus, residual plant

Table 3.1 Types of lignocellulosics and their uses (Howard *et al*, 2003; Haq *et al*, 2016).

Lignocellulosic material	Residues	Uses
1. Grain Harvesting (Wheat, Rice, Corn)	Straws, cobs, stalks, husks	Burnt as fuel
2. Fruit and Vegetable Harvesting	Seeds, peels, husks, stones, rejected whole fruits and vegetables	Oil extraction
3. Sugarcane and Other Sugar Products	Bagasse	Burnt as fuel
4. Oil and Oil Seed Plants Nut, Cotton, Soybean, etc.	Shells, husks, lints, fiber	Burnt as fuel
5. Forestry (Paper and Pulp)	Wood residuals, barks, leaves	Burnt as fuel
6. Pulp and Paper Mills	Fiber waste, sulphite liquor	Reused in pulp and board industry as fuel
7. Lignocellulose Waste from Communities	Old newspapers, paper, cardboard	Small percent recycled, others burnt

biomass generated as agro-industrial byproducts is a potential raw material for the fermentation and production of different value-added products such as biofuels and intermediate chemical compounds. Therefore, it is classified as high-value biotechnological material due to innate composition and properties (Howard *et al*, 2003).

3.3 Degradation of Lignocellulose

The typical composition of Lignocellulose encompasses lignin, hemicelluloses, and cellulose and it varies with the type of plant, plant portion, and components. The details are enclosed in Table 3.2. However, cellulose forms the major component of lignocellulose. Cellulose is made up of linear chains of β- (1-4) glucose units with an average degree of polymerization

Table 3.2 Lignocellulose contents of common agricultural residues and waste (Howard *et al*, 2003; Haq *et al*, 2016).

Lignocellulosic materials	Cellulose (%)	Hemicellulose (%)	Lignin (%)
Hardwood Stem	40-55	24-40	18-25
Softwood Stem	45-50	25-30	25-40
Wheat Straw	85-99	50	15
Rice Straw	32.1	24	18
Leaves	15-20	80-85	0
Cotton Seed Hair	80-95	5-20	0
Newspaper	40-55	25-40	18-30
Waste Paper from Chemical Pulps	60-70	10-20	5-10
Fresh Bagasse	33.4	30	18.9
Swine Waste	6	28	NA
Nut Shell	25-30	25-30	30-40

of about 10000 units. It is insoluble and composed of highly crystalline and non-crystalline (amorphous) regions forming a structure with high tensile strength and resistance from enzymatic hydrolysis (Walker and Wilson, 1991; Lugani *et al*, 2020).

Hemicellulose and heteropolysaccharides soluble in alkali are associated with the cellulose of the plant cell wall and its major components include D-xylose, D-mannose, D-glucose, D-galactose, L-arabinose, D-glucouronic acid, etc. (Adams and Castagne, 1952, Yadav *et al*, 2011).

In general, lignin contains three aromatic alcohols: coniferyl alcohol, sinapyl, and p-coumaryl. Lignin is linked to both hemicelluloses and cellulose, forming a physical seal around the latter two components (Howard *et al.*, 2003). Below Table 3.3 pre-treatment processes of lignocellulosic materials.

Table 3.3 Pre-treatment processes of lignocellulosic materials (Taherzadeh et al, 2008; Lamichhane et al, 2021).

Pre-treatment methods	Processes	Applied application
Physical Pre-treatment	Milling	Ethanol
	Irradiation	Ethanol and Biogas
	Others	Ethanol and Biogas
	Hydrothermal	
	High Pressure Steaming	
	Pyrolysis	
	Explosion	Ethanol and Biogas
	Alkali	Ethanol and Biogas
	Acid	Ethanol and Biogas
Chemical and Physicochemical Pre-treatments	Gas	Ethanol and Biogas
	Solvent Extraction of Lignin	Ethanol
Biological Pre-treatments	Fungi and Actinomycetes	Ethanol and Biogas

3.4 Bioprocessing of Lignocellulosic Materials

Bioconversion of lignocellulose into beneficial high-value products requires multi-step processes, as listed below:

- ✓ Pre-treatment (mechanical, chemical, or biological) (Grethlien and Converse, 1991)
- ✓ Hydrolysis of the polymers to produce readily metabolizable molecules (e.g., Hexose or pentose sugars)
- ✓ Bio-utilization of these molecules to support microbial growth or to produce chemical products
- ✓ Separation and purification

3.4.1 Fermentation

Cellulose hydrolysis leads to the production of sugar, which is further utilized for the production of ethanol. *S. cerevisiae* converts hexose sugars

Table 3.4 Various raw materials for ethanol production.

Raw material	Pretreatment and saccharification	Fermentation conditions	Microorganism
Sugarcane Bagasse	Dilute acid hydrolysis	Batch	C. shehatae NCIM3501
Wheat Straw	Dilute acid, Enzymatic hydrolysis	SSF, SHF	E. coli FBR5
Rice Straw	Auto hydrolysis	Batch	C. shehatae NCIM3501
Sorghum Straw	Steam explosion, Enzymatic	SSF	Kluyveromyces marxianus CECT10875
Corn Stover	Steam, Enzymatic	Fed-batch	S. cereviseae TMB3400
Barley Husk	Steam, Enzymatic	SSF	S. cereviseae
Sun Flower Stalk	Steam, Enzymatic	Batch	S. cereviseae var ellipsoideus
Sugarcane Leaves	Alkaline H_2O_2	SSF	S. cereviseae NRRL-Y-132
Wheat Bran	Dilute acid, Enzymatic hydrolysis	Batch	S. cereviseae
Ground Nut Shell	Acid hydrolysis	Batch	S. cereviseae
Alfalfa Fibers	Liquid hot water	SSF, SHF	C. shehatae FPL-702
Aspen	Acid hydrolysis	Continuous, Immobilized cells	P. stipitis R
Saw Dust	Acid hydrolysis	Batch, Continuous up-flow reactors	Clostridium Thermosaccharolyticum ATCC-31925
Pine	Acid hydrolysis	Continuous stirred tank reactor, Immobilized cells	P. stipitis NRRL-1724

* Chandel *et al.*, 2007; Lugani *et al*, 2020

to ethanol, whereas *Pichia stipis, Candia shehatae,* and *Pachysolan tannophilus* yeasts have the capacity to convert both C5 and C6 sugars. Various research trials proved that thermotolerant yeasts are better suited for ethanol production at an industrial scale. The present strategy is targeted to develop recombinant yeast, which can metabolize all forms of sugar at the lowest cost. Above mentioned Table 3.4 various raw materials for ethanol production.

Raw Materials: Dried Rice straw can be collected from local rice fields.

3.4.2 Microorganisms for Ethanol Production

Baker's Yeast or *Saccharomyces cerevisiae* can be used. This is a commercial food grade preparation, purchased from the local market. *Saccharomyces cerevisiae* culture can be maintained on YEPDA (1% Yeast extract, 2% peptone, 2% dextrose, 2% agar).

3.4.3 Pretreatment of Lignocellulosic Material and Preparation of Extracts

The husk can be washed repeatedly with water to remove water soluble impurities like dust particles. The heavy insoluble impurities which settle down at the bottom of the container will be also removed and rejected. After that, dry the husk at 80-90^{0}C overnight.

3.4.3.1 Analytical Method

- ✓ The total sugar concentration is estimated by the method of (Scott and Melvin, 1972).
- ✓ The total reducing sugar concentration is measured by (Miller *et al.*, 1972)'s method.

3.4.3.2 Estimation of Total Cellulose Contents

- ✓ Total cellulose content is measured by (Updegraph *et al.*, 1969)'s method

3.4.3.3 Estimation of Hemicelluloses Content

- ✓ The total hemicelluloses concentration is measured by the method of (Bailey, 1957).

3.4.4 Preparation of Inoculum

Saccharomyces cerevisiae from slants is subcultured on Petriplates containing nearly 15 ml of YEPDA (Yeast Extract, Peptone, Dextrose, Agar) medium and incubated at 30^0C for 24 hours. The cells are aseptically harvested in YEPD broth and incubated at 30^0C for 24 hours at 150 r.p.m. 10% inoculums containing nearly 10^6 to 10^7 cell/ml are generally used to inoculate the fermentation in cultures in the flasks for production of alcohol.

3.4.5 Estimation of Ethanol

The ethanol content of the fermented extract is estimated calorimetrically after distillation by the method described by Caputi *et al*; 1968.

3.5 AI Technologies

AI has been used to predict the properties of biomass feedstock using the data from either proximate or ultimate analysis. The proximate analysis data is generally less expensive and more timesaving than the ultimate analysis data. Therefore, there is a need to develop an application for predicting the ultimate analysis data based on the proximate analysis data. Various scientists compared the performance of AI with traditional empirical correlations and concluded the superior performance of AI, although the performance of AI is subject to the combinations of training datasets, AI techniques, training algorithms, and other statistical analysis methods (Xing, Luo, Wang, & Fan, 2019). Artificial neural networks (ANN), multiple linear regression, statistical regression, and multiple nonlinear regression models are the most popular methods for predicting the quality of feedstock. Blend composition, temperature, mixing speed, and mixing time are typical input variables and the output variables are viscosity, flash point, oxidation stability, density, methane fraction, higher heating values, and cetane number. Further, at the production stage, different biomass conversion pathways and technologies are assessed to identify similarities and differences for broader and more effective AI applications. AI studies can be organized for quality and yield optimization, quality prediction, estimation of yield to select the best method or process conditions, and efficiency. Bioenergy has three types of end-uses: electricity, heat, and work (e.g., the chemical energy in biofuel is converted to kinetic energy by a vehicle engine that drives a vehicle to move). AI technologies for the end-use help us predict correlations between the environment and energy to develop a sustainable model for bioethanol production. Figure 3.1 described the conversion pathway for making ethanol from cellulosic biomass.

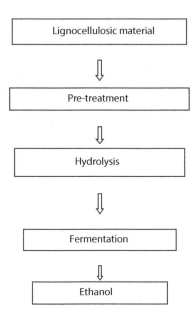

Figure 3.1 Conversion pathway for making ethanol from cellulosic biomass.

3.6 Conclusions

In the present study, different pre-treatment methods such as steam, acid, and alkali can be applied to study different aspects (% hemicelluloses, % lignin, and % cellulose) of rice husk. The utilization of rice husk for the production of bioethanol can be done by using acid and alkali extracts. Analytical tests can also be performed to estimate total sugar, reducing sugar, and xylose concentration. Fermentable sugar can be estimated and a comparison can be drawn between acid and alkali treated rice husks.

AI technologies and machine studies help us draw the future path of research in biofuels with a sustainable approach incorporating various aspects of quality of feed stock, process methodologies, process efficiencies, product quality, and economics.

References

Adams G. A. and Castagne A. E. (1952) Purification and composition of a polyuronide hemicelluloses isolated from wheat straw. *National Research Lab*, Canada. (198): 36-38.

A. Dubey, S. Narang, A. Srivastav, A. Kumar, V. Díaz, Woodhead Publishing, Science Direct, Artificial Intelligence for Renewable Energy Systems. Paperback ISBN: 9780323903967

A. Dubey, S. Narang, A. Srivastav, A. Kumar, V. Díaz, Woodhead Publishing, Science Direct, a Visualization Techniques for Climate Change with Machine Learning and Artificial Intelligence. ISBN: 9780323997140

Amutha R. and Gunasekaran P. (2001) Production of ethanol from liquefied cassava starch using co-immobilized cells of *Zymomonas mobilis* and *Saccharomyces diastaticus*. *Journal of Bioscience & Bioengineering*, 92: 560-564.

Arunachalam V. S., Tongia R., Bhardwaj A. (2007) Scoping technology options for India's oil security: Part 1- ethanol for petrol. *Current Science*, 92: 8.

Bailey R. W. (1957) The reaction of pentoses with anthrone. *Plant Chem. Lab.*, New Zealand. 68:21-28

Caputi A., Chu N. and Pi D. (1968) Spectrometric determination of ethanol in wine. American Society for Ecology & Viticulture.

Chandel A., Shah P. and Kim L. (2007) Economics and environmental impact of bioethanol production technologies: an appraisal. *Biotechnology and Molecular Biology Review*, 2(1): 14-32.

Dubey, A.K., Kumar, A., Ramirez, I.S., Marquez, F.P.G. (2022). A Review of Intelligent Systems for the Prediction of Wind Energy Using Machine Learning. In: Xu, J., Altiparmak, F., Hassan, M.H.A., García Márquez, F.P., Hajiyev, A. (eds) Proceedings of the Sixteenth International Conference on Management Science and Engineering Management – Volume 1. ICMSEM 2022. Lecture Notes on Data Engineering and Communications Technologies, vol 144. Springer, Cham. https://doi.org/10.1007/978-3-031-10388-9_35

Grethlein H. E. and Converse A. O. (1991) Common aspects of acid prehydrolysis and steam explosion for preheating word. *Bioresource Technology*,36: 77-82.

Haq F., Ali H., Shuaib M., Badshah M., Hassan, S. W. and Munis M. F. H. (2016) Recent progress in bioethanol production from lignocellulosic materials: A review. *International Journal of Green Energy*, 13 (14):1413-1441.

Kang Q., Appels J., B., Dewil, R. and Tan, T. (2014) Energy efficient production of cassava- based bioethanol. *Advances in Bioscience and Biotechnology*, 5 (12):925-939.

Howard R. L., Abotsi E., Jansen van Rensburg E. L. and Howard S. (2003). Lignocellulosic biotechnology: issues of bioconversion and enzyme production. *African Journal of Biotechnology*,12: 602-619.

Lamichhane G., Acharya A., Poudel D. K., Aryal B., Gyawali N. and Niraula P. (2021) Recent advances in bioethanol production from Lignocellulosic biomass. *International Journal of Green Energy*, 18 (7): 731-744.

Lugani Y., Rai R., Prabhu A. A., Maan P., Hans M., Kumar V., Kumar S.Chandel A. K. and Sengar R. S. (2020) Recent advances in bioethanol production from lignocellulosics: a comprehensive review with a focus on enzyme engineering and designer biocatalysts. *Biofuel Research Journal*, 28: 1267-1295.

Miller G. I. (1972). *Anal.Chem*. 31:426

Kumar, A., Dubey, A.K., Ramírez, I.S., Muñoz del Río, A., Márquez, F.P.G. (2022). A Review and Analysis of Forecasting of Photovoltaic Power Generation Using Machine Learning. In: Xu, J., Altiparmak, F., Hassan, M.H.A., García Márquez, F.P., Hajiyev, A. (eds) Proceedings of the Sixteenth International Conference on Management Science and Engineering Management – Volume 1. ICMSEM 2022. Lecture Notes on Data Engineering and Communications Technologies, vol 144. Springer, Cham. https://doi.org/10.1007/978-3-031-10388-9_36

Rathore, P.S., Chatterjee, J.M., Kumar, A. et al. Energy-efficient cluster head selection through relay approach for WSN. J Supercomput 77, 7649–7675 (2021). https://doi.org/10.1007/s11227-020-03593-4

Scott T. A. and Melvin, E. H. (1953). *Anal. Chem.* 25:1656

Taherzadeh J. M., Teri M. and Karimi K. (2008). Pre-treatment of lignocellulosic wastes to improve ethanol and biogas production: A review. *International Journal of Molecular Sciences*, 2:1422-0067.

Updehgraph K. and David M. (1969). Semi micro determination of cellulose in biological materials. *Analytical Biochem. Engg.* 32: 420-424.

Yadav K.S., Naseeruddin S., Prashanthi G,S., Sateesh L. and Rao L. V. (2011) Fermentation of concentrate rice straw hydrolysate using co-culture of *Saccharomyces cerevisiae* and *Pichia stipites*. *Bioresource Technology*, 102(11): 6473-6478.

Walker L. P., Wilson D. B. (1991). Enzymatic hydrolysis of cellulose: an overview. *Bioresource Technology,* 36:13-14.

Xing Y., Zheng Z., Sun Y. and M.A. Alikhani (2021). A review on machine learning application in biodiesel production studies. *International Journal of Chemical Engineering.* https://doi.org/10.1155/2021/2154258

Xing, J., Luo, K., Wang, H., & Fan, J. (2019). Estimating biomass major chemical constituents from ultimate analysis using a random forest model. *Bioresource Technology*, 288;121541. https://doi.org/10.1016/j.biortech.2019.121541

4

A Novel-Based Smart Home Energy Management System for Energy Consumption Prediction Using a Machine Learning Algorithm

N. Deepa[1*], Devi T.[1], S. Rakesh Kumar[2] and N. Gayathri[2]

[1]*Department of Computer Science & Engineering, Saveetha School of Engineering, Saveetha Institute of Medical and Technical Sciences, Saveetha University, Chennai, India*
[2]*Department of Computing Science and Engineering, GITAM (Deemed to be University), Bangalore, India*

Abstract

Energy is produced from various sources from electric resources for multi-consumption. Smart homes, smart business enterprises, and education systems are evolving using smart systems. Households and the smart environment are utilizing more energy consumption for sophistication and easy access. Not only smart homes, but NH roads, Artificial Intelligence systems, and other applications are also accessed via electricity and energy management systems. Linear regression and support vector machine algorithms are used to predict energy consumption in smart systems. The likelihood in the existing model analyzed features based on power and resource utilizations to train and test parameters. A prediction system using machine learning algorithms produced more accurate results for achieving energy resource utilizations where the cost was higher and not focused in existing applications. A novel approach for smart home energy management systems for energy consumption prediction is proposed using an enhanced Bayesian linear regression machine learning algorithm (EBLRML). The residual sum of squares shows the major linear model difference from the existing application of ML techniques, whereas the coefficient based on the correlation is lower and energy consumption is calculated to achieve the best fit model.

Corresponding author: deepa23narayanan@gmail.com

Abhishek Kumar, Pramod Singh Rathore, Ashutosh Kumar Dubey, Arun Lal Srivastav, T. Ananth Kumar and Vishal Dutt (eds.) Sustainable Management of Electronic Waste, (79–96) © 2024 Scrivener Publishing LLC

***Keywords*:** Consumption of energy, linear regression, machine learning, prediction

4.1 Introduction

In the modern world electricity plays a major role where frequent usage of energy resources are high. Buildings, lifts, escalators, and other services for intelligent systems are interconnected with Artificial Intelligence applications [1]. Energy consumption and computation of electricity in latest technologies for smart home energy management systems is handled by machine learning algorithms. Due to pandemic conditions, most of the human population spent their time indoors, which in turn contributed to the consumption of energy in the case of buildings [2]. This occurred due to lack of awareness combined with the carelessness of the consumer. The digital world is revolutionary considering buildings that use applications which depend on automation methods. Technologies involved in automation are shown in Figure 4.1. These buildings, with improved methods, help in (i) satisfying the demands of people living in the respective buildings, (ii) improving the level of comfort, and (iii) improving customer safety [3]. The major development in digital products occurred due to revolution in the digital world. Another factor to be considered in this field is customer satisfaction. The consumers of energy (primary) contribute up to 40% while considering sectors such as buildings. In order to reduce the

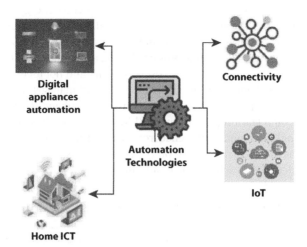

Figure 4.1 Automation methods.

consumption of energy, it is important to concentrate on buildings where residents reside. This, in turn, can help us save energy for the entire world. On the other hand, this job of minimizing the consumption of energy is not easy work because the research towards development of less energy-consuming buildings is a tedious job.

Countries in the development stage are also not able to reduce the consumption of energy. It is therefore necessary to find a better solution for saving energy [4]. AI-related methods are also used by researchers across the world to reduce this consumption rate and thereby help in controlling the operations of digital appliances [5]. Several organizations working towards the development of residential buildings are introducing various technologies such as IoT for improving the environment. IoT or any other AI related techniques can be used to control the devices available in a house and this creates a smart environment for the residents to live in [6]. Analysis based on predictions can be useful in developing various applications in the field of computer science. Additionally, optimization plays a major role in the selection of parameters that are best suited from the available options. Applications of optimization techniques include fields such as mathematics, along with computer science and operations research [13-15]. The fitness function can either be maximized or minimized based on the input parameters and cost estimation is performed with the help of the best suited options. Predicting the level of consumption of energy can be useful in the analysis process while installing the devices in the smart environment [16].

Figure 4.2 artificial intelligence vs. machine learning. This process consisting of consumption of energy can be predicted using various methods;

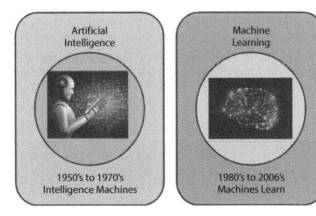

Figure 4.2 Artificial intelligence vs. machine learning.

one significant method among these is machine learning techniques. Emerging technologies such as machine learning play a major role in the field of computer science as well as statistics [17]. Based on experiences, computers can learn and the problem of building them is being handled by machine learning with the help of algorithms. New approaches have developed in the machine learning domain with the help of available online data in addition to increased computational power.

Figure 4.3 represents the lifecycle of ML. The first phase in the life cycle is data gathering where the sources of data are identified and the collection process is carried over. This process is done with the help of mobile devices and the Internet [18]. A dataset is prepared in this phase. Preparing data is the next phase, where data is ordered and is again divided as, (i) exploration of data and (ii) pre-processing of data. This step is followed by wrangling, where the data is converted to a desirable format. Variables are selected in this step and data is cleaned for removal of invalid or missing data or noise. filtering methods help in cleaning data. Data is then subjected to the analysis phase, where techniques are selected and models are built for reviewing the results [19]. Later, the model is trained for improving the output of the model. The model is tested to provide accurate results and deployed in the systems. The differences among AI and ML are presented in Table 4.1.

4.1.1 Bayesian Linear Regression

Vector pairs are used in linear regression. The major functions of the Bayesian regression include (i) estimation of parameters in regression, (ii) parameter distribution, and (iii) standard deviation computation. Mean values for specific variables are computed based on the other combinations.

The objectives of the proposed work are as follows:

(i) Various Machine learning algorithm comparisons are carried out based on prediction algorithms, such as linear regression and SVM, which are applied in finding the accurate results of energy consumption via smart systems; and (ii) To identify the feature-based less cost estimation using an energy consumption dataset that can predict the exact feature selection

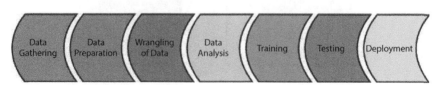

Figure 4.3 Lifecycle of ML.

Table 4.1 Artificial intelligence vs. machine learning.

Artificial intelligence	Machine learning
Behaviors of humans can be simulated by machines using artificial intelligence	Learning from past data occurs in an automatic manner without the need of programs in machine learning
Complex problems can be solved by smart systems by the use of AI	Machines have the ability to learn from old data in order to produce output accurately
Intelligent systems can be created for performing human tasks	Tasks can be done by the machines based on the data provided in order to generate results accurately
The chances available for success can be maximized with AI	Accurate results along with patterns can be considered in ML
The scope of AI is wide	The scope of ML is limited

algorithm proposed as an Enhanced Bayesian Linear Regression Machine Learning Algorithm (EBLRML).

The chapter organization is as follows: Section 4.2 presents a survey on the existing systems in the field of energy saving, Section 4.3 elaborates on the proposed work, Section 4.4 discusses the results obtained, and Section 4.5 concludes the chapter.

4.2 Literature Review

The performance of buildings is estimated and significant for energy efficiency in the case of buildings where residents live. The existing works for effective management of energy have been discussed. Time series information can be estimated and predicted using ANN techniques, along with fuzzy logic. Data in multivariant form can be predicted using other algorithms in AI on both a long term, as well as short term basis. Human mind simulations are the basis for the computational models of ANN [7]. Approaches that are data-driven are considered in the case of neural networks. The results obtained after analysis are mainly dependent on the data that is available.

Energy management in residential buildings can be handled in an optimized way with the help of strategies such as demand response. This, in turn, reduces the cost incurred for the consumption of energy. Various challenges in the system of managing energy are (i) uncertainty in the behavior of users, (ii) usage of various equipment, and (iii) status of working for each device. The proposed work deals with a forecasting model dependent on the optimization techniques for performing scheduling processes. A deep reinforcement technique is utilized along with neural networks for the prediction of the temperature in the outdoors [8].

The major consumers of energy are from sectors, such as buildings, which are significant across the globe. Management of energy is not proper due to certain problems such as strategies for implementation and so on. Usage of techniques for managing buildings in a smart way is necessary for providing suitable living conditions to individuals. Various analyses are performed to identify the aim, as well as challenges, along with an attribute measurement. Additionally, risks related to the health of an individual related to satisfaction as well as comfort need to be addressed. Work that can be done additionally in the future deals with the consumption of energy in an effective way, challenges related to health, and attacks in cyber security [9].

IoT plays a major role in recent smart homes in order to replace old meters with smart ones. The readings from the meter are digitized, including the collection of data. Wireless data transmission takes place, reducing the work done in a manual way. Theft of energy may happen in networks of smart homes. Identification of these attacks is difficult for the existing models as they need several devices for installation. The system to detect energy theft is proposed using machine learning techniques. The phases of the proposed model include modules for making decisions such as (i) the model for performing prediction of consumption level (forecasting based on multimodal), (ii) the model for making decisions in the primary stage (filtration process), and (iii) the model for making decisions in the secondary stage (theft-based decisions are taken). The accuracy obtained in the proposed model is 99%, thereby enhancing the smart home's security [10].

Management of energy is done with the proposed framework depending upon reinforcement learning in order to improve the demand response. The problems pertaining to scheduling are dealt with by the Markov methods with timing in a discrete way. In order to handle these problems, a model is developed depending upon neural networks in combination with algorithms such as Q-learning. Electricity bills can be reduced because of the proposed model and the results are generated after simulation with devices [11].

In the case of smart homes, cost can be minimized considering the energy consumption. The design of an algorithm for performing scheduling

Table 4.2 Survey on existing works.

S. no	Research title and author name	Problem defined	Technique and algorithm applied	Feature limitations
1.	An Intelligent Home Energy Management System to Improve Demand Response, Yusuf Ozturk [et al.], "IEEE Transactions on Smart Grid, Volume: 4 Issue: 2, 2013	Prediction is developed for the user's choice based on their interest in social features and applications to find the demand based on the customer's smart home	An algorithm for learning features based on neural-fuzzy method	Users limited TOU pricing features were used, which is less in certain entities
2.	Mixed Deep Reinforcement Learning Considering Discrete-continuous Hybrid Action Space for Smart Home Energy Management, Chao Huang [et al.], Journal of Modern Power Systems and Clean Energy, Volume: 10, Issue: 3, May 2022	Minimize the cost range based on customers' usage; Thermal range from the dataset was compared with the proposed algorithm	Unique gradient descent based on deep learning technique defined as MDRL algorithm	Household and smart appliances cost from the dataset were applied and less constraints were utilized

(*Continued*)

Table 4.2 Survey on existing works. (*Continued*)

S. no	Research title and author name	Problem defined	Technique and algorithm applied	Feature limitations
3.	IoT Task Management Mechanism Based on Predictive Optimization for Efficient Energy Consumption in Smart Residential Buildings, Imran [*et al.*], Energy and Buildings Volume 257, 15 February 2022, 111762	Prediction based on smart energy system and optimization for consuming energy from multiplex residential places and problems	Focusing on reducing consumption of energy management systems from smart homes using scheduling tasks and managing along with optimization and prediction which can lower energy consumption.	Basic applications were used based on IoT devices where AI-based techniques such as machine learning and block chain with future enhancements for this proposed idea
4.	Deep reinforcement learning for home energy management system control, Paulo Lissa [*et al.*], Energy and AI, Volume 3, March 2021, 100043	Energy which is used for comfortable access and energy resources with less utilization were predicted using deep reinforcement technique	Based on unique house and residence demand, the electric based energy consumption is predicted using a deep learning algorithm called the deep reinforcement technique	Optimization metrics were limited and not measured based on the difference of mean, median, and correlation coefficient which shows less accuracy

in an optimized way is difficult due to various reasons such as generation of output, demand of power, and temperature outdoors. To cope with these problems, an effective algorithm is proposed based on the Markov process. The experimental results represent the algorithm's robustness [12].

Table 4.2 represents the problem defined with the algorithm used and the limitations of the features [20-24].

4.3 Proposed Work

To understand the existing problem identified in smart homes or smart energy consuming units, analysis based on feature selection is presented. Exploratory processes that can focus on existing algorithm performance shows the variances accordingly. The real world produces enormous amounts of energy through electricity and appliances from home-based activities. The overall dataset is collected with prediction household usage and various units for finding the particular grades and status. When there is an existing analysis for the feature choosing procedure, generally the appliance state is identified based on dimension changes which decrease. There are certain limitations where the exploratory data analysis does the modeling with a direct relation between the original and detailed variable. Based on the observation from the dataset, the target feature that is independently described as a response feature is taken as an important feature. When the classes of the feature are illustrated from the supervised learning, a classifier based on the Bayesian rule is evaluated. In terms of regression, the task is accomplished as a result variable. Since existing data is based on dependent terms, the nominal variable alone is considered from the dataset for analysis, which is declared as integers. The proposed system enhances the process of Bayesian linear regression to produce an accurate prediction in the assumption value of electric energy resources.

4.3.1 Enhanced Bayesian Linear Regression Machine Learning Algorithm (EBLRML)

The following are the steps involved in the proposed model to get the best fit model:

Step 1: Identify the current feature and processes for selecting the right features, which are termed as relevant features.

Step 2: Based on a higher or lower value, from the application consumption level correlation as its coefficient value is given.

>   View(smart1)
>   smartcon = smart1$X.8
>   smartenergy = smart1$temperature
>   cor(smartcon,smartenergy)
>   Correlation coefficient : 0.921132
>   temperature - 0.393211
>   humidity - 0.192231
>   Appliance State -0.248791

Step 3: Based on the low value, the correlation varies, and a simultaneous process will occur. Once the variable is selected, encoding is implemented for categorizing the features.

Step 4: In the proposed system, the Bayes line is highlighted to understand metrics such as Mean Perfect Error (MPE) for finding the perfect value that are of a variation between the true rate form prediction.

EBLRML recognizes the model that compares the existing root square value based on the mean difference. Figure 4.4 represents the proposed system which has taken an input from the Kaggle dataset for electrical energy consumption. The consumed price is predicted based on the unit price, which has a higher level of energy consumption, i.e., a higher or lower value, etc. The raw data is analyzed and preprocessed for identifying

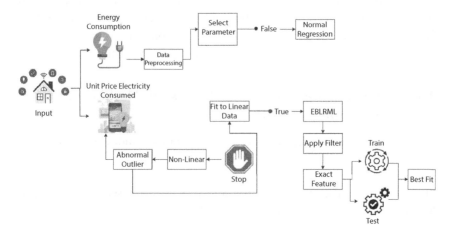

Figure 4.4 Proposed system.

feature selection, which shows the regression point to know the true and false rate. Based on the mean perfect rate (MPR), the normal regression is studied and calculated. Once there is a training process model that is recognized, the enhanced model EBLRML takes the positive value for fitting to linear data condition. This will fix the abnormal outlier can vary from that linearity. When dimension reduction is outperformed, non-linear data is segregated into a separate set of features which are removed from the number of features. Then, the specific feature which was termed for exact mapping features is applied with a filter to remove the noise and gives the best fit of prediction.

4.4 Results and Discussion

Interpreting the proposed model linear value is carried out based on experimental results. The environment used here is a jupyter notebook, along with performance metrics that can show the better results. The proposed algorithm follows regression concepts, such as application consumption, and is validated based on the level of performance. Also, the frequency range, as shown in Table 4.3, represents the percent of continuous truth ranging from fit models that enhance the unit price based on smart home usage. When the valid range differs, the broadcast information received by the consumer can cumulatively update the original model value.

Figure 4.5 shows the correlation coefficient of the proposed model. The distribution of any values in the selected feature provides the encoded plotting as shown below.

When the matrix based on the number of samples was reduced, the model feature was recognized from the Bayesian distribution. Figure 4.6

Table 4.3 Appliance consumption detail.

		Frequency	Percent	Valid percent	Cumulative percent
Valid		2	4.8	4.8	4.8
	Appliance Level Consumption	1	2.4	2.4	7.1
	Higher	20	47.6	47.6	54.8
	Lower	19	45.2	45.2	100.0
	Total	42	100.0	100.0	

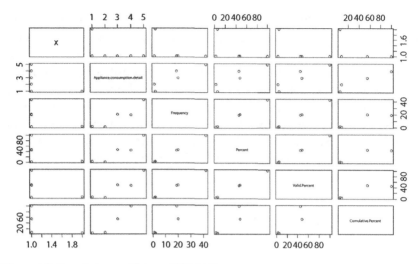

Figure 4.5 Correlation coefficient of EBLRML.

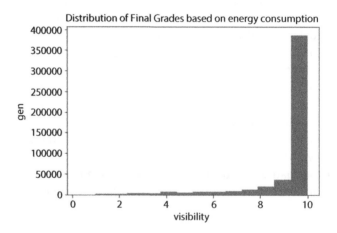

Figure 4.6 Distribution of final grade based on energy consumption.

shows the distribution of final grades that denotes the usage based on the various distributions in energy access from multiple users who are all managing the electric resource based on the required timeline.

When the frequency redirects the model fit that accomplished the role of the feature selection, the mean of normal regression and the model with SD (standard deviation) are probability ranges which can normalize the number of mean values.

Smart Home Energy Prediction with ML 91

The inference used in the jupyter notebook has applied the python modeling package pymc3 for building the right model. The number of samples is shown in Figure 4.7. Various appliance usage and consumption level difference based on electricity resource is shown in Figure 4.8.

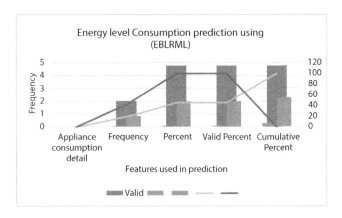

Figure 4.7 Energy level consumption prediction using EBLRML.

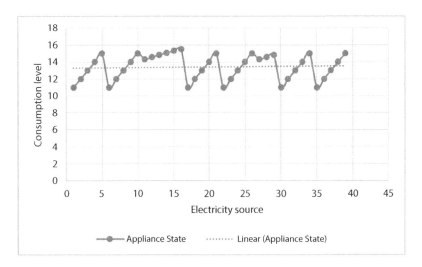

Figure 4.8 Appliance usage and consumption level difference based on electricity resource.

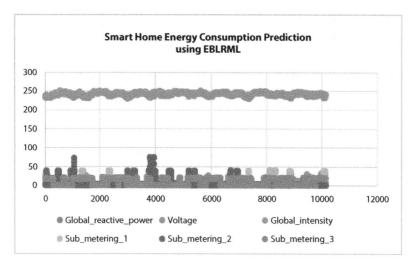

Figure 4.9 Smart home energy consumption prediction using EBLRML.

As the model based on response and its combination are interpreted in the fit model, application consumption levels using the matrix difference are also calculated. Using the below matrix equation, the overall model is formulated:

$$A = \beta^T B + \varepsilon \qquad \text{Equation (4.1)},$$

where the equation shows the regular linear model from the coefficient which is described as beta. Based on the sum of all squares from the electric consumption, the known value is denoted as B here and the response variable is involved in regression analysis.

The distribution of the points calculated by the responses is recognized by the amount of information that is identified from the dataset, including temperature, humidity, etc. The probability of true and false linear models which handle the gaussian distribution shown in Figure 4.9 are in the matrix computation. The squares of sigma that are differentiated as SD are also applicable for multiparameter models. The number of priority and the positive values likelihood are matrices calculated and proportionally distributed from the multiple regression that are normalized. Thus, the sum of the square difference shows the consumption prediction accurately using the enhanced model.

4.5 Conclusion

The existing Bayesian models has many feature selections to build from the training and testing model, which has less limited data that fails in understanding the overall model. From the problem statement, the energy consumption which is taken from the smart home management and resource allocation based on frequency and uncertainty is also recognized. The proposed model estimated the beginning state that gathered the appliance state and consumption level. As the Bayesian line according to the regression is suited for linear models, using the frequency and its counter, the analysis is done and the MPR value can vary and the standard deviation can also minimize the response value. Thus, the proposed model using the Machine Learning algorithm shows a more accurate prediction for knowing the consumption level, which is more accurate.

References

[1] A.-D. Pham, N.-T. Ngo, T.T.H. Truong, N.-T. Huynh, N.-S. Truong, Predicting energy consumption in multiple buildings using machine learning for improving energy efficiency and sustainability, J. Clean. Prod. 260 (2020) 121082.

[2] C.-F. Chen, G.Z.D. Rubens, X. Xu, J. Li, Coronavirus comes home? energy use, home energy management, and the social-psychological factors of covid-19, Energy Res. Soc. Sci. 68 (2020), https://doi.org/10.1016/j.erss.2020.101688 101688.

[3] M. Berry, M. Gibson, A. Nelson, I. Richardson, How smart is smart? Smart homes and sustainability, Steering Sustain. Urbanizing World (2016) 239–251.

[4] Y. Himeur, A. Alsalemi, A. Al-Kababji, F. Bensaali, A. Amira, Data fusion strategies for energy efficiency in buildings: Overview, challenges and novel orientations, Inf. Fusion 64 (2020) 99–120.

[5] A. Alsalemi, C. Sardianos, F. Bensaali, I. Varlamis, A. Amira, G. Dimitrakopoulos, The role of micro-moments: A survey of habitual behavior change and recommender systems for energy saving, IEEE Syst. J. 13 (3) (2019) 3376–3387, https://doi.org/10.1109/jsyst.2019.2899832.

[6] P.T. Akkasaligar, S. Biradar, R. Pujari, Internet of things based smart secure home system, in: International Conference on Intelligent Data Communication Technologies and Internet of Things, Springer, 2019, pp. 348–355.

[7] Dayu Wang, Daojun Zhong, Alireza Souri, Energy management solutions in the Internet of Things applications: Technical analysis and new research directions, Cognitive Systems Research, Volume 67, 2021, Pages 33-49.

[8] Mifeng Ren, Xiangfei Liu, Zhile Yang, Jianhua Zhang, Yuanjun Guo, Yanbing Jia, A novel forecasting based scheduling method for household energy management system based on deep reinforcement learning, Sustainable Cities and Society, Volume 76, 2022.

[9] Muhammad Saidu Aliero, Kashif Naseer Qureshi, Muhammad Fermi Pasha, Gwanggil Jeon, Smart Home Energy Management Systems in Internet of Things networks for green cities demands and services, Environmental Technology & Innovation, Volume 22, 2021.
[10] X. Xu, Y. Jia, Y. Xu, Z. Xu, S. Chai and C. S. Lai, "A Multi-Agent Reinforcement Learning-Based Data-Driven Method for Home Energy Management," in IEEE Transactions on Smart Grid, vol. 11, no. 4, pp. 3201-3211, July 2020, doi: 10.1109/TSG.2020.2971427.
[11] L. Yu *et al.*, "Deep Reinforcement Learning for Smart Home Energy Management," in IEEE Internet of Things Journal, vol. 7, no. 4, pp. 2751-2762, April 2020, doi: 10.1109/JIOT.2019.2957289.
[12] Kumar, P. S., Kumar, A., Agrawal, R., & Rathore, P. S. (2022). Designing a Smart Cart Application with Zigbee and RFID Protocols. Recent Advances in Computer Science and Communications (Formerly: Recent Patents on Computer Science), 15(2), 196-206.
[13] Devi, T., Ramachandran, A., Deepa, N., "A Biometric Approach for Electronic Healthcare Database System using SAML - A Touchfree Technology", Proceedings of the 2nd International Conference on Electronics and Sustainable Communication Systems, ICESC 2021, 2021, pp. 174–178
[14] Deepa, N., Sathya Priya. J, Devi, T., "A Behavioural Biometric system for Recognition and Authenticity via Internet in detecting attacks to provide Information Security Using multiple security interceptive biometric scanners (MSIBS)", Proceedings of the 2nd International Conference on Electronics and Sustainable Communication Systems, ICESC 2021, 2021, pp. 911–915
[15] Devi, T., Priya, J.S., Deepa, N., Framework for detecting the patients affected by COVID-19 at early stages using Internet of Things along with Machine Learning approaches with improved Accuracy, 2022 International Conference on Computer Communication and Informatics, ICCCI 2022, 2022
[16] Deepa, N., Sathya Priya, J., Devi, T., Towards applying internet of things and machine learning for the risk prediction of COVID-19 in pandemic situation using Naive Bayes classifier for improving accuracy, Materials Today: Proceedings, 2022
[17] Kumar, A., Dubey, A.K., Ramírez, I.S., Muñoz del Río, A., Márquez, F.P.G. (2022). A Review and Analysis of Forecasting of Photovoltaic Power Generation Using Machine Learning. In: Xu, J., Altiparmak, F., Hassan, M.H.A., García Márquez, F., Hajiyev, A. (eds) Proceedings of the Sixteenth International Conference on Management Science and Engineering Management – Volume 1. ICMSEM 2022. Lecture Notes on Data Engineering and Communications Technologies, vol 144. Springer, Cham. https://doi.org/10.1007/978-3-031-10388-9_36
[18] Dubey, A.K., Kumar, A., Ramirez, I.S., Marquez, F.P.G. (2022). A Review of Intelligent Systems for the Prediction of Wind Energy Using Machine Learning. In: Xu, J., Altiparmak, F., Hassan, M.H.A., García Márquez, F.P.,

Hajiyev, A. (eds) Proceedings of the Sixteenth International Conference on Management Science and Engineering Management – Volume 1. ICMSEM 2022. Lecture Notes on Data Engineering and Communications Technologies, vol 144. Springer, Cham. https://doi.org/10.1007/978-3-031-10388-9_35

[19] Rathore, P.S., Chatterjee, J.M., Kumar, A. *et al.* Energy-efficient cluster head selection through relay approach for WSN. J Supercomput 77, 7649–7675 (2021). https://doi.org/10.1007/s11227-020-03593-4

[20] A. Dubey, S. Narang, A. Srivastav, A. Kumar, V. Díaz, Woodhead Publishing, Science Direct, Artificial Intelligence for Renewable Energy Systems. Paperback ISBN: 9780323903967

[21] A. Dubey, S. Narang, A. Srivastav, A. Kumar, V. Díaz, Woodhead Publishing, Science Direct, a Visualization Techniques for Climate Change with Machine Learning and Artificial Intelligence. ISBN: 9780323997140

[22] Deepa, N., Udayakumar, N., Devi, T., Management of Traffic in Smart Cities Using Optical Character Recognition for Notifying Users, 2022 International Conference on Computer Communication and Informatics, ICCCI 2022, 2022

[23] Deepa, N., Devi, T., Gayathri, N., Kumar, S.R., Decentralised Healthcare Management System Using Blockchain to Secure Sensitive Medical Data for Users, EAI/Springer Innovations in Communication and Computing, 2022, pp. 265–282

[24] Devi, T., Alice, K., Deepa, N., Traffic management in smart cities using support vector machine for predicting the accuracy during peak traffic conditions, Materials Today: Proceedings, 2022, 62, pp. 4980–4984.

5

AI-Based Weather Forecasting System for Smart Agriculture System Using a Recurrent Neural Networks (RNN) Algorithm

Devi T.[1]*, N. Deepa[1], N. Gayathri[2] and S. Rakesh Kumar[2]

[1]*Department of Computer Science & Engineering, Saveetha School of Engineering, Saveetha Institute of Medical and Technical Sciences, Saveetha University, Chennai, India*
[2]*Department of Computing Science and Engineering, GITAM (Deemed to be University), Bangalore, India*

Abstract

The survival of human life depends on the interconnectedness of agriculture and economy. Precision in agriculture needs to be monitored for productivity and an increase in the growth of farming. It is the quality of food that determines the price of any crop based on its growth. It is mainly soil and its types that may bring risks, such as uncontrolled pesticides, water scarcity, farmer negligence, bad weather, low rainfall, and many others. The use of artificial intelligence today improves all the limitations that are outlined in the problem statements above. Machine learning algorithms have been used to predict crop cultivation, optimization metrics, irrigation levels, and so on, but very little has been done to predict weather and forecast agriculture. By using advanced sensors connected to a network, it is possible to identify bad weather that may threaten agriculture itself at an earlier stage. A live farm's features can be used to improve crop exploitation earlier in the growing cycle. One of the major terms to be focused on is soil fertility improvement and automation of smart agriculture systems. In the proposed system, AI-based weather forecasting is introduced to overcome manual operations in agriculture to improve soil quality, temperature variation, and weather conditions. One of the important features of smart agriculture management systems is monitoring

Corresponding author: devi.janu@gmail.com

Abhishek Kumar, Pramod Singh Rathore, Ashutosh Kumar Dubey, Arun Lal Srivastav, T. Ananth Kumar and Vishal Dutt (eds.) Sustainable Management of Electronic Waste, (97–112) © 2024 Scrivener Publishing LLC

weather conditions on a timely basis. Combining AI-based weather broadcasts with a smart agricultural management Recurrent Neural Network (RNN), one of the machine learning algorithms, is proposed for the prediction of weather conditions and their sequential transformation. In order to improve the weather updates based on timelines, an enhanced neural network pattern is used where the number of hidden points helps to produce accurate data. Compared to all existing Machine Learning algorithms, this algorithm has an accuracy rate of 98.76%.

Keywords: Artificial intelligence, recurrent neural network, sensors, weather forecasting

5.1 Introduction

In several countries, sectors such as agriculture play a major role in improving the economy. Advancements in this field also provide a major contribution towards the economy of a country, leading to development in a sustainable way. Monitoring systems nowadays are implemented to work in a smart way along with systems for sensing conditions in an automatic way to help farmers. These systems help them in monitoring the land as well as allow required actions to be taken in order to increase productivity. The application of various technologies including AI along with deep learning and IoT leads to smart agriculture. Due to the increase in population, the production of food also should be increased. Making the availability of food globally with this increasing population becomes a challenge while using these technologies.

Deep learning can be used in case of image processing as well as analysis of data. The results generated by deep learning models can be utilized for several domains such as agriculture. Various activities in agriculture are managed using the data obtained with the help of different sources. These systems in AI are different based on their capability to interpret data, which in turn helps the farmers make better decisions at a specific time. Recording of data is done utilizing the nodes of IoT, such as sensors, followed by usage of any deep learning algorithm for processing and usage of actuators for decisions. Figure 5.1 represents the technologies incorporated in recent years for real time management of agriculture.

These agricultural methods based on AI are also used for scheduling several resources like fertilizer or water, which in turn reduces pollution as well as expenditure necessary for production. This is the main reason for also increasing production. Contamination of the environment can be reduced as AI helps in detecting diseases in plants in the early phases. Water, along with fertilizers, is necessary for plant growth and should be

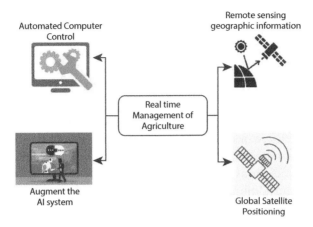

Figure 5.1 Technologies involved in real-time management of agriculture.

available continuously because its absence results in stress in a biotic way. AI helps in deciding the correct time for utilizing a specific resource and helps in predicting the future. The proposed work makes use of AI along with deep learning techniques. Parameters pertaining to agriculture can also be monitored by IoT and for processing these algorithms, deep learning is used. Difference between machine learning and deep learning is described in Table 5.1.

Various methods are incorporated for improving the agricultural industry. Systems based on computer vision, prediction of diseases, or productivity can be considered as a few examples. Both computer vision and deep

Table 5.1 Difference between machine learning and deep learning.

Characteristics	Machine learning	Deep learning
Requirement of data	Data larger in size is necessary	Data in smaller sizes can be trained
Generation of output	Accuracy rate is higher	Accuracy rate is smaller
Time taken for training	Time taken for training is more	Time taken for training is less
Dependency on hardware	Need of GPR for training	CPU is necessary
Tuning of hyperparameters	Tuning is performed in several modes	Capabilities of tuning are limited

learning are combined in platforms such as AirSurf for measurement purposes. Systems based on artificial intelligence can also be employed for predicting the growth of various grains such as maize, wheat, and so on.

5.1.1 Deep Learning

One AI subset is deep learning which has a neural network consisting of several layers. These networks are similar to the human brain and its activities and can handle huge amounts of data. If a neural network consists of a single layer, predictions can be done in an approximate way. The layers that are available and additionally hidden help in accuracy refinement. Another subset of AI is machine learning, which requires the learning process to be done by the systems, thereby eliminating the need for programming. The process starts by making observations like interactions done face to face, preparation of features related to data, and improving the results in the future. Several algorithms in machine learning are combined to build deep learning using various transformations for modeling data abstraction. The main advantage of using deep learning is extracting a feature based on raw data in an automatic manner. The components or features in the lower level are composed to form higher levels.

The standard model in deep learning for agriculture is RNN, which stands for Recurrent Neural Network. The performance of the model is good during applications, as shown in Figure 5.2. Compared to other traditional networks, sequential data of the network is utilized as it is critical for various applications. The sequence of data contains information that is valuable. Understanding the context is more significant for understanding the sentence containing the word.

This makes ANN a short-term memory block with layers such as input along with hidden and output ones. Forecasting the weather is done with RNN as dependency is on time series data. Data pertaining to weather helps in sectors such as agriculture, where parameters like accuracy are considered for crop cultivation. Information technology plays a major role in forecasting the weather and so on, as shown in Figure 5.3. These algorithms are used in the monitoring of parameters along with observations made from any place across the world. Researchers have shown interest in the study of RNN for forecasting weather to carry over into smart agriculture.

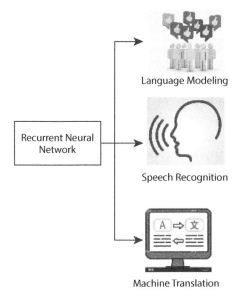

Figure 5.2 Applications of RNN (Recurrent Neural Networks).

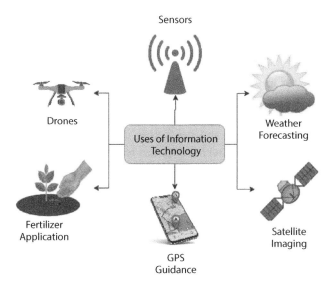

Figure 5.3 Usage of information technology.

The objectives of the research work are given as follows:

(i) Using satellite data, detect the critical images that show the variation in rainfall using Artificial Intelligence sensor data;
(ii) Applying Recurrent Neural Networks (RNNs) to analyze the features by reducing the dimension and captioning the area that exhibits rainfall and weather changes such as humidity, sunny, rainy, cloudy, etc. that can forecast the time changes for smart agriculture management systems;
(iii) The collective database of the Kaggle forum is used for analyzing the hazards that can affect the growth of cultivation using IOT-based sensors, cameras, and others that can track weather changes. To analyze performance, weather and smart agriculture parameters are categorized for finding accuracy, sensitivity, and a confusion matrix.

The chapter is organized as follows: Section 5.2 discusses the existing systems, Section 5.3 presents the proposed system, Section 5.4 discusses the results obtained, and Section 5.5 concludes the chapter.

5.2 Literature Review

The literature review section deals with the existing work done in the field of agriculture for improving the cultivation of crops. AI plays a major role in smart agriculture in ensuring the sustainability of this sector. Techniques of AI are used for the management of soil along with irrigation, forecast of weather, growth of plants, and predicting disease and are considered to be important domains. AI techniques are reviewed in terms of their performance. The research work focuses on highlighting the usage of artificial intelligence along with research directions moving towards the future. Compared to traditional algorithms in machine learning, algorithms in the domain of deep learning have generated better results while handling huge amounts of data. It also enables decisions to be made in an intelligent way that is similar to the decisions made by human beings [1][2].

Plants' growth can be analyzed with the usage of smart farming, along with an analysis of the parameters used for optimizing their growth. This, in turn, helps in supporting the activity of the farmers. IoT products like sensors are employed for the measurement of data along with processing, helping in bridging the gap between the cyber and physical world. The proposed work emphasizes designing systems for performing smart farming using the smart platform for prediction by incorporating techniques in AI. The major phases in this work are: (i) collection of data by using sensors incorporated in the irrigation fields, (ii) cleaning data and storing it, and (iii) processing the prediction using AI techniques [3].

In countries such as Asia, the predominant food is rice. This impacts the development of these nations in an economic way. Hence, it is necessary to maintain sustainability among the cultivation of paddies as well as the demand of consumers. This yield and demand are dependent on various factors including rainfall, lifestyle of humans, and humidity. Harvest prediction along with demand becomes a complex process. The proposed work is developed in the following phases: (i) module for prediction of harvest and (ii) module for prediction of demand. The algorithms employed for prediction are RNN along with LSTM. The experimental results denote that the results are accurate with lesser time [4][5][6].

Sectors including agriculture and industry play a significant role in forecasting the weather in an accurate manner based on the conditions of the weather. Natural disasters are also forecasted. Determining the parameters of weather with the right values is called forecasting weather. Figure 5.4

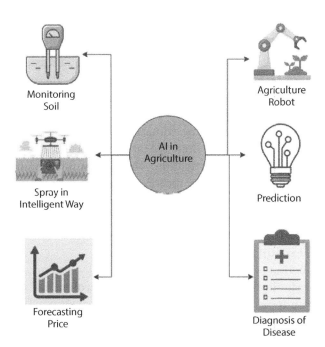

Figure 5.4 Application of AI techniques in agriculture.

shows the application of AI techniques in agriculture. From data based on the climate of a particular region, the training process LSTM (Long Short-Term Memory) is utilized. For predicting the condition of weather, training of LSTM with various parameters such as temperature, precipitation, and so on is carried out. Followed by this, predicting the weather is carried out [7][8][9][10][11].

For activities such as agriculture, freshwater is also utilized for making various products in order to ensure the survival of humans. Smart agriculture also aims to manage the amount of water for fields for every stage of developing the crops. In various spots, to monitor these works, sensors are utilized for finding the moisture level in order to identify the retained water content. The data related to moisture content in soil is not received by the smart agricultural systems because of connectivity issues or failure of the sensors. Techniques in deep learning are employed for predicting the moisture level in soil and with this information it is easy to manage irrigation facilities [12]. The issues related to connectivity can also be handled by fog computing where the computing resources are provided to the edge nodes for processing and analyzing the requests from the sensors. Deep Neural Networks are built for prediction of moisture content in soil with the data from KNN. The model for evaluation denotes that the models for prediction can achieve effective results for saving water used in irrigation [13][14].

The demand for resources such as food and water for the increasing population has helped in precision agriculture. The necessity of water as well as land for farmers is there to meet the demand. There is a necessity for a solution for the farmers to use as the resources are limited. Precision irrigation stands as a solution for yielding more profits using limited resources. While considering the climatic conditions, few of the models are not suitable for research. Both IoT and deep learning are combined in the proposed system for irrigation. The system is based on feedback and provides better functionality for forecasting the weather of any region in specific time. LSTM is the deep learning algorithm used for predicting moisture level content during the period of irrigation. The results obtained from simulation provide a deep insight into the usage of water in the areas of farming [15][16][17].

The production of agriculture can be increased by systems such as smart agriculture. Monitoring of data consisting of environmental conditions is obtained from these systems employing smart agriculture. The algorithm used in the proposed work is backpropagation and its comparison is done with other existing algorithms. LSTM provides better prediction compared to backpropagation for the application of smart agriculture [18][19][20].

5.3 Proposed Work

One of the important terms in agriculture management today is smart agriculture management. A growing number of AI-based devices are being used to measure soil fertility, crop growth, rainfall conditions, etc. Currently, machine learning algorithms are based on one-to-one and single neurons, which sometimes produce inaccurate machine translation.

In our proposed system, one of the benefits of RNN will be its ability to be applied in AI-based weather tracking, which will provide real-time data for understanding crop health. It is described as precision agriculture that may occasionally face risks in growth due to adverse weather conditions.

5.3.1 Dataset Description

In this instance, weather status and agriculture field status are collected from different sources, such as Kaggle and our own database. In this analysis, independent features such as weather, uncertainty that leads to risk, etc. are taken into account. It is essential to extract the right location of crops from the input image, transform it to the expected report, and load the exact production. Generally, the measurement used for crops is one hectogram per hectare defined as hg/ha. These agriculture fields and crops are mostly

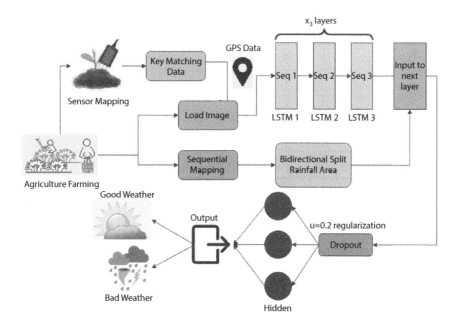

Figure 5.5 Proposed system for smart agriculture.

expected for good production but may fail due to less maintenance and water scarcity. Also, there are many gases present in farming which are monitored through sensors and cameras to avoid risks like soil diseases. Many kilotons of measures in soil may cause productivity to get low. To understand crop cultivation and its risks in various conditions and weather such as rain, temperature, and much more, research is focused on Recurrent Neural network (RNN). Figure 5.5 shows the proposed system for smart agriculture.

In Table 5.2, a summary based on the analysis based on the factors that affected it is clearly shown. Agricultural dependency is based on climatic differences where temperature and rainfall are keys to predicting the exact weather. To overcome the existing limitations based on the distributive variables that are captured before and after cultivation, these factors are used. The optimized value is motivated through the iteration, as shown in Table 5.3. Algorithms are used as epochs for exact prediction.

Table 5.2 Summary of analysis based on factors that affected it.

Name of the factors	Measures in rainfall (in mm)	Observation in temperature (in Celsius)
Value of Factor Mean	81.56	20.89
Standard Deviation (SD)	31.22	0.93
Minimum (min)	14.22	18.92
Testing (25%)	69.33	22.44
Training (75%)	109.55	25.01
Maximum (max)	209.73	29.29

Table 5.3 Statistical summary of RNN.

Group statistics						
		Algorithms	N	Mean	Std. deviation (SD)	Std. error mean (SME)
Accuracy		RNN	70	67.67	8.02	0.959
		SVM	70	89.9	5.27	0.63
Loss		RNN	70	56.16	4.714	0.563

5.3.2 Recurrent Neural Network (RNN)

Feed forward neural networks are used in all new applications to share sequential information based on sensors' timeline data. This input data frequently sends an image to the next line, such as t1, t2, …, tn, where tm describes the data center along with its variable dimensions. To analyze the image type, dimension can be carried along with the image captions that are evolved with time stamps. The following are some of the steps applied in smart agriculture systems using RNN:

Step 1: Data is captured from satellite images which were taken as live photos or stored images from the camera for predicting weather conditions.

Step 2: Sensor mapping datapoints are identified based on preprocessed datapoints that can connect with sequential mapping data.

Step 3: As data changes dynamically, the value on datapoints need to be in a sequential update for framing the right location. There are hidden layers based on the internal layers, described as a 1,2,3 LSTM formation.

Step 4: In comparison with LSTM, layer-wise image data is passed to the next phase as a dropout layer where u =0.2 to reduce the dimensions of the cropped image.

Step 5: Next, regularization is carried out based on a hidden state, forget gate, and other sophisticated gates which can predict the correct values to fix the model properly.

Step 6: As the RNN activation function helps to forecast the selected location alone, for example, tasks are classified based on tasks such as farming, growing, visualizing critical defects, and water conditions for prediction. Wherever the hidden state observed changes in temperature, humidity, and cloudy nature, changes in the weather condition are considered impacts that are high risk.

5.4 Results and Discussion

Exploring and reviewing based on target variables is handled with an RNN algorithm. A log based on its changes is shown in Figure 5.6, which transforms the distribution to normal terms.

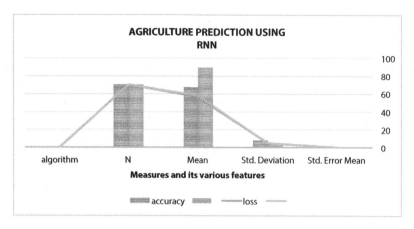

Figure 5.6 Prediction of smart agriculture data.

By analysis, we can verify the factors show a response feature as standard deviation sigma, representing the value as 534000.54 hg per ha. Also. the coefficient for correlation is with the probability rate which shows the plot value shown in Figure 5.7.

The next step is feature selection. The training variable is taken from the dataset and is categorical, so the proposed model can handle it. A similar feature selection process is used in the testing phase to maximize the quality of updates based on data sets. Table 5.3 represents the statistical summary of two parameters, as accuracy and loss, which shows the number

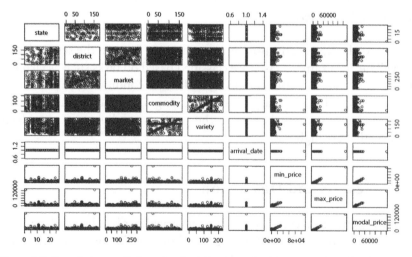

Figure 5.7 Correlation coefficient plot for smart agriculture management system.

Table 5.4 Algorithm RNN to predict exact farming.

Create nodes loop (N)
N = input feature (at)
Repeated iterations i = a0,a1,a2 to recurrent (R)
Memory block (M)
M1,M2.. Mn prioritize division
Action -> input state
Activation(af)
If right side move to output
not
forget activate
predict right input
output(at)

of samples N =70. Also, the comparison algorithm is a recurrent neural network and support vector machine to show the prediction distribution based on factors like Mean, SD, and SME. Table 5.4 shows the algorithm RNN to predict exact farming.

Let us consider the algorithm to understand the working principle of RNN.

A neural network as an image from an agriculture crop has many hidden layers that can obtain important features for viewing results. According to the hidden layer, the data accessed is sorted, which is similar to other machine learning algorithms.

The complete input is split into numbered of blocks, LSTM1, LSTM2, etc., for filling the current memory which can reproduce the feedback recurrent loop to show the input. If there is a delay or no response variable based on time variation, then the data flow is to be identified. This recurrent loop allows the input from a selected neural network which was prioritized from selected features. A memory gate is an initial input that can store the first stage image. Then, the action carries to the current gate to handle the flow which can travel to the next phase.

Since the activation function rectified linear unit (ReLu) is one of the major cell states that achieves memory, it will be stored based on the ReLU filtering to regulate the image to an exact size. At last, the forget gate is activated only if loss of previous information, such as the wrong side, can group, recognize, and predict the selected output using LSTM. Thus, the backpropagation is carried out simultaneously based on the time changes from weather data that are broadcast from AI devices. Also, random weight and bias are adjusted accordingly so that backpropagation time can avoid relapse and fluctuation. From the result, an accurate prediction can be achieved by adjusting the epoch and training percentage based on a gradual decrease or loss function.

5.5 Conclusion

Forecasting of weather is done with AI methods, which eradicates the usage of manual agricultural systems. The proposed work focuses on the improvement of quality of soil and variations in temperature. Changes in weather need to be monitored on a regular interval basis. An RNN algorithm is also employed for predicting the weather along with its transformations. Enhanced neural networks are incorporated to generate accurate information with hidden points. Feature analysis can be done using RNN, where the dimensions can be reduced along with area captioning. This helps in depicting the changes in the weather including temperature, humidity, and so on. Changes in time can be forecasted for management of smart agriculture. The dataset obtained from Kaggle helps in analysis of hazards which affect crop cultivation. Sensors or cameras are employed for this tracking work. The parameters used are accuracy, along with a confusion matrix. The data obtained using satellites is utilized and the images can be detected where the variation is depicted. This occurs in the case of rainfall with the help of data collected from the sensors used in AI.

References

[1] Ridany, H., Latif, R., Saddik, A. (2022). Deep Learning in Smart Farming: A Survey. In: Elhoseny, M., Yuan, X., Krit, Sd. (eds) Distributed Sensing and Intelligent Systems. Studies in Distributed Intelligence. Springer, Cham. https://doi.org/10.1007/978-3-030-64258-7_16

[2] F. K. Shaikh, M. A. Memon, N. A. Mahoto, S. Zeadally and J. Nebhen, "Artificial Intelligence Best Practices in Smart Agriculture," in IEEE Micro, vol. 42, no. 1, pp. 17-24, 1 Jan.-Feb. 2022, doi: 10.1109/MM.2021.3121279.

[3] A. Dahane, R. Benameur, B. Kechar and A. Benyamina, "An IoT Based Smart Farming System Using Machine Learning," 2020 International Symposium on Networks, Computers and Communications (ISNCC), 2020, pp. 1-6, doi: 10.1109/ISNCC49221.2020.9297341.

[4] M. R. S. Muthusinghe, P. S.T., W. A. N. D. Weerakkody, A. M. H. Saranga and W. H. Rankothge, "Towards Smart Farming: Accurate Prediction of Paddy Harvest and Rice Demand," 2018 IEEE Region 10 Humanitarian Technology Conference (R10-HTC), 2018, pp. 1-6, doi: 10.1109/R10-HTC.2018.8629843.

[5] D. N. Fente and D. Kumar Singh, "Weather Forecasting Using Artificial Neural Network," 2018 Second International Conference on Inventive Communication and Computational Technologies (ICICCT), 2018, pp. 1757-1761, doi: 10.1109/ICICCT.2018.8473167.

[6] Matheus Cordeiro, Catherine Markert, Sayonara S. Araújo, Nádia G.S. Campos, Rubens S. Gondim, Ticiana L. Coelho da Silva, Atslands R. da Rocha, Towards Smart Farming: Fog-enabled intelligent irrigation system using deep neural networks, Future Generation Computer Systems, Volume 129, 2022, Pages 115-124.

[7] P. K. Kashyap, S. Kumar, A. Jaiswal, M. Prasad and A. H. Gandomi, "Towards Precision Agriculture: IoT-Enabled Intelligent Irrigation Systems Using Deep Learning Neural Network," in IEEE Sensors Journal, vol. 21, no. 16, pp. 17479-17491, 15 Aug.15, 2021, doi: 10.1109/JSEN.2021.3069266.

[8] P. S. Budi Cahyo Suryo, I. Wayan Mustika, O. Wahyunggoro and H. S. Wasisto, "Improved Time Series Prediction Using LSTM Neural Network for Smart Agriculture Application," 2019 5th International Conference on Science and Technology (ICST), 2019, pp. 1-4, doi: 10.1109/ICST47872.2019.9166401.

[9] Devi, T., Ramachandran, A., Deepa, N., "A Biometric Approach for Electronic Healthcare Database System using SAML - A Touchfree Technology", Proceedings of the 2nd International Conference on Electronics and Sustainable Communication Systems, ICESC 2021, 2021, pp. 174–178

[10] Deepa, N., Sathya Priya.J, Devi, T., "A Behavioural Biometric system for Recognition and Authenticity via Internet in detecting attacks to provide Information Security Using multiple security interceptive biometric scanners (MSIBS)", Proceedings of the 2nd International Conference on Electronics and Sustainable Communication Systems, ICESC 2021, 2021, pp. 911–915.

[11] Devi, T., Priya, J.S., Deepa, N., Framework for detecting the patients affected by COVID-19 at early stages using Internet of Things along with Machine Learning approaches with improved Accuracy, 2022 International Conference on Computer Communication and Informatics, ICCCI 2022, 2022

[12] Deepa, N., Sathya Priya, J., Devi, T., Towards applying internet of things and machine learning for the risk prediction of COVID-19 in pandemic situation using Naive Bayes classifier for improving accuracy, Materials Today: Proceedings, 2022

[13] Deepa, N., Udayakumar, N., Devi, T., Management of Traffic in Smart Cities Using Optical Character Recognition for Notifying Users, 2022 International Conference on Computer Communication and Informatics, ICCCI 2022, 2022

[14] Kumar, A., Dubey, A.K., Ramírez, I.S., Muñoz del Río, A., Márquez, F.P.G. (2022). A Review and Analysis of Forecasting of Photovoltaic Power Generation Using Machine Learning. In: Xu, J., Altiparmak, F., Hassan, M.H.A., García Márquez, F.P., Hajiyev, A. (eds) Proceedings of the Sixteenth International Conference on Management Science and Engineering

Management – Volume 1. ICMSEM 2022. Lecture Notes on Data Engineering and Communications Technologies, vol 144. Springer, Cham. https://doi.org/10.1007/978-3-031-10388-9_36

[15] Dubey, A.K., Kumar, A., Ramirez, I.S., Marquez, F.P.G. (2022). A Review of Intelligent Systems for the Prediction of Wind Energy Using Machine Learning. In: Xu, J., Altiparmak, F., Hassan, M.H.A., García Márquez, F.P., Hajiyev, A. (eds) Proceedings of the Sixteenth International Conference on Management Science and Engineering Management – Volume 1. ICMSEM 2022. Lecture Notes on Data Engineering and Communications Technologies, vol 144. Springer, Cham. https://doi.org/10.1007/978-3-031-10388-9_35

[16] Rathore, P.S., Chatterjee, J.M., Kumar, A. *et al.* Energy-efficient cluster head selection through relay approach for WSN. J Supercomput 77, 7649–7675 (2021). https://doi.org/10.1007/s11227-020-03593-4

[17] A. Dubey, S. Narang, A. Srivastav, A. Kumar, V. Díaz, Woodhead Publishing, Science Direct, Artificial Intelligence for Renewable Energy Systems. Paperback ISBN: 9780323903967

[18] A. Dubey, S. Narang, A. Srivastav, A. Kumar, V. Díaz, Woodhead Publishing, Science Direct, a Visualization Techniques for Climate Change with Machine Learning and Artificial Intelligence. ISBN: 9780323997140

[19] Deepa, N., Devi, T., Gayathri, N., Kumar, S.R., Decentralised Healthcare Management System Using Blockchain to Secure Sensitive Medical Data for Users, EAI/Springer Innovations in Communication and Computing, 2022, pp. 265–282.

[20] Devi, T., Alice, K., Deepa, N., Traffic management in smart cities using support vector machine for predicting the accuracy during peak traffic conditions, Materials Today: Proceedings, 2022, 62, pp. 4980–4984.

6

Comprehensive Review of IoT-Based Green Energy Monitoring Systems

Aishwarya V.

Department of Electrical and Electronics Engineering, Toc H Institute of Science and Technology, Kerala, India

Abstract

The rapid growth of the human population has had a direct impact on energy consumption. Energy consumption has risen exponentially, putting pressure on utilities to increase energy production. One of the significant concerns of the current day is energy conservation, and energy monitoring systems have been developed for optimization of the increasing demand for energy and its consumption. Energy management systems help to decrease current consumption, prevent energy wastage, and enable the optimized utilization of available resources. The deployment of the Internet of Things (IoT) for energy conservation and optimization methods has led to a new, disruptive technology called the Internet of Energy (IoE). This paper presents a critical review of recent IoT-based green energy monitoring systems.

Keywords: Cloud, data acquisition, energy consumers, energy efficiency, energy monitoring systems, Internet of Things, renewable energy

6.1 Introduction

The exponential increase in the world's population has made energy conservation imperative. Energy use is directly impacted by this population boom. In order to satisfy rising energy demands, industrial utilities are advised to increase their power generation [1]. Recently, energy monitoring systems (EMS) have been developed to optimize the demand and

---

Email: aishwarya@tistcochin.edu.in

Abhishek Kumar, Pramod Singh Rathore, Ashutosh Kumar Dubey, Arun Lal Srivastav, T. Ananth Kumar and Vishal Dutt (eds.) Sustainable Management of Electronic Waste, (113–144) © 2024 Scrivener Publishing LLC

consumption of energy [2]. These systems optimize consumption and prevent energy wastage, ensuring the efficient utilization of the available resources. According to the Renewables-Global Energy Review 2020, the International Energy Agency's flagship study, the consumption of renewable energy increased across all sectors by 1.5% in the first quarter of 2020 compared to the same period in 2019. Renewable energy now makes up 28% more of the total power generated in 2020 than it did in 2010 [1, 3]. Green energy solutions that are inexpensive and environmentally benign are gaining popularity quickly. One of the many ways to reduce environmental pollution is to choose green energy technology [1]. The Internet of Things (IoT) has evolved as the technology of the decade and helps achieve reduced losses, high efficiency, and cost minimization. It is an ongoing development that can interconnect objects, computing devices, mechanical and digital machines, etc. everywhere, every day and transfer data between them via a network without requiring any interaction between humans and computers [4, 5]. It allows the detection and remote control of objects through the available network infrastructure, ensuring improved efficiency, accuracy, and affordability [5]. This paper presents a detailed review of recent articles on IoT-based green energy monitoring systems.

6.2 IoT-Based Green Energy Monitoring Systems

Conventional energy-monitoring systems employing serial-ports for data transmission have limitations when high rates of data sampling are required [6–9]. Forero *et al.* [10] designed an automatic monitoring and data acquisition system (DAS) for solar photovoltaic (PV) panels based on virtual instrumentation. It employed a data acquisition card (DAC), modular field-point devices, and LabVIEW as the graphical user interface (GUI) to collect, store, and display the various electrical parameters and system and environmental variables. This system overcomes the limitation of data transmission during high-frequency sampling [10].

Hou and Gao [11] designed a solar-energy-based wireless sensor network monitoring system for a massively large greenhouse. This system consumes very little energy by employing a high-efficiency microcontroller unit, a network monitoring chip, and multi-level energy memory. By integrating both energy transfer and energy management, the energy collected by the batteries is efficiently utilized. Low energy consumption, economic efficiency, and huge internet capacity are the advantages of this green-house monitoring system [11]. Nkoloma *et al.* [12] explained a wireless-remote tracking system for a solar PV plant established in a Malawi primary school. A solar PV plant was implemented at the Malawi plant and a center for energy management was established at Malawi Polytechnic. The electricity generated at the remote plant was accessed at the management center via wireless sensor networks and text messages were transmitted through mobile networks. The intelligent devices at the center receive the messages via FrontlineSMS and disseminate instantaneous remote-energy generation data and performance analysis reports via the internet. The presented system is cost-effective and highly efficient as it uses open resources and facilitates the reproducibility of the solutions [12].

Web-aided software for distant control and observation of a solar PV plant is illustrated in [13]. The software is designed on a web-based server using HTML, JAVA, and JSP. The web-based foundation enabled the accurate collection, storage, and dissemination of required data to the stakeholders irrespective of their geographical location. The proposed system facilitates faster identification, repair, and maintenance of large solar PV plants. The use of JAVA for developing the software eliminates the need for the selection of a particular platform during installation, as it is widely accepted on a majority of the platforms.

A monitoring system for the energy-range extender of electric and hybrid electric vehicles based on solar energy is proposed in [14]. Here, the system detects a drop in the short-circuit current in the PV panel, isolates the partially-shaded panels, and thus increases the degree of efficiency of the energy-range extender system by up to 60%. Also, the pay-back time of the solar panel is minimized and the data obtained is useful for the vehicle driver.

A real-time surveillance system for a stand-alone PV system is introduced [15]. Data is acquired and processed, and the system performance is analyzed and displayed using virtual instrumentation in LabVIEW. This data is shared via the internet for the stakeholders to view and analyze the daily yield of all acquired parameters. The system is designed so that it facilitates all the acquired and analyzed data files to be downloaded for further offline analysis and these files comply with the standards prescribed by IEC 61724.

Kabalci *et al.* [16] implemented an instant renewable energy monitoring system comprising of PV panels and a wind turbine. The generated voltages of the hybrid system are combined on a dc-bus that charges the battery bank. The system parameters are measured through sensors, are later processed using an 18F4450 microcontroller, and transmitted by serial communication bus to a personal computer for real-time analysis and storage for future use. The limitation of this system is the lack of user controls that allow the stakeholders to analyze daily, weekly, or monthly graphical data based on their requirements [16].

Amruta and Sathish [17] developed a solar-powered water quality surveillance system based on wireless networks. The system comprises of distributed sensor nodes along with a base station. The system acquires information relating to the quality of water like pH, oxygen level, turbidity, etc. from the sensor nodes and communicates it to the base station via the ZigBee wireless network. The received data is represented graphically and analyzed through simulation studies at the base station. The advantages of the system include reduced power consumption, flexibility in employing remote inconvenient sites, zero carbon emissions, etc. [17].

Woyte et al. [18] presented a systematic linear regression analysis for the monitored data of a solar PV installation. The proposed method is a simple and systematic technique for mathematical analysis and visual representation of the monitored parameters. With a selected range of parameters, the energy flow in a grid-tied solar plant can be analyzed. These variables, when combined, aid in the analysis of the energy-conversion process in a solar PV installation [18].

An online surveillance and control system for a solar photovoltaic plant based on the Android platform was created by Jiju et al. [19]. This system employs an Android-based tablet, which uses its Bluetooth facility for data transfer with the hardware. It utilizes its inherent display and touch-screen interface and, thus, eliminates the requirement for a separate internet modem, keypad, and graphical display in the power-conditioning hardware. The proposed system achieves a cost saving of Rs.17900/- relative to the traditional online monitoring system [19].

A TinyOS-based solar PV power-plant monitoring system design was proposed by Ye and Wang. The proposed system overcomes the limitations of existing systems like uneven solar PV system distribution, long-term management of field operations by technicians, and technical difficulty in some remote monitoring systems. In this method, an IoT-based supervisory system that collects the energy output data and environmental parameters from the solar PV plant is employed. Wireless sensor nodes used in the plant collect the data and send it to a distant computer via wireless sensor networks and ARM gateways. This information is then analyzed to check the proper functioning of the plant [20]. This system's main drawback is that it is only appropriate for small-scale photovoltaic plants.

Gaurav et al. [21] created a weather-surveillance device for hybrid energy sources based on an 8051 micro-controller and GSM. It automatically senses weather data like humidity, temperature, insolation of the PV panel, and the turbine speed and transmits data as a Short Message Service

(SMS) via the Global System for Mobile Communications (GSM) transmitter to ground-station. The system is cheap, compact and light-weight, floatable and recoverable using a hydrogen balloon, optimised for the maximum saving of power, has a long operational life, and can be used even during extreme weather conditions. The main limitation of this method is the unexpected delay arising in the delivery of SMS [21].

The problems like an unexpected delay in the delivery of messages in the SMS mode of communication and the requirement of repeating stations every 100m in Global Packet Radio Service (GPRS)-based communication, etc., are overcome by using the voice channel of the GSM cellular network for transfer of data. In this technique, the data from the remote standalone PV plant is transmitted as analog signals of different frequencies and shapes through a mobile phone by initiating a call. Upon automatic reception of the call at the central station, the data is received. The proposed method is economical, easy to implement relative to other methods, and has very high efficiency with a maximum error of below 1% in the received data [22].

Moon *et al.* [23] presented an intelligent solar PV conditioning system comprising of a Maximum Power Point Tracking (MPPT) controller and a ZigBee communication module. This single integrated intelligent controller was specifically developed to control multiple PV plants located at remote locations. This host-controller communicates with the PV plants, retrieves information from them via the ZigBee module, and utilizes this information for the MPPT control of each PV plant remotely. This system achieves drastic cost reduction and 99% MPPT performance even under full insolation and unbalanced variable insolation conditions [23].

Oliveira *et al.* [24] demonstrated an IoT-based EMS for residential environments. The EMS is developed on the foundation of IEEE 802.15.4 and PLC HomePlug Green PHY. 6LoWPAN is used to communicate between IoT devices and between IoT devices and the Internet. A demonstrator comprising of gateways and IoT nodes has been developed based on the COTS chipset. All the software in the system is open-source and freeware. The only limitation of this work is the lack of security and semantic interoperability mechanisms [24].

Gad and Gad [25] constructed a novel temperature DAS based on a micro-controller and low-cost sensors for solar energy applications. It comprises of a power supply, a master control board, an Arduino board, and a sensor-terminal unit. While the sensor-terminal unit interconnects the sensors to the Arduino board, the Arduino board measures and stores the sensor voltages and the Master control board monitors and controls the DAS. This system's advantages include flexibility and ease in changing the sensor type and method of data recording; flexibility during dynamic conditions; suitability for large and remote solar PV plants; and rapid development [25].

The existing remote EMS configurations do not focus on energy conservation. Hence, the energy wastage and cost involved are high. Jayapandian et al. [26] described the monitoring and supervisory control of the electrical appliances in a solar-powered house via cloud computing. Every appliance is monitored and data is gathered via the internet and controlled independently through mobile phones. The excess or unused power generated in one house can be rented to another house connected to the same EMS. Thus, power wastage is reduced, which results in cost savings. This system gives a cost report of energy consumed on a daily, weekly, and yearly basis [26].

Rezk et al. [27] created a DAS for surveillance of meteorological data and solar PV insolation using a hardware-software integrated system and Kosmos-3M for collection, storage, and analysis of the images taken by satellites for forecasting cloudy and sunny weather. The proposed system is cheap and is useful for analyzing the climatic aspects of solar insolation after long-duration data collection [27].

Visconti and Cavalera [28] designed an intelligent EMS for remote monitoring of robbery, improper functioning, and energy optimization of the solar MPPTs. The energy production and solar PV system efficiency are estimated from the acquired environmental parameters, insolation, and temperature. An anti-theft alarm in the system detects emergencies like loss of electrical continuity and drastic changes in the string voltage of the PV plant. The two operational modes in the intelligent system make it automatically adaptable to differentiate between dawn and dusk [28].

Habaebi et al. [29] developed an IoT-based EMS using Wi-Fi communication for the measurement of energy consumption in residential buildings. The system facilitates the deployment of an IoT-based EMS using Wi-Fi communication for the measurement of energy consumption in residential buildings. The system facilitates consumers' ability to collect, monitor, analyze, control, and manage their daily energy consumption. The deep-sleep method is implemented in the data monitoring process to reduce the consumption of energy. Wi-Fi communication is preferred owing to the ease of penetrating the markets. The proposed system has the lowest power dissipation of less than 0.1W/s [29].

Shrihariprasath and Rathinasabhapathy [30] implemented a smart IoT-based EMS for solar PV systems employed in greenhouses. It solves the problems in management and maintenance and minimizes the mean repair time. Communication to a remote desktop computer is achieved via the internet, GPRS network, embedded-system gateways, etc. The system allows data to be monitored, gathered, preserved, and analysed to check the functionality of the solar PV system [30].

A smart remote surveillance and supervisory control system for a solar photovoltaic power generation system based on the IoT was constructed by Adhya *et al.* [31]. This cost-effective system facilitates enhanced performance, fault-detection, preventive maintenance, real-time surveillance and control, and historical data analysis of the solar PV plant. This smart-monitoring system solves the problems of reduced automaticity and inefficient real-time monitoring in prevalent PV-monitoring systems. The only limitation of the proposed system is its applicability to small-scale solar PV systems only. The IoT-enabled system achieved high efficiency by employing advanced wireless modules that consume reduced power [31].

Rastogi *et al.* [32] described the concept of smart IoT-enabled energy meters, different methods of communication, PLC-based communication, optimization of cost, and fraud detection and security. The advantages of these smart meters include reduced proneness to error, remote data transfer, rural and remote sites that can be monitored easily without physical presence, and easy detection of tampering with energy meters. Energy consumption is reduced by programming these smart meters with different home appliances [32].

Touati *et al.* [33] designed a customized green energy wireless monitoring and recording system for a solar PV plant comprising of a Buck-Boost DC/DC converter enhanced with an MPPT. It collects atmospheric parameters, dust, and wireless radio signals and stores them in a monitoring station based on LabVIEW. This system is cost-effective, can be employed anywhere, and sends an alarm signal whenever the efficiency drops below a particular threshold [33].

Akkas and Sokullu [34] developed a greenhouse surveillance system based on IoT using a wireless sensor network comprising of MicaZ motes. The MicaZ nodes precisely measure various greenhouse parameters like humidity, pressure, temperature, and light and transfer the acquired data to a distant computer or mobile of the user via the internet. This system overcomes the limitations of large-scale greenhouse automation systems like the need for proper maintenance, expensive cabling, and difficulty involved in relocation once the installation is completed. The advantages of this system include precise data collection, flexible sensor position, and storage of historical data, which allow us to understand the most recent trends in agriculture and help plan for future agricultural cycles. The only limitations of this system are unidirectional communication and the lack of actuators that facilitate bidirectional control of the greenhouse [34].

Prathibha *et al.* [35] produced an IoT-based intelligent agriculture surveillance system. In this smart monitoring system, the environmental factors, soil pH, soil moisture, etc. are sensed and the acquired data is transferred to the mobile phones of farmers via Wi-Fi communication. A camera is integrated within the system to take photographs of the field and transfer them as MMS to the mobile phones of the farmers. A wireless field monitoring system eliminates the requirement for human resources for field monitoring and facilitates viewing the precise changes in agricultural yields. The system is economical, highly efficient, and has reduced power consumption [35].

Rajkumar *et al.* [36] described a wireless IoT-based irrigation system for agricultural fields. The control techniques used in this system are automated through a micro-controller, which enables the farmers to switch ON/OFF the motor pump depending on the dampness of the soil. This intelligent irrigation system has the advantages of reducing manpower, providing the right quantity of water automatically, reducing power consumption, optimizing the use of resources, reducing water logging and water shortages, enhancing the quality of the environment, and improving irrigation [36].

An automated monitoring and control system was implemented by Hazarika and Chaudhury [37] to overcome the limitations of manual monitoring methods. The manual monitoring method suffers from the need for human resources and difficulty in cleaning, maintaining, and cooling large solar PV plants spanning large areas. The proposed system employs an AT89S52 microcontroller that acquires and displays the various parameters of the photovoltaic panel and the ambient temperature in real-time. This circuit monitors the health of PV plants, sends an alarm signal when the PV panel parameters exceed the standard reference values, monitors the status of the entire PV plant, and provides control signals necessary for enhancing the performance of the PV plant during normal and long-duration operation [37].

Kabalci and Kabalci [38] proposed a novel smart-grid monitoring system for green energy sources. In this system, the power lines, in addition to carrying the electricity generated, are also employed to transfer the energy-consumption details of the loads connected to the microgrid. It is accomplished using the Power Line Communication (PLC) infrastructure, which is controlled by modems placed at different locations. The proposed method eliminates the installation cost of remote EMS relative to its existing counterparts like SCADA, Wireless Sensor Networks-based systems, Ethernet-based systems, etc., and facilitates the easy augmentation of additional load plants for further expansion of the microgrid [38].

Kim et al. [39] introduced daily forecasting models for the solar PV system based on weather forecast data. The system predicts the magnitude of solar insolation and the loss adjustment factor from the temperature and cloud data obtained from the weather prediction data. This predictive model is embedded in the current EMS to enhance its performance. This model has the advantages of easy usability due to automatic tuning of parameters; eliminates the need for a large quantity of historical data for predictive modeling; has simple computational logic and hence facilitates easy implementation; and indicates degradation in the photovoltaic system performance over the regression models and Artificial Neural Network (ANN) models [39].

Gupta et al. [40] constructed a remote real-time surveillance system for accessing PV plants remotely from anywhere. The system monitors the PV panel output easily and any interruptions in the output are detected spontaneously. The system hardware comprises of a RaspberryPi (RPi) microcontroller, an Analog-to-Digital converter, sensors, and a DC voltage transducer. The reported system is cost-effective and suitable for PV plants located in distant areas with difficult accessibility. The data collected can be stored, retrieved, and analysed at any point in time. The collected data yields information about historical energy generation and the degradation of solar PV cells. Easy transmission, stability, reliable operation, accessibility to data from anywhere, graphical presentation of the collected data, and load optimization are a few of the advantages of this system [40].

Patil et al. [5] developed a flask-framework-based RPi microcontroller that employs solar EMS. Daily energy consumption is monitored and the corresponding data is acquired through smart metering. The Flask-based platform is lightweight, flexible, and employs Python programming, which yields a simple and versatile template for web-portal development. The RPi has an in-built Wi-Fi facility in it so that the data can be displayed and stored in the cloud. This enables the users to access the data through the cloud for further analysis, weather forecasting, and estimating battery usage. The advantages of the proposed EMS include cost-effectiveness, high efficiency of 95%, and helping to achieve energy savings [5].

Kekre and Gawre [41] implemented an IoT-based remote surveillance system for a PV system. The hardware infrastructure comprises of a remote monitoring unit, a GPRS module, and a cheap Arduino Uno microcontroller that gathers and sends real-time data via the Internet to a distant computer at the receiving end. This gives the users first-hand real-time information about the installation, its maintenance, intelligent fault identification, and a periodic record of data. The system is very flexible and allows the facility to add more devices and sensors for possible future expansion of the solar PV plant [41].

Spanias [42] presented an IoT-based high-efficiency solar energy management system for an 18 kW solar PV plant. It is programmable and has the advantages of delivering data analytics of the plant via mobile phone, reducing transients in the inverter, fault identification and rectification, and power optimization under various shading conditions [42].

López-Lapeña and Pallas-Areny [43] presented a low-power solar energy harvester for the quantification of solar energy radiation. This harvester measures the open-circuit voltage and current of the solar panels at the Maximum Power Point (MPP) to calculate the solar irradiance and compensate for the temperature drift. This energy harvester consumes less power, incurs less expense, and has a high power efficiency and daily solar irradiance error compared to existing pyranometer-based radiation measurement systems [43].

An intelligent traceability system for tracking and record-keeping crops in a greenhouse was implemented by Gonzalez-Amarillo et al. [44]. Parameters like luminosity, temperature, water consumption, and humidity are tracked and provide details regarding the overall use of water, plant growth patterns, and timeline of the harvest of the crop produce. The system facilitates the internal tracking of the greenhouse via an automatic irrigation system and temperature control and ensures complete tracking of the crops from the seedling stage to the end crop product. The IoT platform of this smart traceability system helps to evaluate the behaviour of a wide variety of crops inherent to a locality [44].

Gunturi and Reddy [45] constructed a smart energy meter based on an IoT platform for domestic electricity consumers. It is lightweight, portable, economical, compact, and technically sound compared to existing portable, hand-held bulky energy meters. This meter acquires and sends the data to the cloud-assigned location, from where the user can obtain the energy consumption of a specific appliance for a particular duration by scanning a unique QR code on the device. Thus, the user can cross-verify the reliability of the energy meter and measure the energy usage of the device [45].

Gusa et al. [46] introduced a smartphone-based real-time PV panel surveillance system. The hardware structure of this real-time monitoring system comprises of an Arduino Atmega 2560 microcontroller; sensors for measuring current, voltage, and temperature; an ADC; a Wi-Fi module; and an Android-based smartphone installed with the Blynk app. The monitored real-time information from the solar panels is acquired and sent to the smartphone through the Wi-Fi module connected to the Atmega 2560 microcontroller. The Blynk app on the smartphone receives the data and displays it in its GUI. The system offers high accuracy and a very low mean error of less than 10% in the monitored solar PV panel output [46].

Badave et al. [47] introduced a wireless IoT-based health-tracking system for a solar PV farm. The system infrastructure comprises of a CC3200 microcontroller having ARM Cortex-M4 as the core, sensors, an ADC, and an onboard Wi-Fi module for data transfer. Sensors acquire the solar PV system parameters and transfer them to a smartphone or remote PC via the onboard Wi-Fi connected to the microcontroller. Real-time PV panel monitoring is attained via the IoT connectivity protocol, Message Queue Telemetry Transfer (MQTT), which transfers the information to the remote destination [47].

Singh and Chandra [48] presented an IoT-based greenhouse monitoring system where parameters like light intensity, temperature, humidity, soil pH, moisture content, etc. are tracked in real-time by sensors using the MQTT protocol. This protocol was chosen owing to its numerous advantages, including lightweight data transmission, reduced consumption of resources, offering flexibility, and ensuring efficient detection of various greenhouse parameters, compared to the Constrained Application Protocol (CoAP) and Extensible Messaging and Presence Protocol (XMPP). The proposed real-time greenhouse monitoring system closely tracks and controls the greenhouse parameters to maintain ambient environmental conditions to maximize crop yields throughout the season and reduce manpower significantly [48].

Sadowski and Spachos [49] constructed an economical IoT-based smart monitoring system powered by solar energy for Precision Agriculture (PA). Nodes powered by solar energy were designed to monitor and acquire the environmental parameters in a field and the gathered data was transferred to a base station for storage and analysis [49]. The nodes comprise of a micro-controller, an energy-harvesting device, a battery, and wireless antennas proven to be a robust and reliable solution for smart precision agriculture [49].

An intelligent greenhouse monitoring system was demonstrated by Danita et al. [50]. The proposed system structure comprises of humidity and temperature sensors, an RPi microcontroller, and an IoT-based platform for intelligent control of the system. For obtaining maximum growth and yield from the crops, the microcontroller maintains the ambient condition inside the greenhouse by actuating the irrigation pump, sliding windows, and turning on a cooling fan based on the sensed values of humidity and temperature. The recorded values of these parameters are stored in the cloud and the user can retrieve them from the web page whenever required [50].

Icaza et al. [51] developed a novel, real-time monitoring and tracking system for the border areas of Ecuador and Columbia through micro-cameras installed in native birds. The system is powered by solar energy, which guarantees its operation. A telemetry system suitable for the existing communication protocol of Ecuador, SIS ECU911, is designed for the system. It is ensured that the micro-cameras embedded in the native birds do not affect their normal growth and coexistence among the different species in the border areas [51].

Márquez and Ramírez [52] presented a condition tracking system utilizing unmanned aerial vehicles (UAV) for solar PV plants. This system checks dust in the solar PV panels by analyzing the emissivity generated on a surface and is characterized by a low value of emissivity when dust accumulates on the surface. For the emissivity analysis of the PV panels, an Arduino micro-controller board connected with radiometric sensors and a thermographic camera is embedded in a UAV. Radiometric information is transferred and analyzed and thermographic data is stored for detailed analysis [52].

Shamshiri et al. [53] introduced an intelligent IoT-based electricity tracking system for a university. This intelligent device interacts with several digital energy meters via the Modbus protocol and effectively tracks the campus-wide electricity consumption. The field data is collected from different sub-metering levels and is communicated to a central hub, which receives, stores, and processes the real-time data through IoT technology. Measures to save electricity are implemented by the energy manager based on the hourly load profile received from the EMS [53].

Khan et al. [54] described an IoT-based real-time EMS for tracking electricity generation in a solar PV plant. The system integrates a multi-function Arduino microcontroller, sensors, an ADC, and a Wi-Fi module to acquire and transmit real-time data of PV panel variables and environmental parameters to a distant computer or smartphone over an IoT platform. The data is received, stored, and analysed to study the electricity generated and identify the various faults on the PV panels. This system is very economical, improves the efficiency of the PV plant, and has a minimum error percentage compared to the existing SCADA-based monitoring and controlling system [54].

Rahman et al. [55] presented an IoT-based intelligent highway lighting system powered by hybrid green energy systems. To overcome the limitations of pure solar PV systems, like high cost and inconsistencies, hybrid green energy sources are employed to power the highway lighting systems. An IoT-based platform is used to control the multiple green energy sources for improving the performance of the battery of the highway lighting system. In addition to solar PV panels, a wind turbine is also employed for utilizing the energy from the aerodynamic losses generated by moving automobiles on the highway, thus enhancing the efficiency of the system and providing an uninterruptible power supply to highway lights at all times. A microcontroller is employed to achieve the utmost effectiveness and system performance [55].

Hassnuddin and Kumar [56] proposed a method of advanced green energy scheduling using IoT in a smart grid. This method is user-independent; that is, it monitors consumer-side pricing and makes its own decisions without the need for consumer intervention. The IoT platform interconnects the consumer and the utility to enable bidirectional communication between them based on electricity consumption and energy cost. The consumer achieves energy savings and economy by taking relevant measures to reduce their energy consumption based on the real-time data analysis available on the web portal [56].

Michael *et al.* [57] produced a water tracking system based on green IoT for a metro city. The system monitors the real-time water consumption and uploads it to the cloud, even in locations with no power supply. A solar PV plant and a battery bank are used to provide continuous power supply in such areas and the water consumption details are uploaded to the cloud via an IoT-based NODEMCU micro-controller. This design achieves the lowest probability of power outages [57].

Agyeman *et al.* [58] created an IoT-based EMS for smart home management. This intelligent meter facilitates the remote and independent control of each household appliance through a mobile application and a web portal by utilizing an Arduino micro-controller and sensors. The system is beneficial for both consumers and the utility in terms of the facility for visualizing energy consumption, managing and controlling the intelligent meter, and a web portal for tracking and administering the system. The proposed smart meter has several advantages, including different payment options for paying the energy bill, sending a message to the consumer if their consumption exceeds 75% of the customizable limit, facilitating bidirectional communication between the utility and consumers, and independent control of appliances, among other things [58].

Al-Turjman *et al.* [2] reviewed the different types of energy monitoring systems using IoT-based ad hoc networks. The advantages of the EMS include improved resource utilization, minimizing energy wastage, and a reduction in energy consumption. They reviewed two types of EMS: indirect feedback systems and direct feedback systems. In an indirect feedback system, the data collected from the EMS is administered by the utility and later sent to the consumers with their monthly bills. In contrast, in a direct feedback system, consumers provide real-time raw data via cloud-based IoT platforms [2].

Sharma et al. [59] presented ambient solar energy harvesting for charging the batteries of the wireless sensor network nodes extensively used in smart agriculture. Thus, by facilitating battery charging via the harvested solar energy, the lifespan of these nodes is extended. It is observed that by using this technique, the lifespan of these nodes raised from 5.75 days to 115.75 days at a duty ratio of 25% and the throughput of the network rose from 100kBits/s to 160 kBits/s [59].

Al-Ali et al. [60] described a smart, solar-powered IoT-based irrigation system for agricultural farms in areas with acute water shortages and frequent power outages. The system architecture comprises of a microcontroller with an onboard Wi-Fi feature and connections to solar PV panels for its operational power. The controller tracks the soil moisture, soil pH, humidity, temperature, etc. and outputs a control signal to actuate the water pump. It also tracks the underground water level to protect the water pump motor from burning due to the shortage in the underground water level. The irrigation system has three operational modes, namely, mobile-monitoring mode, local-control mode, and fuzzy-logic-based control mode. The advantages of this include very high accessibility, the ability to control the pump via smartphone or web-based PC, and the ability to scale up during future expansion of the farm [60].

Jumaat et al. [61] developed a dual-axis solar tracker with an IoT-based surveillance system employing an Arduino microcontroller. The system hardware comprises of Light Dependent Resistors (LDRs) to detect the sunlight intensity, servo motors to turn the solar panel according to sunlight intensity, and a Wi-Fi module to enable bidirectional interaction between the IoT monitoring device and the solar PV system. IoT monitoring acquires, stores, and displays the solar panel parameters in a graphical format on a web portal for further analysis [61].

Karbhari and Nema [62] introduced a novel method of monitoring the health of a solar PV plant employing IoT and Machine Learning techniques. The power generation of the PV panel is tracked and increased using this technique. The power generation efficiency of the solar panel is increased by aligning the panel according to the sun's direction remotely using an IoT-enabled MPPT system. A Wide Area Network (WAN) based web server is deployed to transfer, store, and retrieve the PV plant information from anywhere on the globe that has access to the internet [62].

Cheddadi et al. [63] implemented a cost-effective smart EMS for solar PV plants. This intelligent solution tracks the PV panel continuously and transmits the PV panel parameters and environmental conditions in real-time via the IoT platform to a distant PC pre-installed with a web app, Grafana. This IoT-enabled PC receives, stores, processes, and displays the information in a graphical format for further processing and analysis. The system also sends an alarm message to the remote user whenever the quality of the PV Panel performance parameters deviates from the set reference values. The proposed EMS is highly cost-effective as it uses low-cost sensors and an open-source IoT platform for data transfer and processing [63].

Krishnamoorthy *et al.* [64] created an IoT-enabled EMS that tracks the energy consumption of the load and saves electricity efficiently. The EMS comprises of an intelligent energy meter that acquires the energy consumption of each load individually and transmits the details via GPRS to a distant web server using IoT technology. The energy consumption details are stored, processed, and displayed on a website so that the remote user can access them from anywhere [64].

Hossain *et al.* [65] demonstrated an IoT-based condition tracking system for a Wind Energy Conversion System (WECS). A WECS experiences a number of faults in the DC-bus capacitor and power converter switches which, if not rectified in the early stages, could lead to an interruption in the power supply. In the proposed monitoring system, a PI-based Space Vector Pulse Width Modulated Controller embedded with a novel algorithm is employed to detect in real-time the location of the power converter fault. With WECS plants being mostly offshore, IoT technology is being embraced to promise the reliability of the WECS. The fault details are monitored remotely in real-time and relevant steps are taken to rectify these faults firsthand. This system is cost-effective and has high efficiency and reliability [65].

Saxena and Dutta [66] presented an IoT-enabled wireless sensor monitoring system for enhancing the productivity of agricultural farms. The system employs a solar energy harvester that is connected to a microcontroller for battery charging of the sensor nodes. It monitors the various environmental conditions and soil parameters and transmits them via an IoT platform to a distant web server. Based on the obtained data, the system manages the scheduled tasks on the farms. This system thus controls the losses and enhances the efficiency, quality, and productivity of crop cultivation [66].

Xia *et al.* [67] implemented an intelligent EMS for solar PV plants. The communication and networking between the sensor nodes are achieved via ZigBee and the information is uploaded to the cloud storage using the 4G-communication network. The server sets up a composite current characteristic and develops a fault-diagnostics model from the uploaded information using a probabilistic neural network to track the health conditions of the solar panel in real-time. The system gathers and stores the health information of the PV panels and provides a graphical display to the remote end-user via smartphone or web portal on a PC. The merits of this system include the requirement for little sampling frequency, eliminating the need for extra sensors, and cost-effectiveness [67].

Avancini *et al.* [68] developed a novel IoT-based intelligent-energy meter with a multi-protocol connection and various interfaces for communication. The mobilized data is wirelessly transmitted through IoT protocols and this data is received in real-time by an IoT-based middleware that provides the users with the required energy information via the internet. The system employs ZigBee, Bluetooth, and 6LoWPAN for data communication. This meter monitors energy consumption in real-time and helps customers save money and energy [68].

Mukta *et al.* [69] proposed a highly efficient and cost-effective IoT-based standalone highway lighting system. It includes an IoT-based network configuration that takes into account real-time automobile status for dynamic and smooth highway illumination. The proposed solution incorporates and unifies the main features, including hybrid green energy generation, demand-based resilient lighting, connected light posts, and distant control and monitoring for rapid repairs and maintenance. The Digi XBee S2 module, 3G module, and WiMAX were employed for establishing short-distance communication between light posts, communication between the leader light post and the Remote System Management Unit, and remote highway applications, respectively. The proposed solution is highly cost-effective and greatly improves the energy efficiency and performance of present-day standalone lighting systems [69].

Muthubalaji *et al.* [70] developed an EMS that employs a hybrid system integrated with an IoT framework. The hybrid system integrates the Seagull Optimization Algorithm (SOA) and Owl Search Algorithm (OSA) for the optimal control of the power and resources of the distribution system via real-time monitoring through an IoT-based platform [70]. The proposed technique is highly efficient and is accountable for satisfying the complete power and supply demands [70]. The IoT-enabled distribution system improved the adaptiveness of the wireless networks and facilitated optimal usage of the available resources [70].

Bhau *et al.* [71] demonstrated an intelligent IoT-based solar EMS. The current and voltage sensors are interfaced to an Arduino board. The Arduino tracks the real-time data and communicates the information to an IoT-enabled desktop PC or tablet via a WiFi module. This IoT-enabled cloud is user-friendly and enables the administrators to visualize the data illustration and conduct surveillance over the internet [71].

Zhang *et al.* [72] introduced an IoT-based green EMS for smart cities powered by deep reinforcement learning. The developed system automatically monitors the energy demands of the users and allocates smart power systems, thereby reducing energy consumption by limiting its unwanted dissipation. The EMS optimally balances the energy availability and demand by maintaining the stable states using the recurrent learning technique. The developed EMS reduces emissions through the optimal and efficient utilization of local power sources and the use of decentralized urban power supplies [72].

Sadeeq and Zeebaree [73] demonstrated an IoT-based EMS for distribution systems. The system combines both IoT-enabled Heating, Ventilation, and Air Conditioning (HVAC) systems and EMS to include both smart energy devices and a smart energy hub. It has been demonstrated that the Internet of Things is the best method for energy conservation and management, as well as for monitoring and controlling distributed systems [73].

Pal *et al.* [74] designed and developed an IoT-enabled intelligent EMS for virtual power plants using a Programmable Logic Controller (PLC). The system deploys IoT-based automation and PLC for the optimal management of energy through the regulation of the various sources and loads of the virtual power plant. Raspberry Pi and PLC were used to transmit the controlling signals to the relay switches and regulate the different sources and loads. The PLC was programmed using the Nod-Red programming package. The developed system efficiently handled the power demand of the virtual power plant [74].

Xiaoyi *et al.* [75] developed an IoT-based green EMS with a Multi-Objective Distributed Dispatching Algorithm (MODDA) for smart cities. The suggested algorithm is used to make decisions regarding when to use the power from the grid and batteries and when to direct power to the load. The suggested algorithm attains a large utility function and obtains reduced energy usage in smart cities [75].

Ramu *et al.* [76] created an intelligent IoT-based surveillance system for solar photovoltaic applications. The proposed system comprises of three layers. The first layer is attributed to the PV system architecture. The second layer is attributed to communication links between the PV system hardware and the remote server via router and internet firewall. An Arduino is deployed to integrate with the web server via an Ethernet cable or a wireless router. The parameters of the PV system are monitored and controlled by Arduino. The final third layer controls the PV system hardware based on the feedback received from the end users from the real-time information available to them. This system greatly enhanced real-time data analysis and effective scheduling of the loads [76].

6.3 Comparative Analysis

From the above reviews, the characteristics of different Energy Systems are represented in Table 6.1. It is observed that most of the EMS discussed in Section 6.2 use a Wi-Fi Wireless Sensor Network, which has replaced the very expensive Ethernet wired connection. The Arduino Uno remains the most sought-after controller for Energy Monitoring Systems owing to its cost-effectiveness, high performance, and independent and open-source software. All the EMS display their real-time results through PCs, laptops, Android-based smartphones, and tablets.

Table 6.2 presents the significant differences in the characteristics of different monitoring methods. It is observed that each of these monitoring methods has its own merits and demerits, hence the type of monitoring method for a particular application is chosen based on the requirements of the consumer.

Table 6.1 Characteristics of different IoT-based energy monitoring systems.

Author and Year	Data transfer mechanism	Controller	Monitoring methods
Hou and Gao, 2010	Wireless Sensor Network	MSP430	PC, Web
Nkoloma et al., 2011	Wireless Sensor Network	Arduino Uno	PC, Web, Smartphone
Kumar, 2011	Wireless Sensor Network	Arduino Uno	PC, Web, Laptop
Schuss et al., 2012	Wireless Sensor Network: LIN Bus, CAN Bus	Arduino Uno	Smartphone
Torres et al., 2012	Wireless Sensor Network: GPRS	Agilent 3490A	PC, Web
Kabalci et al., 2013	Wired: Universal Serial Bus	PIC18F4450	PC

(*Continued*)

Table 6.1 Characteristics of different IoT-based energy monitoring systems. (*Continued*)

Author and Year	Data transfer mechanism	Controller	Monitoring methods
Amrutha *et al.*, 2013	Underwater Wireless Sensor Network: ZigBee	ARM 7	PC, Web
Jiju *et al.*, 2014	Wireless Sensor Network: Bluetooth	PIC18F4550	Smartphone, Tablet
Ye and Wang, 2014	Wireless Sensor Network: GPRS	CC2530	PC
Gaurav *et al.*, 2014	Wireless Sensor Network: GSM-based SMS	8051 MUC	Smartphone
Tejwani *et al.*, 2014	Wireless Sensor Network: GSM Voice Channel	Arduino Uno	Smartphone
Moon *et al.*, 2015	Wireless Sensor Network: ZigBee	TMS320F28335	PC, Laptop
Oliviera *et al.*, 2015	Wireless Sensor Network: 6LoWPAN, Wired Network: PLC	MSP430	PC, Smartphone
Gad and Gad, 2015	Wired Network: Ethernet	PIC18F46K20	PC, Web
Rezk *et al.*, 2015	Satellite Communication	Kosmos-3M	PC, Laptop, Web
Visconti and Cavalera, 2015	Wireless Sensor Network: GPRS, GSM	CS097	PC, Laptop
Habaebi *et al.*, 2016	Wireless Sensor Network: Wi-Fi	Atmega328	PC, Laptop

(*Continued*)

Table 6.1 Characteristics of different IoT-based energy monitoring systems. (*Continued*)

Author and Year	Data transfer mechanism	Controller	Monitoring methods
Adhya et al., 2016	Wireless Sensor Network: GPRS	PIC18F46K22	Laptop, PC
Rastogi et al., 2016	Wired Network: PLC	Arduino Uno	Laptop, PC
Shrihariprasath and Rathinasabapathy, 2016	Wireless Sensor Network: GPRS	Arduino Uno R3	Laptop, PC
Touati et al., 2016	Wireless Sensor Network: ZigBee	Atmega328	PC, Web, Laptop
Akkas and Sokullu, 2017	Wireless Sensor Network: MicaZ Motes	MIB520	PC, Web, Smartphone
Prathibha et al., 2017	Wireless Sensor Network: Wi-Fi, GPRS	CC3200	PC, Web, Smartphone
Rajkumar et al., 2017	Wireless Sensor Network: Wi-Fi	Atmega328	PC, Web, Smartphone
Gupta et al., 2017	Wireless Sensor Network: Wi-Fi	Raspberry Pi	PC, Laptop, Web
Patil et al., 2017	Wired Network: LAN	Raspberry Pi	PC, Laptop, Web
Hazarika and Choudhury, 2017	Wireless Sensor Network: Wi-Fi	AT89S52	PC, Laptop, Web
Kabalci and Kabalci, 2017	Wired Network: PLC	PIC18F4450	PC, Laptop, Web
Kim et al., 2017	Wireless Sensor Network: Wi-Fi	Arduino Uno	PC, Laptop, Web

(*Continued*)

Table 6.1 Characteristics of different IoT-based energy monitoring systems. (*Continued*)

Author and Year	Data transfer mechanism	Controller	Monitoring methods
Kekre and Gawre, 2017	Wireless Sensor Network: GPRS, GSM	Arduino Uno	PC, Laptop, Web
Gonzalez-Amarillo *et al.*, 2018	Wireless Sensor Network: Wi-Fi	RaspberryPi3	PC, Laptop, Web
Gunturi and Reddy, 2018	Wireless Sensor Network: Wi-Fi	MSP430F6736	PC, Laptop, Web
Danita *et al.*, 2018	Wireless Sensor Network: Wi-Fi	RaspberryPi3	PC, Laptop, Web
Gusa *et al.*, 2018	Wireless Sensor Network: Wi-Fi	Arduino Uno	Smartphone
Badave *et al.*, 2018	Wireless Sensor Network: Wi-Fi	CC3200	PC, Laptop, Web, Tablet
Singh and Chandra, 2018	Wireless Sensor Network: MQTT	Raspberry Pi	PC, Laptop, Web, Tablet
Sadowski and Spachos, 2019	Wireless Sensor Network: ZigBee	Arduino Uno Rev3	PC, Laptop, Web
Marques and Ramirez, 2019	Wireless Sensor Network: Wi-Fi	Arduino DUE	PC, Laptop, Web
Shamshiri *et al.*, 2019	Wireless Sensor Network: Modbus	PIC18F4450	PC, Laptop, Web
Khan *et al.*, 2019	Wireless Sensor Network: Wi-Fi	Arduino Uno	PC, Laptop, Web
Rahman *et al.*, 2019	Wireless Sensor Network: Wi-Fi	Arduino Uno R3	PC, Laptop, Web
Hassnuddin and Kumar, 2019	Wired Network: Ethernet	Arduino Uno	PC, Laptop

(*Continued*)

Table 6.1 Characteristics of different IoT-based energy monitoring systems. (*Continued*)

Author and Year	Data transfer mechanism	Controller	Monitoring methods
Michael et al., 2019	Wireless Sensor Network: Wi-Fi	Node-MCU ESP8266	PC, Laptop, Web
Agyeman et al., 2019	Wired Network: Ethernet, Wireless Sensor Network: Wi-Fi	Arduino Uno R3	PC, Laptop Smartphone, Tablet
Sharma et al., 2019	Wireless Sensor Network: Wi-Fi	Arduino Uno	PC, Laptop, Web
Al-Ali et al., 2020	Wired Network: Ethernet, Wireless Sensor Network: Wi-Fi	MyRIO	PC, Laptop Smartphone, Tablet
Saxena and Dutta, 2020	Wireless Sensor Network: Wi-Fi	Arduino Uno	PC, Laptop, Web
Xia et al., 2020	Wireless Sensor Network:4G Communication	CC2530	PC, Smartphone
Hossain et al., 2020	Wireless Sensor Network: Wi-Fi	Arduino Uno	PC
Cheddadi et al., 2020	Wireless Sensor Network: Wi-Fi	ESP32	PC, Laptop, Web
Said et al., 2020	Wireless Sensor Network: Wi-Fi	Arduino Uno	PC, Laptop
Karbhari et al., 2020	Wireless Sensor Network: Wi-Fi	Atmega328P	PC, Laptop, Web
Krishnamoorthy et al., 2020	Wireless Sensor Network: GPRS	Arduino Uno	PC
Xia et al., 2021	Wireless Sensor Network: ZigBee	Arduino Uno	PC, Web

(*Continued*)

Table 6.1 Characteristics of different IoT-based energy monitoring systems. (*Continued*)

Author and Year	Data transfer mechanism	Controller	Monitoring methods
Avancini et al., 2021	Wireless Sensor Network: ZigBee, Bluetooth	Arduino Uno	PC, Web
Mukta et al., 2021	Wireless Sensor Network: DigiXBee, WiMax	Arduino Uno	PC, Web
Bhau et al., 2021	Wireless Sensor Network: Wi-Fi	Arduino Uno	PC, Web
Pal et al., 2021	Wired Network: PLC	Raspberry Pi	PC, Web

Table 6.2 Comparison of different monitoring methods.

Characteristic	Types	Differences
Data Transfer Mechanism	Wired Wireless PLC	Excellent bandwidth, low error, expensive Lower bandwidth, poor service quality, expensive Cheap, low error, excellent bandwidth
Controller	DSP MUC	A highly integrated device, expensive, depends on software Cheap, high-performance, independent of software
Software Language	LabVIEW MATLAB JAVA C, TURBO C++	One operator with a laptop is needed to run the (LabVIEW, MATLAB, and JAVA) software, expensive, requirement of grid power supply Cheap, independent of PC and grid power supply

(*Continued*)

Table 6.2 Comparison of different monitoring methods. (*Continued*)

Characteristic	Types	Differences
Monitoring Methods	PC Web	Difficult to access the monitored data from remote PV plants Users can access the data from anywhere at any time
Parameters	Temperature Solar Irradiance Solar Panel Output Voltage Solar Panel Output Current Wind Velocity Soil pH Humidity Pressure	-

6.4 Conclusion

The application of the Internet of Things (IoT) in energy conservation and optimization methods has led to a new, disruptive technology called the Internet of Energy (IoE). This paper presents a detailed review of the recent trends in IoT-based green Energy Monitoring Systems (EMS). The various characteristics of EMS, like data transfer mechanism, controller, and monitoring methods, are presented in Table 6.1. It is shown that most of the EMS discussed in Section 6.2 use Wi-Fi Wireless Sensor Networks, which have replaced the very expensive Ethernet wired connection. The Arduino Uno remains the most sought-after controller for Energy Monitoring Systems owing to its cost-effectiveness, high performance, and independent and open-source software. All the EMS display their real-time results through PCs, laptops, Android-based smartphones, and tablets. The significant differences in the characteristics of different EMS are described in Table 6.2. It is observed that each of these monitoring methods has its own merits and demerits, hence the type of EMS for a particular application is chosen based on its requirements.

References

[1] Aishwarya V, Gnana Sheela K. Topologies and Comparative Analysis of Reduced Switch Multilevel Inverters for Renewable Energy Applications. In: Bhaskar MS, Gupta N, Padmanaban S, Holm-Nielsen JB, Subramaniam U, editors. Power Electron. Green Energy Convers., John Wiley & Sons, Ltd; 2022, p. 221–64. https://doi.org/https://doi.org/10.1002/9781119786511.ch7.

[2] Al-Turjman F, Altrjman C, Din S, Paul A. Energy monitoring in IoT-based ad hoc networks: An overview. Comput Electr Eng 2019;76:133–42. https://doi.org/10.1016/j.compeleceng.2019.03.013.

[3] Renewables – Global Energy Review 2020 – Analysis - IEA. 2020.

[4] Ibrahim DiM. Internet of Things Technology based on LoRaWAN Revolution. 2019 10th Int. Conf. Inf. Commun. Syst. ICICS 2019, Institute of Electrical and Electronics Engineers Inc.; 2019, p. 234–7. https://doi.org/10.1109/IACS.2019.8809176.

[5] Patil SM, Vijayalashmi M, Tapaskar R. IoT based solar energy monitoring system. 2017 Int. Conf. Energy, Commun. Data Anal. Soft Comput. ICECDS 2017, Institute of Electrical and Electronics Engineers Inc.; 2018, p. 1574–9. https://doi.org/10.1109/ICECDS.2017.8389711.

[6] Blaesser G. PV system measurements and monitoring the European experience. Sol Energy Mater Sol Cells 1997;47:167–76. https://doi.org/10.1016/S0927-0248(97)80008-6.

[7] Wilshaw AR, Pearsall NM, Hill R. Installation and operation of the first city centre PV monitoring station in the United Kingdom. Sol. Energy, vol. 59, Pergamon; 1997, p. 19–26. https://doi.org/10.1016/S0038-092X(96)00123-5.

[8] Mukaro R, Carelse XF, Olumekor L. First performance analysis of a silicon-cell microcontroller-based solar radiation monitoring system. Sol Energy 1998;63:313–21. https://doi.org/10.1016/S0038-092X(98)00072-3.

[9] Koutroulis E, Kalaitzakis K. Development of an integrated data-acquisition system for renewable energy sources systems monitoring. Renew Energy 2003;28:139–52. https://doi.org/10.1016/S0960-1481(01)00197-5.

[10] Forero N, Hernández J, Gordillo G. Development of a monitoring system for a PV solar plant. Energy Convers Manag 2006;47:2329–36. https://doi.org/10.1016/j.enconman.2005.11.012.

[11] Hou J, Gao Y. Greenhouse wireless sensor network monitoring system design based on solar energy. Int. Conf. Challenges Environ. Sci. Comput. Eng. CESCE 2010, vol. 2, 2010, p. 475–9. https://doi.org/10.1109/CESCE.2010.274.

[12] Nkoloma Mayamiko, Zennaro Marco, Bagula Antoine. SM^2: Solar monitoring system in Malawi - IEEE Conference Publication. Proc. ITU Kaleidosc. 2011 Fully Networked Human? - Innov. Futur. Networks Serv., Cape Town, South Africa: IEEE; 2011.

[13] Kumar BA. Solar Power Systems Web Monitoring 2011.

[14] Schuss C, Eichberger B, Rahkonen T. A monitoring system for the use of solar energy in electric and hybrid electric vehicles. 2012 IEEE I2MTC - Int. Instrum. Meas. Technol. Conf. Proc., 2012, p. 524–7. https://doi.org/10.1109/I2MTC.2012.6229214.

[15] Torres M, Muñoz FJ, Muñoz J V., Rus C. Online Monitoring System for Stand-Alone Photovoltaic Applications—Analysis of System Performance From Monitored Data. J Sol Energy Eng 2012;134. https://doi.org/10.1115/1.4005448.

[16] Kabalci E, Gorgun A, Kabalci Y. Design and implementation of a renewable energy monitoring system. Int. Conf. Power Eng. Energy Electr. Drives, 2013, p. 1071–5. https://doi.org/10.1109/PowerEng.2013.6635759.

[17] Amruta MK, Satish MT. Solar powered water quality monitoring system using wireless sensor network. Proc. - 2013 IEEE Int. Multi Conf. Autom. Comput. Control. Commun. Compress. Sensing, iMac4s 2013, 2013, p. 281–5. https://doi.org/10.1109/iMac4s.2013.6526423.

[18] Woyte A, Richter M, Moser D, Mau S, Reich N, Jahn U. Monitoring of photovoltaic systems: good practices and systematic analysis. Proc. 28th Eur. Photovolt. Sol. Energy Conf., 2013, p. 3686–94.

[19] Jiju K, Ramesh P, Brijesh P, Sreekumari B. Development of Android based on-line monitoring and control system for Renewable Energy Sources. I4CT 2014 - 1st Int. Conf. Comput. Commun. Control Technol. Proc., Institute of Electrical and Electronics Engineers Inc.; 2014, p. 372–5. https://doi.org/10.1109/I4CT.2014.6914208.

[20] Ye J, Wang W. Research and design of solar photovoltaic power generation monitoring system based on TinyOS. Proc. 9th Int. Conf. Comput. Sci. Educ. ICCCSE 2014, Institute of Electrical and Electronics Engineers Inc.; 2014, p. 1020–3. https://doi.org/10.1109/ICCSE.2014.6926617.

[21] Gaurav D, Mittal D, Vaidya B, Mathew J. A GSM based low cost weather monitoring system for solar and wind energy generation. 5th Int. Conf. Appl. Digit. Inf. Web Technol. ICADIWT 2014, IEEE Computer Society; 2014, p. 1–7. https://doi.org/10.1109/ICADIWT.2014.6814689.

[22] Tejwani R, Kumar G, Solanki C. Remote monitoring for solar photovoltaic systems in rural application using GSM voice channel. Energy Procedia, vol. 57, Elsevier Ltd; 2014, p. 1526–35. https://doi.org/10.1016/j.egypro.2014.10.145.

[23] Moon S, Yoon SG, Park JH. A New Low-Cost Centralized MPPT Controller System for Multiply Distributed Photovoltaic Power Conditioning Modules. IEEE Trans Smart Grid 2015;6:2649–58. https://doi.org/10.1109/TSG.2015.2439037.

[24] Oliveira LML, Reis J, Rodrigues JJPC, De Sousa AF. IOT based solution for home power energy monitoring and actuating. Proceeding - 2015 IEEE Int. Conf. Ind. Informatics, INDIN 2015, Institute of Electrical and Electronics Engineers Inc.; 2015, p. 988–92. https://doi.org/10.1109/INDIN.2015.7281869.

[25] Gad HE, Gad HE. Development of a new temperature data acquisition system for solar energy applications. Renew Energy 2015;74:337–43. https://doi.org/10.1016/j.renene.2014.08.006.

[26] Jayapandian N, Rahman AMJMZ, Poornima U, Padmavathy P. Efficient online solar energy monitoring and electricity sharing in home using cloud system. IC-GET 2015 - Proc. 2015 Online Int. Conf. Green Eng. Technol., Institute of Electrical and Electronics Engineers Inc.; 2016. https://doi.org/10.1109/GET.2015.7453775.

[27] Rezk H, Tyukhov I, Raupov A. Experimental implementation of meteorological data and photovoltaic solar radiation monitoring system. Int Trans Electr Energy Syst 2015;25:3573–85. https://doi.org/10.1002/etep.2053.

[28] Visconti P, Cavalera G. Intelligent system for monitoring and control of photovoltaic plants and for optimization of solar energy production. 2015 IEEE 15th Int. Conf. Environ. Electr. Eng. EEEIC 2015 - Conf. Proc., Institute of Electrical and Electronics Engineers Inc.; 2015, p. 1933–8. https://doi.org/10.1109/EEEIC.2015.7165468.

[29] Habaebi MH, Ashraf QM, Azman AA Bin, Rafiqul Islam M. Development of Wi-Fi based home energy monitoring system for green internet of things. J Electron Sci Technol 2016;14:249–56. https://doi.org/10.11989/JEST.1674-862X.604051.

[30] Shrihariprasath B, Rathinasabapathy V. A smart IoT system for monitoring solar PV power conditioning unit. IEEE WCTFTR 2016 - Proc. 2016 World Conf. Futur. Trends Res. Innov. Soc. Welf., Institute of Electrical and Electronics Engineers Inc.; 2016. https://doi.org/10.1109/STARTUP.2016.7583930.

[31] Adhya S, Saha D, Das A, Jana J, Saha H. An IoT based smart solar photovoltaic remote monitoring and control unit. 2016 2nd Int. Conf. Control. Instrumentation, Energy Commun. CIEC 2016, Institute of Electrical and Electronics Engineers Inc.; 2016, p. 432–6. https://doi.org/10.1109/CIEC.2016.7513793.

[32] Rastogi S, Sharma M, Varshney P. Internet of Things based smart electricity meters. Int J Comput Appl 2016;133:13–6.

[33] Touati F, Al-Hitmi MA, Chowdhury NA, Hamad JA, San Pedro Gonzales AJR. Investigation of solar PV performance under Doha weather using a customized measurement and monitoring system. Renew Energy 2016;89:564–77. https://doi.org/10.1016/j.renene.2015.12.046.

[34] Akkaş MA, Sokullu R. An IoT-based greenhouse monitoring system with Micaz motes. Procedia Comput. Sci., vol. 113, Elsevier B.V.; 2017, p. 603–8. https://doi.org/10.1016/j.procs.2017.08.300.

[35] Prathibha SR, Hongal A, Jyothi MP. IOT Based Monitoring System in Smart Agriculture. Proc. - 2017 Int. Conf. Recent Adv. Electron. Commun. Technol. ICRAECT 2017, Institute of Electrical and Electronics Engineers Inc.; 2017, p. 81–4. https://doi.org/10.1109/ICRAECT.2017.52.

[36] Rajkumar MN, Abinaya S, Kumar VV. Intelligent irrigation system - An IOT based approach. IEEE Int. Conf. Innov. Green Energy Healthc. Technol. - 2017, IGEHT 2017, Institute of Electrical and Electronics Engineers Inc.; 2017. https://doi.org/10.1109/IGEHT.2017.8094057.

[37] Hazarika K, Kumar Choudhury P. Automatic monitoring of solar photovoltaic (SPV) module. Mater. Today Proc., vol. 4, Elsevier Ltd; 2017, p. 12606–9. https://doi.org/10.1016/j.matpr.2017.10.069.

[38] Kabalci Y, Kabalci E. Modeling and analysis of a smart grid monitoring system for renewable energy sources. Sol Energy 2017;153:262–75. https://doi.org/10.1016/j.solener.2017.05.063.

[39] Kim JG, Kim DH, Yoo WS, Lee JY, Kim YB. Daily prediction of solar power generation based on weather forecast information in Korea. IET Renew Power Gener 2017;11:1268–73. https://doi.org/10.1049/iet-rpg.2016.0698.

[40] Gupta A, Jain R, Joshi R, Saxena R. Real time remote solar monitoring system. Proc. - 2017 3rd Int. Conf. Adv. Comput. Commun. Autom. (Fall), ICACCA 2017, vol. 2018- January, Institute of Electrical and Electronics Engineers Inc.; 2018, p. 1–5. https://doi.org/10.1109/ICACCAF.2017.8344723.

[41] Kekre A, Gawre SK. Solar photovoltaic remote monitoring system using IOT. Int. Conf. Recent Innov. Signal Process. Embed. Syst. RISE 2017, vol. 2018- January, Institute of Electrical and Electronics Engineers Inc.; 2018, p. 619–23. https://doi.org/10.1109/RISE.2017.8378227.

[42] Spanias AS. Solar energy management as an Internet of Things (IoT) application. 2017 8th Int. Conf. Information, Intell. Syst. Appl. IISA 2017, vol. 2018- January, Institute of Electrical and Electronics Engineers Inc.; 2018, p. 1–4. https://doi.org/10.1109/IISA.2017.8316460.

[43] López-Lapeña O, Pallas-Areny R. Solar energy radiation measurement with a low-power solar energy harvester. Comput Electron Agric 2018;151:150–5. https://doi.org/10.1016/j.compag.2018.06.011.

[44] Gonzalez-Amarillo CA, Corrales-Munoz JC, Mendoza-Moreno MA, Gonzalez Amarillo AM, Hussein AF, Arunkumar N, et al. An IoT-Based traceability system for greenhouse seedling crops. IEEE Access 2018;6:67528–35. https://doi.org/10.1109/ACCESS.2018.2877293.

[45] Gunturi SKS, Reddy MSK. IoT Based Domestic Energy Monitoring Device. 2018 3rd Int. Conf. Converg. Technol. I2CT 2018, Institute of Electrical and Electronics Engineers Inc.; 2018. https://doi.org/10.1109/I2CT.2018.8529756.

[46] Gusa RF, Sunanda W, Dinata I, Handayani TP. Monitoring System for Solar Panel Using Smartphone Based on Microcontroller. Proc. - 2018 2nd Int. Conf. Green Energy Appl. ICGEA 2018, Institute of Electrical and Electronics Engineers Inc.; 2018, p. 79–82. https://doi.org/10.1109/ICGEA.2018.8356281.

[47] Badave PM, Karthikeyan B, Badave SM, Mahajan SB, Sanjeevikumar P, Gill GS. Health monitoring system of solar photovoltaic panel: An internet of things application. Lect. Notes Electr. Eng., vol. 435, Springer Verlag; 2018, p. 347–55. https://doi.org/10.1007/978-981-10-4286-7_34.

[48] Singh TA, Chandra J. IOT based Green House monitoring system. J Comput Sci 2018;14:639–44. https://doi.org/10.3844/jcssp.2018.639.644.

[49] Sadowski S, Spachos P. Solar-Powered Smart Agricultural Monitoring System Using Internet of Things Devices. 2018 IEEE 9th Annu. Inf. Technol. Electron. Mob. Commun. Conf. IEMCON 2018, Institute of Electrical and Electronics Engineers Inc.; 2019, p. 18–23. https://doi.org/10.1109/IEMCON.2018.8614981.

[50] Danita M, Mathew B, Shereen N, Sharon N, Paul JJ. IoT Based Automated Greenhouse Monitoring System. Proc. 2nd Int. Conf. Intell. Comput. Control Syst. ICICCS 2018, Institute of Electrical and Electronics Engineers Inc.; 2019, p. 1933–7. https://doi.org/10.1109/ICCONS.2018.8662911.

[51] Icaza D, Ordoez A, Ochoa E, Juela C, Blandin P, Icaza F. Monitoring of illegal activities in the border area between ecuador and Colombia using telecommunications networks and microcameras supplied by solar energy installed in native birds. IEEE ICA-ACCA 2018 - IEEE Int. Conf. Autom. Congr. Chil. Assoc. Autom. Control Towar. an Ind. 4.0 - Proc., Institute of Electrical and Electronics Engineers Inc.; 2019. https://doi.org/10.1109/ICA-ACCA.2018.8609782.

[52] García Márquez FP, Segovia Ramírez I. Condition monitoring system for solar power plants with radiometric and thermographic sensors embedded in unmanned aerial vehicles. Meas J Int Meas Confed 2019;139:152–62. https://doi.org/10.1016/j.measurement.2019.02.045.

[53] Shamshiri M, Gan CK, Baharin KA, Azman MAM. IoT-based electricity energy monitoring system at Universiti Teknikal Malaysia Melaka. Bull Electr Eng Informatics 2019;8:683–9. https://doi.org/10.11591/eei.v8i2.1281.

[54] Khan MS, Sharma H, Haque A. IoT Enabled Real-Time Energy Monitoring for Photovoltaic Systems. Proc. Int. Conf. Mach. Learn. Big Data, Cloud Parallel Comput. Trends, Prespectives Prospect. Com. 2019, Institute of Electrical and Electronics Engineers Inc.; 2019, p. 323–7. https://doi.org/10.1109/COMITCon.2019.8862246.

[55] Rahman MA, Mukta MY, Yousuf A, Asyhari AT, Bhuiyan MZA, Yaakub CY. IoT based hybrid green energy driven highway lighting system. Proc. - IEEE 17th Int. Conf. Dependable, Auton. Secur. Comput. IEEE 17th Int. Conf. Pervasive Intell. Comput. IEEE 5th Int. Conf. Cloud Big Data Comput. 4th Cyber Sci., 2019, p. 587–94. https://doi.org/10.1109/DASC/PiCom/CBDCom/CyberSciTech.2019.00114.

[56] Hassnuddin M, Kumar S. Advance Green Energy Scheduling In Smart Grid Using IOT. 2019 Int. Conf. Recent Adv. Energy-Efficient Comput. Commun. ICRAECC 2019, Institute of Electrical and Electronics Engineers Inc.; 2019. https://doi.org/10.1109/ICRAECC43874.2019.8995043.

[57] Michael E, Agajo J, Osanaiye OA, Oyinbo AM. Design of a green iot based water monitoring system for metropolitan city. 2019 15th Int. Conf. Electron. Comput. Comput. ICECCO 2019, Institute of Electrical and Electronics Engineers Inc.; 2019. https://doi.org/10.1109/ICECCO48375.2019.9043294.

[58] Agyeman MO, Al-Waisi Z, Hoxha I. Design and implementation of an iot-based energy monitoring system for managing smart homes. 2019 4th Int. Conf. Fog Mob. Edge Comput. FMEC 2019, Institute of Electrical and Electronics Engineers Inc.; 2019, p. 253–8. https://doi.org/10.1109/FMEC.2019.8795363.

[59] Sharma H, Haque A, Jaffery ZA. Maximization of wireless sensor network lifetime using solar energy harvesting for smart agriculture monitoring. Ad Hoc Networks 2019;94:101966. https://doi.org/10.1016/j.adhoc.2019.101966.

[60] Al-Ali AR, Nabulsi A Al, Mukhopadhyay S, Awal MS, Fernandes S, Ailabouni K. IoT-solar energy powered smart farm irrigation system. J Electron Sci Technol 2020;30:1–14. https://doi.org/10.1016/J.JNLEST.2020.100017.

[61] Jumaat SA, Said MNAM, Jawa CRA. Dual axis solar tracker with IoT monitoring system using arduino. Int J Power Electron Drive Syst 2020;11:451–8. https://doi.org/10.11591/IJPEDS.V11.I1.PP451-458.

[62] Karbhari G V, Nema P. IoT & Machine Learning Paradigm for Next Generation Solar Power Plant Monitoring System. Int J Adv Sci Technol 2020;29:6894–902.

[63] Cheddadi Y, Cheddadi H, Cheddadi F, Errahimi F, Es-sbai N. Design and implementation of an intelligent low-cost IoT solution for energy monitoring of photovoltaic stations. SN Appl Sci 2020;2:1–11. https://doi.org/10.1007/s42452-020-2997-4.

[64] RK, LSP, SA, CG. Design and Implementation of IoT based Energy Management System with Data Acquisition, Institute of Electrical and Electronics Engineers (IEEE); 2020, p. 1–5. https://doi.org/10.1109/icsss49621.2020.9201997.

[65] Hossain ML, Abu-Siada A, Muyeen SM, Hasan MM, Rahman MM. Industrial IoT based condition monitoring for wind energy conversion system. CSEE J Power Energy Syst 2020:1–12. https://doi.org/10.17775/cseejpes.2020.00680.

[66] Saxena M, Dutta S. Improved the efficiency of IoT in agriculture by introduction optimum energy harvesting in WSN. 2020 Int. Conf. Innov. Trends Inf. Technol. ICITIIT 2020, Kottayam, India: Institute of Electrical and Electronics Engineers Inc.; 2020. https://doi.org/10.1109/ICITIIT49094.2020.9071549.

[67] Xia K, Ni J, Ye Yanhong, Xu P, Wang Y. A real-time monitoring system based on ZigBee and 4G communications for photovoltaic generation. CSEE J Power Energy Syst 2020;6:52–63. https://doi.org/10.17775/cseejpes.2019.01610.

[68] Avancini DB, Rodrigues JJPC, Rabêlo RAL, Das AK, Kozlov S, Solic P. A new IoT-based smart energy meter for smart grids. Int J Energy Res 2021;45:189–202. https://doi.org/10.1002/er.5177.

[69] Mukta MY, Rahman MA, Asyhari AT, Alam Bhuiyan MZ. IoT for energy efficient green highway lighting systems: Challenges and issues. J Netw Comput Appl 2020;158:102575. https://doi.org/10.1016/j.jnca.2020.102575.

[70] Muthubalaji S, Srinivasan S, Lakshmanan M. IoT based energy management in smart energy system: A hybrid SO2SA technique. Int J Numer Model Electron Networks, Devices Fields 2021;34:e2893. https://doi.org/10.1002/jnm.2893.

[71] Bhau GV, Deshmukh RG, Kumar TR, Chowdhury S, Sesharao Y, Abilmazhinov Y. IoT based solar energy monitoring system. Mater Today Proc 2021. https://doi.org/10.1016/j.matpr.2021.07.364.
[72] Zhang X, Manogaran G, Muthu BA. IoT enabled integrated system for green energy into smart cities. Sustain Energy Technol Assessments 2021;46:101208. https://doi.org/10.1016/J.SETA.2021.101208.
[73] Sadeeq MAM, Zeebaree S. Energy management for internet of things via distributed systems. J Appl Sci Technol Trends 2021;2:59–71.
[74] Pal P, Parvathy AK, Devabalaji KR, Antony SJ, Ocheme S, Babu TS, *et al.* IoT-Based Real Time Energy Management of Virtual Power Plant Using PLC for Transactive Energy Framework. IEEE Access 2021;9:97643–60.
[75] Xiaoyi Z, Dongling W, Yuming Z, Manokaran KB, Benny Antony A. IoT driven framework based efficient green energy management in smart cities using multi-objective distributed dispatching algorithm. Environ Impact Assess Rev 2021;88:106567. https://doi.org/10.1016/J.EIAR.2021.106567.
[76] Ramu SK, Irudayaraj GCR, Elango R. An IoT-Based Smart Monitoring Scheme for Solar PV Applications. In: S.L. Tripathi PAA and US, editor. Electr. Electron. Devices, Circuits, Mater., Wiley; 2021; p. 211–33. https://doi.org/10.1002/9781119755104.ch12.
[77] Kumar, A., Dubey, A.K., Ramírez, I.S., Muñoz del Río, A., Márquez, F.P.G. (2022). A Review and Analysis of Forecasting of Photovoltaic Power Generation Using Machine Learning. In: Xu, J., Altiparmak, F., Hassan, M.H.A., García Márquez, F.P., Hajiyev, A. (eds) Proceedings of the Sixteenth International Conference on Management Science and Engineering Management – Volume 1. ICMSEM 2022. Lecture Notes on Data Engineering and Communications Technologies, vol 144. Springer, Cham. https://doi.org/10.1007/978-3-031-10388-9_36
[78] Dubey, A.K., Kumar, A., Ramirez, I.S., Marquez, F.P.G. (2022). A Review of Intelligent Systems for the Prediction of Wind Energy Using Machine Learning. In: Xu, J., Altiparmak, F., Hassan, M.H.A., García Márquez, F.P., Hajiyev, A. (eds) Proceedings of the Sixteenth International Conference on Management Science and Engineering Management – Volume 1. ICMSEM 2022. Lecture Notes on Data Engineering and Communications Technologies, vol 144. Springer, Cham. https://doi.org/10.1007/978 3 031 10388 9_35
[79] Rathore, P.S., Chatterjee, J.M., Kumar, A. *et al.* Energy-efficient cluster head selection through relay approach for WSN. J Supercomput 77, 7649–7675 (2021). https://doi.org/10.1007/s11227-020-03593-4
[80] A.Dubey, S.Narang, A.Srivastav, A.Kumar, V.Díaz, Woodhead Publishing, Science Direct,Artificial Intelligence for Renewable Energy Systems. Paperback ISBN: 9780323903967
[81] A.Dubey, S.Narang, A.Srivastav, A.Kumar, V.Díaz, Woodhead Publishing, Science Direct, a Visualization Techniques for Climate Change with Machine Learning and Artificial Intelligence. ISBN: 9780323997140

7
The Contribution of Renewable Energy with Artificial Intelligence to Accomplish Organizational Development Goals and Its Impacts

K. M. Baalamurugan* and Aanchal Phutela

School of Computing Science and Engineering, Galgotias University, Noida, India

Abstract

The electricity sector is progressively dependent on Variable Renewable Energy sources (VRE), which are likely to grow their proportion of energy generation further. A significant impediment to widespread adoption of these energy sources is their high integration costs. Artificial Intelligence (AI) explanations and data-intensive expertise are currently being deployed in various sectors of the electrical significance chain and, given the future intelligent grid's increasing complexity and data generation capacity, have the probability to add substantial value to the system. Conversely, many uncertainties or an absence of understanding about the impact frequently impedes people who make decisions' willingness to invest in AI and data-intensive technologies, including those in the energy sector. We give a comprehensive review of how artificial intelligence can significantly reduce integration costs using a cost-effective model for variable renewable integration costs from research first. They examine several application scenarios and address the difficulties inherent in determining the value produced by AI solutions in the power industry.

Keywords: Renewable energy, energy utilization, impact analysis, elicited expert, artificial intelligence, sustainable development goals

---

*Corresponding author: baalaresearch@outlook.com

Abhishek Kumar, Pramod Singh Rathore, Ashutosh Kumar Dubey, Arun Lal Srivastav, T. Ananth Kumar and Vishal Dutt (eds.) Sustainable Management of Electronic Waste, (145–166) © 2024 Scrivener Publishing LLC

7.1 Introduction

As the climate is changing, discussions and related legislation reveal that decarbonization has become a major problem. To meet the growing need for energy in a sustainable manner, there is an urgent need for more environmentally friendly and affordable energy. Electricity, largely as a result of renewable energy sources (RES), has the potential to develop a critical green and moderately inexpensive fuel in the future. Between 2000 and 2019, global electricity generation via VRE (a subset of RES), including wind and solar photovoltaic (PV), climbed from 32 TWh to 1.857 TWh. Artificial intelligence and information technologies are already being deployed in several sectors of the electricity value chain and require the potential to add substantial value to the system. As with other sectors and marketplaces, this offers the ability to be a driver of value conception and, contingent on the location of value pools, could potentially impact the attractiveness of geographies, enterprises, and new entrants to the power sector. Adoption of AI and data-intensive technologies ought to be assisted by better defining their potential for value generation, as uncertainty or an absence of consideration of this frequently impedes or even prevents policymakers from committing to investment in these technologies [1]. One of the objectives of this assessment is to provide provision for such interpretations and to assist in identifying future research possibilities.

7.1.1 AI will Balance Millions of Assets on the Grid

AI systems can let decentralized energy sources transfer any excess electricity they create to the grid and utilities can distribute that energy to where it is needed. Energy storage in industrial facilities, offices, residences, and cars can be used to store excess energy while demand is low and AI uses that power when generation is insufficient or impossible. Additionally, a lot of cooperation and forecasting, including optimization, is required to keep the grid stable. To put it another way, imagine a conductor maintaining the ensemble in rhythm as AI assembles a symphony in real time. An automation system has the potential to be a game-changer because of these factors. A more resilient and adaptable grid may be built by replacing an infrastructure-heavy system with one based on AI, which enables forecasting and control in seconds rather than days. Decentralized energy resources necessitate that utilities, policymakers, and regulatory agencies begin to consider how they want to play a role in their development. Coordinated and efficient administration is essential to the patchwork of distributed energy sources. This is a great opportunity for utilities to take the lead because they

AI-POWERED RENEWABLE ENERGY FOR ORGANIZATIONAL DEVELOPMENT 147

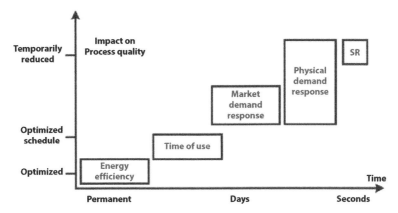

Figure 7.1 Predicted assets in AI growth of redict in year wise.

confront a decreasing pool of customers purchasing electricity due to rooftop solar panels and the like. Predicted assets in AI growth of redict in year wise is shown in Figure 7.1. Bloembergen predicts that by 2050, Europe's average power plant will shrink from 800 megawatts to 562 megawatts and that eventually, this will drop even further to 32 megawatts [2][3].

Services providers will have to choose between collaborating with software firms and establishing themselves as independent software firms. A paradigm change is needed from the traditional model of big capital investments in a few large energy producing assets to demand management of an exponentially rising number of privately held assets, while also safeguarding customer data and privacy and guaranteeing grid security. National and municipal administrations must also speed up and rethink their approaches to spending on infrastructure, particularly in the areas of energy generation and distribution.

Using infrastructure-based approaches to grid stability maintenance will take years of planning and construction and billions of dollars. Crisis disruption will endure to increase the frequency and severity of life-threatening weather events around the world, therefore there is no time to waste due to the current economic climate and anticipated lengthy political battles in the US.

Investing in centralized networks is a mistake; instead, governments should plan for a grid where individuals and buildings generate their own electricity, which is then managed in real-time by software.

Public subsidies for renewable energy generation and incentives for increased distributed energy generation in homes and businesses should be considered. We also need worldwide oversight of AI software to assure interoperability, transparency, and equal access throughout the energy environment [4].

The events in Texas have exposed how vulnerable our current electric networks are to attack. For the same reasons that we shouldn't strive to improve internal combustion engines, we also cannot afford to replace an outdated electrical infrastructure. An all-electric world can only be achieved through the use of AI and software.

Astonishing advancements in technology are taking place all the time to the point where many of us are just unable to keep up with it all. Artificial intelligence (AI) is poised to transform our environment in profound ways in the next several decades.

In the renewable energy area, there is no difference. Artificial Intelligence (AI) has the ability to learn from large amounts of data. As a result, it is able to make adjustments to maximize energy efficiency, conversion, and distribution. Such methods enable the accurate forecasting of temperature as well as load. Among a plethora of other uses, it also helps reduce the likelihood of such an electrical surge [5].

7.2 AI Contributions in Conventional and Renewable Energy

Solar, wind, and water-based renewable energy are generated utilizing technology that ensures that energy stores are replenished over time. Conventional methods of generating electricity have a significant impact on the climate. Renewable energy is a viable option for making the world a safer and more energy efficient place because of its low carbon emissions. Several sorts of renewable energy resources have been exploited in the last several years, including wind, sun, geothermal, biomass, and hydropower. Renewable energy is exploding as costs are driven down by technological innovation. About 20% of the world's energy demands are presently fulfilled by renewable sources. They are also responsible for around a quarter of the world's power generation. For a variety of reasons, it makes sense to rely on renewable energy. Relying on resources that are both secure and local saves money in the long run by lowering dependence on non-renewable sources of energy. In addition to creating new jobs in the renewable energy industry, this will lower your energy costs. By vending your excess energy back to your energy provider, you may be able to make money.

7.3 AI-Based Technology in Renewable Energy

If the load would extract enough power for renewable sources to maintain high energy conversion efficiency, a sustainable energy application would be

AI-Powered Renewable Energy for Organizational Development 149

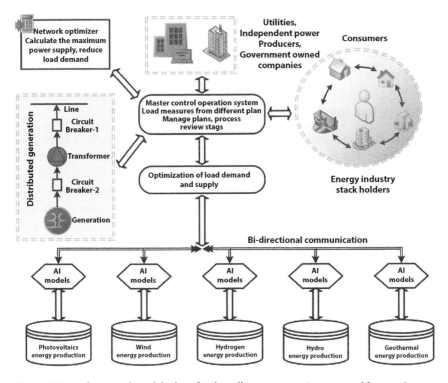

Figure 7.2 Architectural model of artificial intelligence on various renewable energies.

a success. The assimilation of renewable energy sources into the grid is constrained by a number of nonlinear interactions between various components. For complex situations that are difficult to foresee, artificial intelligence (AI) technology can help. Renewable energy systems modelling, identification, optimization, forecasting, and control are increasingly being solved using AI methods (e.g., neural networks, intelligent optimization algorithms, fuzzy logic) instead of conventional ways [6]. Figure 7.2 shows the architectural model of artificial intelligence on various renewable energies.

Incorporating breakthrough AI-based technologies into renewable energy systems will result in even more efficiency gains. Researchers in electrical engineering and artificial intelligence (AI) are now able to work together to propose and progress new technologies that will advance:

- Modeling and parameters approximation
- Net load predicting
- Line loss prophecies
- Retaining system dependability
- Energy effectiveness

- Renewable energy processes
- Incorporating hybrid solar and battery storage systems
- Predicting equipment failure
- Decision processing on behalf of grid operators

7.3.1 Challenges of the Renewable Energy Sector

The weather's variability is a substantial issue in the production of renewable energy. The production of electricity from renewable energy sources like solar and wind is highly dependent on the state of the weather. The output current can be disrupted by rapid changes in climate, even though we have effective systems for weather forecasting. Renewable energy's supply chain is susceptible to such flaws. As a result, it must be flexible enough to adapt to new circumstances. A second reason for optimism is the recent progress being made in energy storage technologies. However, they haven't been properly tested to their fullest extent. Renewable energy's popularity will only grow in the years to come. As a result, the developing technologies of AI, IoT, and machine learning should be prioritized by renewable energy companies looking to boost production and close the gap. AI technology can be used to make data-driven decisions by huge customers of renewable energy, such as factories, supermarkets, railways, and offices [7].

7.3.2 Artificial Intelligence in Wind Farming

Predicting Wind Speeds: WSF is becoming an increasingly significant part of electric power and energy system planning and operation as wind power's penetration into modern grids grows rapidly. According to NWP, a mathematical model of the atmosphere and oceans is used to predict future weather conditions based on existing weather patterns. Artificial Intelligence (AI) algorithms are trained on industry data to generate accurate projections, which helps to inform the supply and demand of power. Artificial Intelligence (AI) is used to monitor and improve energy efficiency. In order to calculate utility savings and recommend smart home investments, AI is applied.

Ingenious Forecasting of Wind Speeds (iWSF): In order to effectively integrate variable generation into the power grid, the wind power forecasting system delivers crucial information that fulfils both the needs of reliable electric grids and energy trading. In a time frame of minutes to several days, grid operators and energy traders need reliable wind power projections. No single weather forecasting (WSF) methodology will work efficiently over different temporal scales, thus specialized methods based on empirical evidence can be combined with numerical weather projections (NWPs) to increase extremely long or relatively brief forecasts, as well as future projections beyond a few hours.

There are many benefits to using renewables, including:

- Creating use of secure and local resources
- Decreasing dependency on non-renewable energy
- Aiding to decrease the fabrication of carbon dioxide and further greenhouse gases
- Generating new works in renewable energy industries
- Decreasing energy bills and in some cases allowing the ability create income by selling excess energy back to an energy provider [8].

7.4 A Look at Some Challenges Faced by the Renewable Energy Industry

The weather's instability is an important issue in the production of renewable energy. The production of electricity from renewable energy sources like solar and wind is highly dependent on the state of the weather. The energy flow can be disrupted by rapid climatic changes, even if we have effective systems for weather forecasting. Renewable energy's supply chain is susceptible to such flaws. As a result, it must be flexible enough to adapt to new circumstances shown in Figure 7.3.

A second reason for optimism is the recent progress being made in energy storage technologies. However, they haven't been properly tested to their

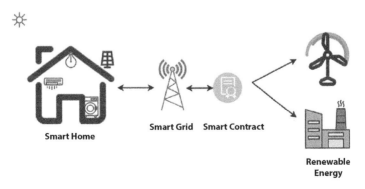

Figure 7.3 AI applied to forecasting wind energy.

Table 7.1 Various algorithm designs for variable objects in AI technology.

Year	Problem description	Design variables	Objects	Algorithms
1993	Use the AI technologies to realize fault detection, control applications, operational optimization, load forecasting, and security assessment	Heating steam temperature, circulating brine temperature, circulating brine flow	The energy cost	ANN
1993	Minimize energy cost	Seawater temperature, steam enthalpy and distillate demand or steam supply	Cost	GRG-QP
1994	Keep the desired distillate production	Seawater temperature, heat exchange coefficients, brine top temperature and brine recirculation flowrate	Distillate production flowrate	ES
2007	Optimize the performance of desalination	Dry and damp bubble temperature of the air, the inlet and outlet cooling water temperature of seawater	Water production ratio	ANN
2007	Solve the large-scale MSF model	Flow rate of distillate, brine flow rate, feed flow rate	Initial guesses and starting guesses	GA/ANN

(Continued)

Table 7.1 Various algorithm designs for variable objects in AI technology. (*Continued*)

Year	Problem description	Design variables	Objects	Algorithms
2013	Optimize feed-forward control system	Feed flow rate	Radiation and distillate flow rate	L-M
2015	Assess the ranking of potential development	Environment and the economy	Quantity and quality of water supply and demand	DSS
2017	Examine the relationships between different variables	The cold feed inlet temperature, hot feed inlet temperature, and feed-in flow rate	Gained output ratio (GOR)	PSO

fullest extent. Renewable energy's popularity will only grow in the years to come. As a result, the developing technologies of AI, IoT, and machine learning should be prioritized by renewable energy companies looking to boost production and close the gap. Artificial intelligence can be used to make data-driven decisions by huge consumers of renewable energy, such as grocery store, businesses, workplaces, and railroads. Various algorithm designs for variable objects in AI technology is shown in Table 7.1.

The use of Artificial Intelligence (AI) in the energy sector is on the rise. For instance, an AI use that could have a long-standing impact is accelerating the expansion of green, low-carbon electricity generation over an ideal energy-storing scenario. Utility transportation and energy transformation can benefit from the usage of AI in contemporary energy technologies such as ML, DL, and sophisticated neural networks. AI is being used by power generation, utility services, and several other energy companies to guarantee a balance between supply and demand that has developed a contest due to the rising stake of RES in an energy industry that is being decentralized, decarbonized, and flooded with emerging technologies [9,10,11]. AI can help the energy sector evaluate and condense a significant volume of data. Other uses include energy storage, stand-alone verifiable and permanent methods (e.g., high grid stability, intuitive operation, real-time metering, voltage regulation), power outages (AI can identify power failures before they happen and save lives, time, and money; AI can also be used to predict

Table 7.2 Comparison statement of artificial intelligence in pros and cons with renewable energy.

Algorithms	Advantages	Disadvantages	Application
ANN, BP	Self-learning function, associative storage function, optimal solutions at high speed	High computing cost	Applicable to optimization and prediction
GA	Large coverage, overall optimization	Low efficiency, computational complexity	Commonly used for optimization
PSO, DSS, ES	Fast convergence speed, self-renewal ability	Easily trapped local solution	More suitable for decision weight

and prevent power outages), and so on. Over the next few years, AI energy applications are predicted to span the whole value chain of the energy industry. Comparison statement of artificial intelligence in pros and cons with renewable energy is shown in Table 7.2.

Modeling, planning, and simulation are methods that can be applied to topics such as eco-friendly and storm prediction, as well as energy systems. It is ideally suitable for forecasting and optimization applications because of artificial neural networks, which have considerable skills in this area.

Investment models are looking for market players and AI-powered entities can interact with markets autonomously. Technology is used to optimize and forecast maintenance plans and happenings, as well as to secure energy infrastructure from cyber and physical threats. A focus on customer-oriented services, such as smart home management, integration of micro-generation and storage with virtual power stations, and energy earning and billing, are all included in this category. AI is aimed at energy planning, ultimatum, and control.

ML approaches are considered to build models for a specified function onto a precise number of data inputs. A series of evolutionary iterations have been used to train the algorithm in order to get the required outcomes. There are three types of machine learning models: supervised, unsupervised, and reinforcement learning. Modeling and categorization are two subcategories of supervised methods. Diagnostics, identity fraud detection, retaining existing customers, and picture classification all benefit from classification models. It is common practice to utilize econometric models to predict market trends, life expectation estimations, weather forecasts, population growth forecasts, and publicizing popularity forecasts. Clustering and dimensional reduction are two examples of unsupervised models. Feature reduction models are utilized for elicitation, structure discovery, huge data presentations, and compression. It is common to employ cluster structures for a variety of purposes, including categorization, marketing, and product suggestions. Artificial intelligence (AI) game theory, instructional strategies, and robot navigation all make use of recurrent neural networks [12,13].

7.5 Higher Computational Power & Intelligent Robotics

Energy effectiveness and analytics, grid control, asset management, grid planning, grid simulation, digital twin, smart metering virtual power plants, and digital substations will all benefit from the Internet of Things (IoT) in infrastructure planning. Self-governing smart homes, linked vehicles, business security, fitness trackers, smart watches, machine-to-machine coupled devices, drones for exploration, and the manufacturing revolution are some of the latest advancements in the energy sector made by the Internet of Things (IoT). The use of IoT is extensive and diverse. For example, customers might be more conscious of smart appliances aimed at home appliances that may be distantly controlled by IoT smart devices such as lighting, cooking, heating, cooling, alarms, and entertainment systems. In the industrial sector, IoT technologies are used to accomplish process control in manufacturing lines and optimization in the profit-making retail, supply chain, and storing sectors. In the energy sector, IoT devices are already in use [14,15].

For solar distillation, it has been discovered that intelligent prediction and control have a lot of room for application. As a result, the desalination system has a straight influence on its energy consumption and throughput output. A beneficial mathematical model that seeks to simulate the edifice and gathering of a biological neural network is ANN, which has greater recompenses in calculating day-to-day global sun radioactivity than other models. Many studies have been done on the usage of the multilayer perceptron (MLP) in solar desalination, radial basis function network (RBF), and recurrent neural network (RNN). The A-B-C assembly of an ANN model shows that the constructed ANN model contains an input neuron layer with diverse input variables, Xi, a hidden neuron layer with a diverse transfer function, Hj, and

C, output neurons. The number of neurons in a hidden layer and the optimum transmission function across layers is found through trial-and-error methods in order to choose the best ANN structure [16,17,18]. One of the most important factors in a neural network's construction is the number of neurons in its buried layer. By selecting the transfer function, the input variables (such as wind speed and feed flow rate) can be accessed from an input layer. The transfer function is used to derive the objective function at the output terminal after the results of the hidden layer have been coupled.

Operational parameter optimization is best solved using a heuristic approach. Many common processes tend to evaluate only a few possibilities based on past experience, current knowledge, and pre-established rules. I don't think this technique is real, and it is difficult to balance local and global optimization. Determining the best way to overcome the drawbacks of conventional desalination technologies, such as artificial intelligence (AI), should be a top goal. The design process for seawater desalination can be broken down into four primary directions based on a general classification of the difficulties. Optimization, prediction, and control are the four pillars of decision making [19]. Figure 7.4 shows the different methods of intelligence prediction based on human models.

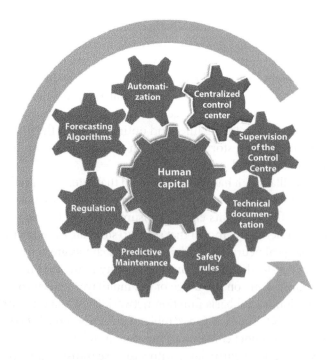

Figure 7.4 Different methods of intelligence prediction based on human models.

7.6 The Impact of Digital Technology on Energy Company Results

Digital technology adoption and firm-level outcomes have been found to be linked in prior research, which is consistent. Information Technology (IT) and information systems are said to improve a company's performance by using digital technologies. Recently, a new stream of research has begun to explore the impact of new digital technologies on corporate performance. As the energy sector undergoes a flurry of changes, renewable energy (RE) is a potent source for imminent global growth. The world cannot abandon the potential of RE in the face of traditional resource depletion, climate change, and growing pollution. Improved usage of Renewable Energy (RE) technologies throughs variable supply, big volumes of data, bidirectional flow of power, and increased requirements for energy storage are important transformations in the sector's main shifts. As renewable energy becomes more popular, more electricity is needed in the generating, transmission, and distribution systems. Load and renewable energy forecasts are crucial to efficient energy management. Real-world power grids were not built to handle variable energy sources or eruptions in loading capacity, such as in the circumstance of renewable energy. Changes are taking place because renewable energy (RE) and new intelligence algorithms are increasingly being used to govern and produce the RE. The new algorithms are faster and more accurate than the old ones [20]. Even the idea of combining two or more algorithms in order to overwhelm the constraints of a distinct algorithm is being considered.

7.6.1 Smart Match of Supply Through Demand

While growing the usage of renewable energy embodies a significant prospect for contemporary society in terms of gating alongside climate change and resource shortages, there will be a risk associated with RE assuming the role of energy sector leader due to its intermittency. Since RE is not reliant on traditional power sources like fossil fuels, the opportunity to not rely on outmoded power sources lies in AI, which might deliver precise forecasting in expressions of RE power supply in order to retort to natural actuation, regulate operations as planned, and supply the required customer demand. It is estimated to boost the effectiveness of RE by automating operations, which will need extensive usage of AI [21,22,23]. The purpose of RE grids is to maximize infeed from a production capacity while taking into account volatility and associated uncertainty costs. Artificial intelligence is cast-off to diagnose energy demand in order to make more accurate adjustments for peaks. It is also critical for handling energy demand and supply from decentralized creation systems (private users who generate energy from renewable sources), which draw energy from the grid when their fabrication falls short of their essentials and feed excess energy back to the grid when their production exceeds their needs. It would facilitate the continual interchange of energy between consumers and providers.

To urge the AI energy sector to modernize and sustain the development of sustainable energy sources, this learning first introduces energy storing technology and the discusses energy storage applications. Energy storage, as a critical and controlled implementation in a grid, plays a critical role in promoting renewable energy immersion, enhancing power grid control capabilities, and ensuring nontoxic and economic grid services. The particulars of AI applications address numerous areas of energy storing and renewable energy integration, including parameter estimates, optimum design, and process control. Lastly, a complete study evaluates the possibility of integrating system challenges and future research areas [24,25].

Lithium-ion battery technology is reasonably developed and has a high energy density, making it one of the most auspicious electrochemical energy storing technologies. Emblematic demonstrative batteries are comprised of lithium iron phosphate batteries, lithium manganite batteries, and ternary lithium batteries.

(a) Lithium iron phosphate batteries, in particular, have been broadly employed in industrialized production due to their comparatively long provision life and low cost. They also provide additional lower business costs and expand wellbeing and performance for large-capacity lithium-ion battery energy storage, which will aid in the promotion of significant profitable requests.

(b) While lead-acid battery technology has a number of advantages, including high safety, low cost, reconditioning possibilities, and high rate, it also has a number of disadvantages, including a low energy density and an undersized cycle life. To address this issue, carbon constituents have been added to lead-acid batteries' adverse electrodes, resulting in the formation of lead-carbon batteries with improved presentation. Nevertheless, lead-carbon batteries cover a high concentration and greatly contribute to environmental pollution. Additionally, no definitive conclusion has been reached concerning the collection and quantification of negative carbon compounds.

(c) The most advanced all-vanadium drift battery is the flow battery. Flow batteries store and release electrical energy electrochemically by reacting with vigorous chemicals in positive and negative electrolyte

AI-Powered Renewable Energy for Organizational Development 159

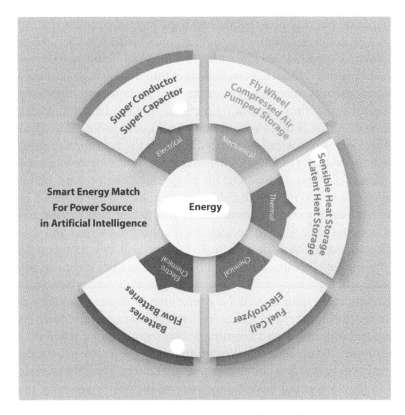

Figure 7.5 Smart energy match for power source in artificial intelligence.

solutions. They explore crucial elements of the connections among cathode materials and electrolyte results in lithium-ion batteries using vanadium ions of various valences as active agents [26,27]. Additionally, flow battery technology provides a number of advantages, including increased safety, increased cycle durations, independent energy/power capacity design, and a low self-discharge rate. Its disadvantages include a poor energy density and high engineering costs. Smart energy match for power source in artificial intelligence is shown in Figure 7.5.

7.6.1.1 Thermal Energy

Sensible heat storage materials store and release heat in response to temperature adjustments. In situations where a heat-storing temperature is not necessary, such as in solar air conditioners, liquid serviceable heat-storing materials like water and solid serviceable heat-storing materials like crushed stone and soil are extensively employed. Their characteristics include a high heat-storing capacity per unit of volume, consistent chemical and physical possessions, and high thermal conductivity. Similar techniques, conversely, cannot be applied on a broad scale because of the high volume required. Sensible materials include liquid metals, molten salts, and organics. Due to its wide temperature range, high heat capacity, and squat viscosity, molten salt is an excellent average- and high-temperature heat transmission and storing material. The fundamental premise of latent heat storing is that two segments for a quantifiable value coexist in stability. Heat is immersed or emitted through the transformation of one phase into another [28,29,30]. This latent heat is the heat immersed per unit mass of material throughout the phase transition process. Heating is the most researched type of heat storage technology at the moment, owing to the fact that the energy storage density for latent heat storage constituents is better than that of sensible heat storing materials and it offers excellent opportunities for practical research and development. Latent heat-storing supplies are classified into low-temperature stage amendment materials and high-temperature stage amendment materials based on their stage amendment temperature. Low-temperature phase-changing materials are mostly employed in industrial leftover heat recovery, solar energy storing and utilization, and warming and air conditioning schemes. Ionic liquids have a tremendous deal of promise to be effective substances for storing latent heat at low and medium temperatures. Typical materials utilized in higher spectral heat storage systems include mixed salts, molten salts, alloys, and metals. They are mostly used in power plants, aerospace systems, and other sectors [31, 32, 33].

7.7 Conclusion

Decomposing connectivity expenses enables the identification of cost-cutting initiatives and value development opportunities. We examined the value-creation power of Artificial Intelligence in the electricity industry's value chain by correlating prevailing AI solutions to particular modules of incorporation costs and examining a number of published use cases. Additionally, we uncovered several guesstimates of the probable value formation of AI for VRE integration cost control in the literature. Finally, we explored the difficulties inherent in developing these estimations. Despite these obstacles, the literature suggests that AI has the potential to provision VRE through lowering their assimilation costs in a variety of methods. Further study into artificial intelligence in the energy industry may help to further enhance these savings. AI solutions are already effectively applied throughout the power sector's value chain, and in numerous cases have been proven to be more effective than formerly utilized elucidations, enhancing the economic sustainability and assumption potential of VRE resources. There will be a rising demand for advanced system optimization approaches in the future, in order to increase the performance of jobs involving a considerably larger number of system components at an even quicker speed.

7.8 Future Work

At the moment, examination of the optimization and incorporation of energy storage systems and renewable energy resources is insufficiently organized and thorough; the underlying principles are insufficiently mature and there are several critical difficulties to overcome. In light of the examinations presented and the potential for further examination, the following findings are presented:

(1) When analyzing working conditions, take into account the influence of several factors on operational characteristics in a congested location. The overfilled region's operation is impacted by elements like climate, commuter peak times, holidays, and weekends.

(2) Developing a comprehensive energy storage system involves not only practical and theoretic research, but also operative assessments in expressions of engineering frugality, ecological influence, and security to decide energy-storing technology's feasibility.

(3) Fine-tune the energy conduction procedure of the power hoard network and the control strategy prototype of the edge energy storing system, for instance, by taking into account the change in impedance of the power supply network by detachment and converting the wayside energy storage system's power control model to a voltage and current dynamic model, etc.

(4) To minimize the size of the solution space in the optimization energy-storing model, a numeral indolent speed control approach is adopted. Consider the constraint on the value collection of verdict variables or experiment with diverse approaches for optimizing driving strategies to determine their effect on the outcomes of energy-convertible optimization.

(5) Adapting the procedure diagram prioritizes energy reduction while also taking into account various security and provision quality limits. Conversely, the construction of the real operation diagram entails considerations such as operative service excellence, managing costs, and so on, which are typically optimized using several objectives.

References

[1] Chandel, S., Shrivastva, R., Sharma, V., Ramasamy, P., 2016. Overview of the initiatives in renewable energy sector under the national action plan on climate change in India. Renew. Sustain. Energy Rev. 54, 866–873.
[2] Appel, F., Ostermeyer-Wiethaup, A., Balmann, A., 2016. Effects of the German Renewable Energy Act on structural change in agriculture–The case of biogas. Utilities Policy 41, 172–182.
[3] Colombo, E., Romeo, F., Mattarolo, L., Barbieri, J., Morazzo, M., 2018. An impact evaluation framework based on sustainable livelihoods for energy development projects: an application to ethiopia. Energy Res. Soc. Sci. 39,78–92.
[4] Cunha, F.B.F., de Miranda Mousinho, M.C.A., Carvalho, L., Fernandes, F., Castro, C., Silva, M.S., et al., 2021. Renewable energy planning policy for the reduction of poverty in Brazil: lessons from Juazeiro. Environ. Dev. Sustain. 23, 9792–9810.
[5] Diachuk, O., Chepeliev, M., Podolets, R., Trypolska, G., Venger, V., Saprykina, T., et al., 2018. Transition of Ukraine to the Renewable Energy by 2050. In: Transition of Ukraine to the Renewable Energy By. p. 2050.
[6] Freire-Gormaly, M., 2018. Experimental characterization of membrane fouling under intermittent operation and its application to the optimization of solar photovoltaic powered reverse osmosis drinking water treatment systems.
[7] Goel, S., Sharma, R., 2016. Feasibility study of hybrid energy system for off-grid rural water supply and sanitation system in Odisha, India. Int. J. Ambient Energy 37, 314–320.
[8] Karmaker, A.K., Ahmed, M.R., Hossain, M.A., Sikder, M.M., 2018. Feasibility assessment & design of hybrid renewable energy based electric vehicle charging station in Bangladesh. Sustain. Cities Soc. 39, 189–202.
[9] Kotrikla, A.M., Lilas, T., Nikitakos, N., 2017. Abatement of air pollution at an aegean island port utilizing shore side electricity and renewable energy. Mar. Policy 75, 238–248.

[10] Onifade, T.T., 2016. Hybrid renewable energy support policy in the power sector: The contracts for difference and capacity market case study. Energy Policy 95, 390–401.

[11] Pino, M., Abarzúa, A.M., Astorga, G., Martel-Cea, A., Cossio-Montecinos, N., Navarro, R.X., et al., 2019. Sedimentary record from Patagonia, southern Chile supports cosmic-impact triggering of biomass burning, climate change, and megafaunal extinctions at 12.8 ka. Sci. Rep. 9, 1–27.

[12] Vyhmeister, E., Muñoz, C.A., Miquel, J.M.B., Moya, J.P., Guerra, C.F., Mayor, L.R., et al., 2017. A combined photovoltaic and novel renewable energy system: An optimized techno-economic analysis for mining industry applications. J. Cleaner Prod. 149, 999–1010.

[13] Rehman, A., 2019. The nexus of electricity access, population growth, economic growth in Pakistan and projection through 2040. Int. J. Energy Sect. Manag..

[14] Semeraro, T., Pomes, A., Del Giudice, C., Negro, D., Aretano, R., 2018. Planning ground based utility scale solar energy as green infrastructure to enhance ecosystem services. Energy Policy 117, 218–227.

[15] Simon, C.A., 2020. Alternative Energy: Political, Economic, and Social Feasibility. Rowman & Littlefield Publishers.

[16] Solaun, K., Cerdá, E., 2019. Climate change impacts on renewable energy generation. A review of quantitative projections. Renew. Sustain. Energy Rev. 116, 109415.

[17] Zvezdov, I.M., 2020. The EU legal and regulatory framework for measuring damage risks to the biodiversity of the marine environment. Environ. Policy: Econ. Perspect. 121–137.

[18] The Need for Virtualization: When and Why Virtualization Took Over Physical Servers, Anand A., Chaudhary A., Arvindhan M. (2021). In: Hura G., Singh A., Siong Hoe L. (eds) Advances in Communication and Computational Technology. Lecture Notes in Electrical Engineering, vol 668. Springer, Singapore. https://doi.org/10.1007/978-981-15-5341-7_102

[19] Srivastava, S.K., 2020. Application of artificial intelligence in renewable energy. In: 2020 International Conference on Computational Performance Evaluation (ComPE). IEEE, pp. 327–331.

[20] Shen, W., Chen, X., Qiu, J., Hayward, J.A., Sayeef, S., Osman, P., et al., 2020. A comprehensive review of variable renewable energy levelized cost of electricity. Renew. Sustain. Energy Rev. 133, 110301. Shields, M.A., Woolf, D.K., Grist, E.P., Kerr, S.A., Jackson, A.

[21] The Modern Way for Virtual Machine Placement and Scalable Technique for Reduction of Carbon in Green Combined Cloud Datacenter, D. Nageswara Rao Arvindhan Muthusamy, Abhineet Anand, T. Vinodh Kannan, ISBN 978-3-030-48140-7 ISBN 978-3-030-48141-4.

[22] Scheming an proficient auto scaling technique for minimizing response time in load balancing on Amazon AWS Cloud, International Conference on Advances in Engineering Science Management & Technology (ICAESMT)-2019, Arvindhan, M and Anand, Abhineet, SSRN: https://ssrn.com/abstract=3390801 or http://dx.doi.org/10.2139/ssrn.3390801

[23] The Modern Way for Virtual Machine Placement and Scalable Technique for Reduction of Carbon in Green Combined Cloud Datacenter,D. Nageswara Rao Arvindhan Muthusamy, Abhineet Anand, T. Vinodh Kannan, ISBN 978-3-030-48140-7 ISBN 978-3-030-48141-4.

[24] Suryakiran, M.N.S., Begum, W., Sudhakar, R., Tiwari, S.K., 2020. Development of wind energy technologies and their impact on environment: A review. Adv. Smart Grid Technol. 51–62.

[25] Mashaly AF, Alazba AA. MLP and MLR models for instantaneous thermal efficiency prediction of solar still under hyper-arid environment. Comput Electron Agric 2016;122:146–55.

[26] Lowitzsch J, Hoicka CE, van Tulder FJ. Renewable energy communities under the 2019 European Clean Energy Package-Governance model for the energy clusters of the future? Renew Sustain Energy Rev 2020;122:109489. https://doi.org/ 10.1016/j.rser.2019.109489.

[27] Baidya S, Nandi C. Green Energy Generation Using Renewable Energy Technologies. In Advances in Greener Energy Technologies 2020 (pp. 259-276). Springer, Singapore.

[28] Khan K, Su C-W, Tao R, Hao L-N. Urbanization and carbon emission: causality evidence from the new industrialized economies. Environ Dev Sustain 2020;22(8): 7193–213.

[29] Marouane Salhaoui, Mounir Arioua, Antonio Guerrero-Gonz´alez, María Socorro García-Cascales. An IoT Control System for Wind Power Generators. In: Springer, Cham; 2018, p. 469–479.

[30] Brynjolfsson E, Rock D, Syverson C. Artificial Intelligence and the Modern Productivity Paradox: A Clash of Expectations and Statistics. Cambridge, MA: National Bureau of Economic Research; 2017.

[31] Kumar, A., Dubey, A.K., Ramírez, I.S., Muñoz del Río, A., Márquez, F.P.G. (2022). A Review and Analysis of Forecasting of Photovoltaic Power Generation Using Machine Learning. In: Xu, J., Altiparmak, F., Hassan, M.H.A., García Márquez, F.P., Hajiyev, A. (eds) Proceedings of the Sixteenth International Conference on Management Science and Engineering Management – Volume 1. ICMSEM 2022. Lecture Notes on Data Engineering and Communications Technologies, vol 144. Springer, Cham. https://doi.org/10.1007/978-3-031-10388-9_36

[32] Dubey, A.K., Kumar, A., Ramirez, I.S., Marquez, F.P.G. (2022). A Review of Intelligent Systems for the Prediction of Wind Energy Using Machine Learning. In: Xu, J., Altiparmak, F., Hassan, M.H.A., García Márquez, F.P., Hajiyev, A. (eds) Proceedings of the Sixteenth International Conference on Management Science and Engineering Management – Volume 1. ICMSEM 2022. Lecture Notes on Data Engineering and Communications Technologies, vol 144. Springer, Cham. https://doi.org/10.1007/978-3-031-10388-9_35

[33] Rathore, P.S., Chatterjee, J.M., Kumar, A. *et al.* Energy-efficient cluster head selection through relay approach for WSN. J Supercomput 77, 7649–7675 (2021). https://doi.org/10.1007/s11227-020-03593-4

8
Current Trends in E-Waste Management

Anjali Sharma[1]*, Devkant Sharma[2], Deepshi Arora[1] and Ajmer Singh Grewal[1]

[1]Guru Gobind Singh College of Pharmacy, Yamunanagar Haryana, India
[2]CH. DeviLal College of Pharmacy, Jagadhari Haryana, India

Abstract

The world's growing reliance on electronic gadgets, including mobile phones, and computer accessories, along with the persistent introduction of newly designed products to the market has resulted in an e-waste problem that is getting worse every year. Electronic trash is a secondary source of valuable and pretentious metal. If we notice, the urban mining of such metals is gaining remarkable recognition due to its economic possibilities, expanded commercial opportunity, source of livelihood, and ultimate achievement of the Agenda for Sustainable Development Goals (SDGs) 2030. At the same time, due to the presence of dangerous chemical components, large amounts of E-waste prove to be an insurmountable task. With its negative effects on human health and the environment, the massive E-waste output has generated multi-dimensional hurdles over present treatment methods. The key issues influencing the entire E-waste value chain in India include a lack of data inventorization, unlawful disposal, and a lack of treatment choices. As a result, this study focuses on strategic interventions that are compliant with existing legislation and are necessary for a long-term E-waste value chain, secure resources, social well-being, reduced environmental consequences, and overall sustainable development. Apart from these, different methods for recycling e-waste are incorporated to maintain the integrity of the environment.

Keywords: E-waste, sustainable development, human health, environment, urban mining

---

Corresponding author: sharma.sharmaa.anjali@gmail.com

Abhishek Kumar, Pramod Singh Rathore, Ashutosh Kumar Dubey, Arun Lal Srivastav, T. Ananth Kumar and Vishal Dutt (eds.) Sustainable Management of Electronic Waste, (167–186) © 2024 Scrivener Publishing LLC

8.1 Introduction

It's challenging to picture a future without cell phones, GPS navigation systems, personal computers, and certain other digital devices because technology has advanced at such a rapid rate in recent decades. Due to the rapidly growing amount of outdated electronics being destroyed, environmentalists, local and state governments, and the United Nations are worried about how to decrease e-waste. Electronic garbage has become a major issue in recent years. The fact that just 30% of India's recoverable trash is currently recycled shows how much potential there is for growth in the country's waste management sector. A few of the many problems causing the nation's waste management inefficiencies include inefficient infrastructure, ineffectual garbage pickup, dumping, recycling policies, and many others.

In 2021, the worldwide e-waste management market held USD 56.56 billion in revenue. By 2030, it is anticipated to reach USD 189.8 billion, expanding at a CAGR of 14.4% (2022-2030). E-waste, or electronic garbage, refers to outdated electrical and electronic equipment. E-waste is a general term that refers to used electronics that may be recycled through resource recovery, restoration, resale, reuse, or disposal. Both wealthy and developing nations are seeing a dramatic increase in the amount of e-waste. The growing need for updating to the most recent technology is the primary factor fueling the growth of the e-waste management industry. The drive to embrace new, advanced technological equipment results in the production of millions of tonnes of e-waste every year in multiple countries.

8.2 What is E-Waste?

E-waste is also known as end-of-life electronics, e-scrap, and electronic garbage. It includes any technological equipment that people desire to throw in the storehouse such as phones, computers, televisions, gaming systems, appliances, office equipment, and even some industrial devices. Many of these items currently end up in landfills as a result of incorrect disposal practices. Those that do make it into the recycling process are handled in a number of ways depending on the components and their final destination. E-waste, according to the Environmental Protection Agency (EPA), is defined as electronic goods and parts that contain reusable materials. The definition focuses on the device's operating components rather than its plastic shell or other insignificant features. While plastic can be recycled in a variety of ways, this occurs outside of the context of e-waste recycling.

The Asia-Pacific region is one of the world's most popular and fastest-growing regions. Additionally, numerous nations in this region experienced fast industrialization supported by foreign direct investments as a result of a labour force that was relatively inexpensive. These conditions benefitted the electrical and electronics industry which had undergone huge changes as a result of greater technological and commercial advancements [1, 2]. Electrical and electronic equipment (EEE) has become indispensable in today's society, raising living standards while simultaneously holding dangerous chemicals that are bad for the environment and human health and worsening the climate crisis. The amount of e-waste has increased as a result of the rise in EEE sales and demand [3-5].

In 2019, the production of e-waste reached over 50 million tonnes (Mt) worldwide. 24.9 million tonnes of electronic garbage were produced in the Asia-Pacific region alone. With predictions that worldwide e-waste creation will reach 74 million tonnes (Mt) by 2030, Global e-waste production surged three times more quickly than world population growth. However, recycling is not developing as quickly. In that same year, only about 13% of the trash was actually recycled. The majority of created e-waste is also dumped in landfills, which is a widespread practice all over the world [6-8]. There are two major problems with current e-waste management processes: (a) there is no effective infrastructure for collection and recycling and (b) there are no systems in place to make EEE producers accountable for end-of-life disposal [9-12].

8.3 How is E-Waste Recycled and Why Do Problems Exist?

The Asian-Pacific area now recycles the majority of e-waste. Developing economies get large quantities of discarded devices and recycle the metal and other materials according to their own guidelines and best practices. The economic realities of recycling companies in these underdeveloped countries make it exceedingly difficult for them to reject cargo once they arrive at their intended destination. They may contain potentially toxic elements or e-waste types that they are unable to process. Incineration, acid baths, and other unethical work practices can have negative consequences for the environment and individuals [13].

8.4 Global E-Waste Management Market Restraints

The high cost of recycling E-waste has been brought on by a lack of facilities for collecting it and by the high cost of processing methods. Many outdated products are disposed of in the garbage or stored in warehouses and storerooms since there isn't a system in place to prevent this. To improve the issue, a system for frequent collection of e-waste must be devised. An insufficient number of waste pickup zones is currently a hindrance to recycling efforts. Customers are also ignorant of these collection locations, which results in waste being unnecessarily disposed of, such as by burning it in traditional ways. This leads to pollution and health problems [14].

8.5 Global E-Waste Management Market Opportunities

Many electronic manufacturers today know the financial benefits of processing and recycling e-waste. During R&D and manufacturing processes, significant amounts of e-waste are produced and firms are taking steps to recover the necessary elements from this trash. After being recycled, mobile phones contain valuable metals, including gold, silver, and palladium, that can be recovered. As a result, many leading cell phone makers have launched their own initiatives to collect old phones from users who want to upgrade their technology. Many governments have also begun to take action by urging makers of electronic items to organize internal e-waste management projects or to outsource these initiatives to external organizations. The improper recycling of this rising e-waste creates environmental damage and health risks. Additionally, some recyclable materials are wasted as a result of this because they are thrown out as rubbish. Increasing awareness of recycling initiatives is now essential to combat these issues.

8.6 Management of E-Waste in India

E-waste generation in the nation is growing three times faster than municipal trash. India is the third-biggest producer of electronic garbage in the world, behind the United States and China, based on the International E-waste Monitor 2020. These three nations produce 53.6 million tonnes of the world's e-waste, or around 38% of the total (Mt). In India, the highly organized informal sector has historically been used for e-waste management's collecting, separation, disassembly, and recycling processes. Recycling in the informal sector is done in an inefficient manner, causing harm to workers' health as well as environmental degradation and the loss of valuable resources. India's approach to e-waste disposal differs from that of the rest of the world. As a result of unauthorized recycling practices, e-waste disposal methods are a major concern. As a result, it is challenging to quantify E-waste in India and there are no tools for controlling the flow of E-waste through the system [15]. Maharashtra produced 19.8% of India's e-waste but recycled only 47,810 tonnes annually (TPA), Tamil Nadu recycled 13% of its E-waste and recycled 52,427 TPA, and Uttar Pradesh recycled 10.1% of its E-waste and recycled 86,130 TPA in 2018, according to an ASSOCHAM-NEC poll. E-waste was generated in West Bengal at a rate of 9.8%, compared to Delhi's 9.5%, Karnataka's 8.9%, Gujarat's 8.8%, and Madhya Pradesh's 7.6%.

In India, there are no large-scale structured E-waste recycling facilities and all recycling takes place in the informal sector (https://cpcbbrms.nic.in). The informal sector handles the majority of tasks such as collection, transportation, sorting, disassembly, recycling, and disposal. The majority of E-waste is collected by rag pickers, who are paid a fee by the client from whom the rubbish is collected. Rag pickers collect all types of garbage, such as paper, books, newspapers, plastic, cardboard, polythene, metals, and E-waste, and sell it to mediators or scrap dealers to make a living. Not only for rag pickers but also for mediators and scrap merchants, this is a substantial source of money. E-waste is usually handled by untrained workers who do not take adequate safety precautions in order to save money [16]. Due to a lack of suitable technology, recycling and disposal processes are not successfully utilized. The "take-back" programme has only been willingly embraced by a limited portion of firms. The volume of waste produced and disposed of annually, as well as the environmental damage it poses, are not well documented in statistics [17]. The literature analysis shows that just 50% of common people are aware of the negative effects that electronic products have on the environment and human health, which shows that India sorely needs to put in place an efficient E-waste treatment system [18].

8.7 New Solutions for E-Waste Excess

How can the world address the issue of e-waste recycling and improve the social and environmental accountability of resource reuse? The National Strategy for Electronics Stewardship of the Environmental Protection Agency serves as the basis for new regulations and strategies for the production and recycling of electronic goods. The changes that have been suggested or implemented include a concentration on a minimal supply chain, increased focus on cyber defense to improve hardware destruction techniques, and a broader emphasis on the importance of e-waste reuse accreditations, workers' rights, and ecological oversight for current service companies [19]. The world will continue to become more dependent on technology and automation in the years to come and people are unlikely to give up their individual electronics. Global economic and environmental authorities must redesign the entire manufacturing-to-consumer pipeline with a focus on long-term sustainability instead of profit alone as the need for recycling e-waste rises.

8.8 Recent Trends in E-Waste Management

In 2016 the Ministry of Environment, Forestry, and Climate Change laid different rules in order to manage e-waste. Globally, there are still many obstacles in the effective treatment of e-waste management, which include a lack of proper infrastructure, improper investments, and knowledge from consumers regarding the dumping of electronic trash. In India, a number of companies and individuals have emerged to assist the unorganized sector in gathering, processing, and recycling electronic waste [20].

8.8.1 Public Health, Environment, and E-Waste

Electronic garbage is presently the component of ordinary municipal solid waste that is rising the fastest in the globe, making its disposal a hazard for both the environment and public health globally. Because most people don't know how to properly dispose of electronic trash, a lot of it is maintained in homes and is referred to as "e-waste" (electrical or electronic garbage). E-waste is any electronics or electrical device that is abandoned, outmoded, or broken. In addition to being complex in form, this growing waste is a rich source of metals, like copper and silver, which may be collected and reintroduced back into the production process. According to associated market research, the global e-waste management market is expected to be worth $49,880 million in 2020. It is anticipated to increase at 14.3% CAGR from 2021 to 2028 and reach $143,870 million. Rare metals have experienced significant price increases as a result of their growing demand and dwindling supply.

8.8.2 Disposal and Management of E-Waste

E-waste reduction involves more than just reducing environmental risks. Lowering e-waste helps preserve resources and lowers the energy needed to produce these products because recycling e-waste items uses substantially less energy than producing new ones.

E-waste can be reduced and managed in a variety of methods such as:

i) **Purchasing Fewer Items**

The purchasing of products we don't need is the most typical source of e-waste. Avoid purchasing new electrical gadgets that the manufacturer will not be able to reuse or dispose of. A sustainable approach to e-waste management is to choose recyclable or long-lasting electronic items.

ii) **Organizing Your Possessions**

You won't know what you have if you don't organize your electronics, wires, connectors, and DVDs. Buying something you believe you need only to discover that you already have a copy is the very last thing you want to do.

iii) **Giving Away or Donating Your E-waste**

Give up anything you don't need in order for another individual to use it. The amount of donations is frequently similar to the worth of the item being sold, making them an excellent strategy to reduce taxes.

iv) **Take Them Back to the Retailer**

A "buy-back programme" is an arrangement between a consumer and a provider that implies the seller has promised to reprice the item purchased at a later date. Few stores provide this service. Find out if the business will take your old camera, PC, or other products before you go shopping for something new.

v) **Market Any Unused Goods**

As when you are finished using your devices, sell them because newer model value quickly decreases. Many businesses will gladly buy your used laptops, fitness bands, cameras, video games, wearables, and many other modern devices. They promote businesses that offer payment in exchange for goods.

vi) **Learn Details About the Recycling Options Available in Your Area**

No matter where you live, look into your local recycling options and tell your family and neighbors about them. For instance, if you reside in the United States, the Nature Conservation Agency's website provides details on nearby alternatives for recycling electronics.

vii) **Consider the Future**

Why not make use of e-waste now, since we won't be able to eliminate it quickly and will eventually need to? Keep them from piling up. Electronic equipment is rapidly becoming obsolete due to technological improvements, so it is advised to sell any outdated electronics to prevent cluttering and damage in the future.

viii) **Data Storage in the Cloud is Now Possible**

There is no requirement to spend money on a sizable server or robust equipment for business or private storage. Without the need for a server, cloud-based data solutions are ideal for preserving and synchronising your data across several devices.

ix) **Do Your Research and Prepare Yourself for Some Anxiety**

Electrical gadgets include hazardous components and appropriate disposal is essential. Educate yourself, your children, and the people around you. We should be more mindful of the dangers of e-waste as a result of these substances.

x) **Maintain What You Have**

Minor adjustments can help you keep what's working for longer. Clean your computer on a regular basis and don't overburden it to lengthen the life of your battery. Electronic product longevity can be considerably aided by regular servicing and maintenance.

xi) **Consider Security Issues while Disposing of Computers and Mobile Devices, and Use Recycling**

Since all of your personally identifiable data is preserved on electronic devices even after you wipe it, there is another reason not to throw away your electronic devices. Recycling facilities can thoroughly clean your device before recycling it, preventing access to the data by cybercriminals.

xii) **Inform Your Kids About E-waste**

Children are our heritage, thus it is advantageous if we can instill in them a commitment to recycling e-waste at a young age.

8.9 E-Waste Regulation in India

Different regulations directly or indirectly apply to toxic and hazardous waste. The Environmental (Protection) Act of 1986 covers an extensive range of environmental issues. Section 6 continuously empowers the Central Government to make rules on a various topics such as:

1. Protocols for processing and measuring hazardous materials
2. Laws related to prohibitions and restrictions related to hazardous materials

In response to the expanding e-waste issue, the Central Government has notified these rules in accordance with its authority under Sections 6, 8, and 25 of the Environmental (Protection) Act, 1986. The 2011 E-waste (Managing and Handling) Rules have been replaced by the 2016 E-waste (Management) Rules. There are 24 rules total, divided into 6 Chapters and 4 Schedules. This Act aims to promote the handling of all types of E-waste while lowering the amount of hazardous waste dumped in landfills and encouraging the recycling of usable E-waste products. The implementation date for these rules is October 1, 2016. The 2016 E-waste (Management) Rules identify and describe all parties involved in the creation of electrical equipment and the disposal of waste created when such equipment reaches the end of its useful life (producer, e-retailers, bulk consumers, dismantlers, collection centres, manufacturers, dealers, and refurbishers). The rule also outlines each stakeholder's responsibilities and functions. The foundation of e-waste management is the Extended Producer Responsibility idea (EPR). The manufacturer of the goods is responsible for the product's whole life span, including grab, recycling, and final disposal, under this environmental protection policy. E-waste management must include the classification and application of EPR [21]. Many academics have utilised EPR to investigate different aspects of e-waste management. The EPR is frequently cited as being one of the best methods for addressing the E-waste issue. However, officials in developing countries are cautious about the execution of EPR due to the unofficial sector's active participation, unlike in the industrialised world [22].

EPR is a regulatory tool that enables businesses to pay for the costs associated with gathering, recycling, and/or properly discarding things that their customers no longer desire [23]. In light of European WEEE Directives, such as the EPR, Favot et al. [24] analysed the Italian community system for the disposal of domestic waste, Waste Electrical and Electronic Equipment (WEEE), in addition to its development over time. The EPR and Producer Responsibility Organization (PRO) projects have been described in detail in European nations, but their potential in India has not yet been adequately examined [25].

Producers have been given specific targets to meet in order to manage 30% of trash created during the first two years of the rule's implementation. This goal has been continuously expanded, and by the seventh year of the rule's adoption, approximately 70% of the E-waste created has been correctly managed. Non-compliance with the aim may result in the cancellation of EPR authorization, preventing the producer from putting items on the market until EPR authorization is re-granted. A planned approach to managing E-waste is required of producers, in addition to a reduction in the number of hazardous substances in their machinery. It is forbidden to use equipment that contains amounts of mercury, lead, polybrominated biphenyls, hexavalent chromium, and polybrominated diphenyl ethers that are more than 0.1 percent cadmium by weight in homogeneous materials.

8.9.1 Treatment of E-Waste

E-waste is a complex mixture of valuable materials that can be recovered and recycled and dangerous materials that need to be properly disposed of. To separate dangerous trash from e-waste, sophisticated techniques that are labour-intensive are needed. The ideal way to handle E-waste that has been suggested uses physical dismantling as the first step in its management. Due to the variability of the substance, recovery of the precious material is still not worthwhile. A significant issue with effective recovery is heterogeneity. Circuit boards also contain other necessary organic substances, such as phenolic resins and isocyanate phosgene acrylic, in addition to these inorganic materials. Glass, copper, steel, aluminium, plastic, printed circuit boards, and other materials are removed from the deconstructed electronic waste. The entire PCB level of e-waste ranges from 3 to 5 percent by weight, necessitating ecologically friendly recycling techniques. Glass, plastics, and the other 95–97% of materials may all be manually disassembled and separated without damaging the environment [26, 27]. At this stage, dangerous components, including batteries, CRT displays, CFC gases, and capacitors, are also separated and eliminated. The rise in recyclable materials in a specialized fraction is made possible by mechanical processing, which is often a large-scale operation. It also further isolates dangerous contaminants. Crushing machines, shredders, magnetic separators, and air separators are typical elements of automated processing systems. In order to sell them as secondary raw materials, the majority of the fractions obtained here are processed. After refinement and the extraction of valuable fractions, the pollutants that are often harmful and unusable are disposed of in facilities designed specifically to handle hazardous waste [28–30].

8.9.2 Amendments in E-Waste Management Rules 2018

A notification dated March 22, 2018, with the number G.S.R. 261(E), modifies the E-waste Management Rules of 2016. The rule change was implemented to formally establish the E-waste recycling market by directing the domestically produced electronic debris to approved dismantlers and recyclers. The collection goals under the Extended Producer Responsibility (EPR) provisions of the Rules have been revised and goals have been established for new producers who have just started their sales operation. Key facets of the E-waste (Management) Amendment Rules, 2018 are as follows:

1. The updated E-waste collection targets under EPR will take effect on October 1, 2017. The weight-based phase-wise collection targets for E-waste are 10% of the waste-generating quantity as specified in the EPR Plan during 2017–18, with an increase of 10% per year by 2023. The aim will be 70% of the amount of waste generated, as mentioned in the EPR Plan starting in 2023.
2. The amount of electronic trash gathered by manufacturers between 1st October, 2016, and 30th September, 2017, up until March 2018, must be taken into account by the new EPR targets.
3. For the new producers, those whose number of years in business is less than the average lifespan of their products, separate E-waste collection targets have been set. Things must live their normal lives in accordance with the regulations set forth occasionally by the Central Pollution Control Board of India (CPCB), 2016.
4. Producer Responsibility Organizations (PROs) must submit a registration application to the CPCB in order to engage in the activities listed in the regulation
5. The government is required by the Reduction of Hazardous Substances (RoHS) rules to cover the cost of sampling and testing in order to carry out the RoHS test. The manufacturer is responsible for paying the test costs if the test results do not support the RoHS specification [31-33].

8.10 Recovery of Resources from E-Waste

Resources are recovered from e-waste through pyrometallurgy and hydrometallurgy methods.

8.10.1 Pyrometallurgy

A traditional method for extracting non-ferrous and precious metals from WEEE is pyrometallurgy. The abundant metal portion is first roasted, disassembled, and smelted as part of the pyrometallurgical procedure to pre-treat the wastes. The extraction of valuable metals and copper from EoL electrical appliances has traditionally been accomplished via a pyrometallurgical process that involves the smelter [34, 35]. The most important step in pyrometallurgy is smelting. Some of the contemporary smelting equipment utilised in pyrometallurgy includes the Mitsubishi continuous smelter, the Outokumpu flash smelter, and the Noranda reactor system [36]. Because e-waste contains both pure metals and alloys, improving the quality of the final metal products presents the biggest challenge in pyrometallurgical operations. Melting in smelters can easily handle pure metallic forms.

Higher temperatures are used in pyrometallurgical processes to volatilize particular metals, which are then condensed and recovered. The majority of industrial recycling procedures use heat treatments and pyrometallurgical techniques. Hydrometallurgical procedures offer comparatively cheap capital costs, minimal environmental impact, totally recoverable leachates, and much less air pollution (e.g., no toxic gases or dusts) than pyrometallurgical operations. It is important to note that the majority of the base and precious metals included in e-waste can be recovered using this method; however, additional processing (downstream) is needed to refine or extract the target metal of interest using electrochemical or hydrometallurgical processes. The steps involved in the Kaldo process are:

(i) Washing with water
(ii) Leaching process with acid
(iii) Reductive smelting of e-waste
(iv) Oxidative blowing followed by refining
(v) Electrolytic refining [37]

Advantages

(i) High rates of recovery for rare metals like gold and silver
(ii) Due to its known furnace design and configuration, it is simple to scale-up and operate
(iii) Treatment capacity is high

8.10.2 Hydrometallurgy

By employing chemicals, hydrometallurgy extracts metals from solid sources. Leaching, in which the metals are dissolved by aqueous reagents at a low pH, and recovery, in which the leached metals are extracted from the polymetallic expectant leaching solution, are the two steps of hydrometallurgy [38]. Hydrometallurgical methods are considered to be effective and well-established and they are developing quickly. Circuit boards, telecom servers, and automated teller machines all contain significant amounts of precious metals, making it more economical to utilize these parts as raw resources in hydrometallurgical operations [39].

8.11 Generation and Management of Mobile Phone Waste

8.11.1 Generation of Mobile Phone Waste

The amount of garbage now in existence and the rate at which it is being produced must be known in order to evaluate mobile phone waste's potential for recycling. Therefore, the estimation of the generation of trash from mobile phones has been the focus of various academics. However, it was noted that rather than a worldwide representation, the majority of this research solely focused on local statistics on trash generation. In the Czech Republic, the generation of garbage from mobile phones was calculated by Polák and Drápalová in 2012 [40]. To estimate the amount of mobile phone garbage that will be produced in the near future, they employed a special methodology that involved calculating the life duration and distribution volume of mobile phones. They estimated the number of abandoned mobile phones and reported that it was only 45,000 in the decade 1990-2000, 6.5 million in the following decade, and is anticipated to be over 26,000,000 in the present decade (2010–2020). They also contrasted their principles with those of nations like Japan, China, the United States, the United Kingdom, and Korea. Rahmani *et al.* (2014) estimated Iranian computer and phone generation trends in the past and the future. They did this by using various models, including the simplified logistic function model and the time-series multiple life span model. They estimated that up until 2014, 39 million mobile phones have entered the trash stream, and they predicted that by 2035, this number will rise to 90 million. In addition, they predicted that it would take 21 years after 2014 for mobile phone trash to reach saturation [41]. Li *et al.* (2015) compared a number of approaches, including the sales and new method, consumption and usage strategy, and market supply method [42]. They discovered that the findings varied greatly depending on the methodology used. In order to obtain reliable estimates of the generation of mobile phone waste, it is crucial to use the best method.

8.11.2 Management of Waste from Mobile Devices

After the EU's RoHS (Restriction of Hazardous Substances) directive and e-waste handling standards were established in 2011, management initiatives for mobile phone garbage began to emerge. Successfully using various components of the mobile phone waste stream has been investigated. The most significant findings from these investigations are discussed in this section. When comparing different mobile phone takeback programmes in the US and Finland, Tanskanen and Butler (2007) came to the conclusion that there is a tremendous opportunity for mobile phone recycling through takeback programmes. They claimed that the effectiveness of such takeback systems is largely dependent on their simplicity of use, compelling messaging, and the correct customer incentive [43]. A thorough analysis of the disposal of used mobile phones in Korea was provided by Jang and Kim (2010). Their study's approach included gathering information about yearly domestic demand for mobile phones using questionnaire surveys, site visits, interviews, and talks, as well as a review of the literature. They talked about the legislative steps implemented by the Korean government for the correct disposal of used mobile phones, including their features and efficacy. One intriguing part of their work is a process used for recycling mobile phones that is broken down into four stages: generation, collecting, reuse/recycling, and treatment. They said that in order to address the growing problem of mobile phone waste, makers, mobile phone companies, customers, and governments all must work together [44]. Remanufacturing used mobile phones may help achieve sustainability, according to research by Rathore *et al.* [45]. A report on sustainable mobile phone waste management in the UK through collection, reuse, and recycling was written by Ongondo and Williams (2011). They evaluated the efficacy of various mobile phone takeback programmes. An essential piece of instruction, the Basel Convention Mobile Phone Partnership Initiative (MPPI), 2012, provides information on how to handle mobile phone waste from collection to recycling [46]. According to Zink *et al.* (2014), who have focused especially on smartphones, the adaptability of modern cell phones offers a new repurposing technique for efficient waste management. Utilizing mobile devices or their components for different purposes is known as repurposing. Zink powered a parking metre with batteries that had been thrown away. Also, they used life cycle analysis to compare it to conventional renovation (LCA). They discovered that recycling is significantly more environmentally favorable than regular renovation [47]. In a related study, Diouf *et al.* (2015) used the batteries from used mobile phones to power lighting equipment. The conclusions drawn from the various studies already mentioned point to the need for effective recycling programmes for used mobile phones, and it should be highlighted that practically all materials may be recovered and recycled, leading to significant savings in terms of fuel, energy, and resources [48].

According to projections, the Asia Pacific will dominate the regional market while growing at a CAGR of 11.57%. As a result of continual innovation and price reductions, new product advancements have had a substantial impact on the actual life lifetime of electronic devices like computers and smartphones. Additionally, the rising per capita earnings in the region's several nations have compelled people to often upgrade their consumer goods purchases. Sales of electronic devices, including computers, refrigerators, and cell phones, have surged over the past 10 years, which has increased the amount of e-waste in the region. Due to rising sales of electrical and electronic equipment and a refusal to fix malfunctioning goods, the amount of e-waste increases every year. Users typically choose not to repair damaged or outdated gadgets since the cost of repairs can occasionally be higher than the cost of a brand-new product. This increases the amount of electronic waste each year. Additionally, a variety of opportunities for effective e-waste management are made available by the flow of e-waste from developed regions to developing countries like India, China, and Pakistan. With a CAGR of 14.66%, Europe will probably contain USD 54 billion in sales. Electronic waste, which includes refrigerators, cell phones, and computers, is one of the waste sources in the EU that is growing the quickest. Due to the rapid increase in electronic and electrical equipment waste, the scarcity of important metals, and the high cost of mining, e-waste recycling is essential. Since there is the least amount of landfill space left among the other continents, Europe has also built strict environmental protection legal frameworks. Due to all of these factors, the market for recycling e-waste in the region is growing. Electronic and electrical equipment makes up a sizable portion of the continent's overall waste. In Europe, stricter regulations have been put in place as a result of rising pollution levels and a lack of available space for garbage disposal. As a result, things have gotten better because there has been a lot of focus on stopping the illegal shipping of e-waste to developing nations. Additionally, a high rate of consumer recycling of electrical and electronic equipment and cooperation between public and private sector organizations have helped to create a favorable environment for the e-waste recycling industry in this area [49].

8.12 Players of the Market

A list of major players in the e-waste management market includes Recycling Management Inc., Financial Environmental Holdings Ltd., International

Electronic Recyclers, Inc., Environmental Hub Holdings Ltd. Company, Sembcorp Industries Ltd., S.A. Veolia Environment, The Mri (Australia) Pty Ltd., UMICORE SA, Tetronics (International) Limited, and TES-AMM.

8.13 Recent Developments

Electronic Recyclers International, Inc. declared in 2022 that it had obtained its compliance certification and had completed the Service Organization Control (SOC) 2 Type 1 evaluation, making ERI the industry's first and only e-waste recycler that was SOC 2 accredited. OneDrumTM, the first mixed consumer battery collection solution on the market, was formally unveiled in 2022 by Electronic Recyclers International, Inc [49].

8.14 Conclusion

In addition to hazardous materials, e-waste also includes valuable resources. Although the public should be aware that e-waste contains many harmful materials, the administration is quite cautious about it. In India, the reuse, reprocessing, and repurposing of e-waste is typically done unofficially, or without official permission. The overall bulk of e-waste is dumped at landfills after the precious material has been collected, however, some are reused in the unauthorized sector. Various environmental and health problems are brought on by it. Therefore, it is essential to bridge the gap between the informal and the formal E-waste sectors in order to efficiently utilise the resource potential of E-waste and establish a sustainable management pattern. The formalization of the informal sector into an ongoing recycling programme is crucial and strongly advised. This substantial volume of E-waste will be turned into useful goods thanks to the effective acquisition, collecting, extracting, and removal of materials facilitated by good E-waste management. Efficient recycling, E-waste management, and strict enforcement of E-waste management regulations are required to lessen the harmful effects of E-waste on human health and the environment, as well as to ensure India's waste management system is profitable.

References

1. https://timesofindia.indiatimes.com/blogs/voices/12-ways-to-reduce-and-control-e-waste-for-2022/

2. Forti V, Balde CP, Kuehr R, Bel G. The Global E-waste Monitor 2020: Quantities, flows and the circular economy potential.
3. Baldé CP, Forti V, Gray V, Kuehr R, Stegmann P. The global e-waste monitor 2017: Quantities, flows and resources. United Nations University, International Telecommunication Union, and International Solid Waste Association; 2017.
4. Kumar B, Bhaskar K. Electronic waste and sustainability: Reflections on a rising global challenge. Markets, Globalization & Development Review. 2016;(1)1-16.
5. Khetriwal DS, Kraeuchi P, Widmer R. Producer responsibility for e-waste management: key issues for consideration–learning from the Swiss experience. Journal of environmental management. 2009 Jan 1;90(1):153-65.
6. Schmidt C.W. Unfair trade: E-waste in Africa. Environ. Health Perspect. 2006;114:232–235.
7. Marinescu C, Cicea C, Ciocoiu CN. The development of a performance assessment method for e-waste management in the European Union. Environmental Impact Assessment Review. 2005;25:436-58.
8. Islam MT, Abdullah AB, Shahir SA, Kalam MA, Masjuki HH, Shumon R, Rashid MH. A public survey on knowledge, awareness, attitude and willingness to pay for WEEE management: Case study in Bangladesh. Journal of cleaner production. 2016 Nov 20;137:728-40.
9. Zuo L, Wang C, Sun Q. Sustaining WEEE collection business in China: The case of online to offline (O2O) development strategies. Waste Management. 2020 Jan 1;101:222-30.
10. Sharma KD, Jain S. Municipal solid waste generation, composition, and management: the global scenario. Social responsibility journal. 2020 Jun 23;16(6):917-48.
11. Andrade DF, Romanelli JP, Pereira-Filho ER. Past and emerging topics related to electronic waste management: top countries, trends, and perspectives. Environmental Science and Pollution Research. 2019 Jun 1;26:17135-51.
12. Schumacher KA, Agbemabiese L. Towards comprehensive e-waste legislation in the United States: design considerations based on quantitative and qualitative assessments. Resources, Conservation and Recycling. 2019 Oct 1;149:605-21.
13. Murthy V, Ramakrishna S. A review on global e-waste management: urban mining towards a sustainable future and circular economy. Sustainability. 2022 Jan 7;14(2):647.
14. Baldé C.P, Brink, S, Forti V, Schalk A, Hopstaken F. The Dutch WEEE Flows 2020. "What happened between 2010 and 2018"; United Nations University (UNU)/United Nations Institute for Training and Research (UNITAR): Tokyo, Japan, 2020.
15. Rajput R, Nigam NA. An overview of E-waste, its management practices and legislations in present Indian context. Journal of Applied and Natural Science. 2021 Feb 5;13(1):34-41.

16. Purushothaman M, Inamdar MG, Muthunarayanan V. Socio-economic impact of the e-waste pollution in India. Materials Today: Proceedings. 2021 Jan 1;37:280-3.
17. Arya S, Kumar S. E-waste in India at a glance: Current trends, regulations, challenges and management strategies. Journal of Cleaner Production. 2020 Oct 20;271:122707.
18. Bhutta MK, Omar A, Yang X. Electronic waste: a growing concern in today's environment. Economics Research International. 2011, 1-8
19. Current Trends In E-Waste Recycling — Digital Responsibility
20. 12 ways to reduce and control e-waste for 2022 (indiatimes.com)
21. Corsini F, Rizzi F, Gusmerotti NM, Frey M. Extended Producer Responsibility and the Evolution of Sustainable Specializations: Evidences From the e-Waste Sector. Business Strategy and the Environment. 2015 Sep;24(6):466-76.
22. Victor D, Agamuthu P. Policy trends of strategic environmental assessment in Asia. Environmental Science & Policy. 2014 Aug 1;41:63-76.
23. Nash J, Bosso C. Extended producer responsibility in the United States: Full speed ahead?. Journal of Industrial Ecology. 2013 Apr;17(2):175-85.
24. Favot M, Veit R, Massarutto A. The evolution of the Italian EPR system for the management of household Waste Electrical and Electronic Equipment (WEEE). Technical and economic performance in the spotlight. Waste management. 2016 Oct 1;56:431-7.
25. Zhao ZY, Yan H, Zuo J, Tian YX, Zillante G. A critical review of factors affecting the wind power generation industry in China. Renewable and Sustainable Energy Reviews. 2013 Mar 1;19:499-508.
26. Ludwig C, Hellweg S, Stucki S, editors. Municipal solid waste management: strategies and technologies for sustainable solutions. Springer Science & Business Media; 2012 Dec 6 320-322.
27. Chatterjee S. Sustainable electronic waste management and recycling process. American Journal of Environmental Engineering. 2012;2(1):23-33.
28. Rajput R, Nigam NA. An overview of E-waste, its management practices and legislations in present Indian context. Journal of Applied and Natural Science. 2021 Feb 5;13(1):34-41.
29. Ravindra K, Mor S. E-waste generation and management practices in Chandigarh, India and economic evaluation for sustainable recycling. Journal of Cleaner Production. 2019 Jun 1;221:286-94.
30. Rajput R, Nigam NA. An overview of E-waste, its management practices and legislations in present Indian context. Journal of Applied and Natural Science. 2021 Feb 5;13(1):34-41.
31. Hong Y, Valix M. Bioleaching of electronic waste using acidophilic sulfur oxidising bacteria. Journal of Cleaner Production. 2014 Feb 15;65:465-72.
32. Tuncuk A, Stazi V, Akcil A, Yazici EY, Deveci H. Aqueous metal recovery techniques from e-scrap: Hydrometallurgy in recycling. Minerals engineering. 2012 Jan 1;25(1):28-37.

33. Jujun R, Yiming Q, Zhenming X. Environment-friendly technology for recovering nonferrous metals from e-waste: Eddy current separation. Resources, conservation and recycling. 2014 Jun 1;87:109-16.
34. Korman T, Bedekovic G, Kujundzic T, Kuhinek D. Impact of physical and mechanical properties of rocks on energy consumption of jaw crusher. Physicochemical Problems of Mineral Processing. 2015;51.
35. Kendrick D, Young B, Mason-Jones AJ, Ilyas N, Achana FA, Cooper NJ, Hubbard SJ, Sutton AJ, Smith S, Wynn P, Mulvaney C. Home safety education and provision of safety equipment for injury prevention. Evidence-based child health: a Cochrane review journal. 2013 May;8(3):761-939.
36. Bu L, Zhang N, Guo S, Zhang X, Li J, Yao J, Wu T, Lu G, Ma JY, Su D, Huang X. Biaxially strained PtPb/Pt core/shell nanoplate boosts oxygen reduction catalysis. Science. 2016 Dec 16;354(6318):1410-4.
37. Hait NC, Allegood J, Maceyka M, Strub GM, Harikumar KB, Singh SK, Luo C, Marmorstein R, Kordula T, Milstien S, Spiegel S. Regulation of histone acetylation in the nucleus by sphingosine-1-phosphate. Science. 2009 Sep 4;325(5945):1254-7.
38. Sethurajan M, Huguenot D, Jain R, Lens PN, Horn HA, Figueiredo LH, van Hullebusch ED. Leaching and selective zinc recovery from acidic leachates of zinc metallurgical leach residues. Journal of Hazardous Materials. 2017 Feb 15;324:71-82.
39. Sethurajan M, van Hullebusch ED, Nancharaiah YV. Biotechnology in the management and resource recovery from metal bearing solid wastes: Recent advances. Journal of environmental management. 2018 Apr 1;211:138-53.
40. Polák M, Drápalová L. Estimation of end of life mobile phones generation: The case study of the Czech Republic. Waste management. 2012 Aug 1;32(8):1583-91.
41. Rahmani M, Nabizadeh R, Yaghmaeian K, Mahvi AH, Yunesian M. Estimation of waste from computers and mobile phones in Iran. Resources, Conservation and Recycling. 2014 Jun 1;87:21-9.
42. Li B, Yang J, Lu B, Song X. Estimation of retired mobile phones generation in China: A comparative study on methodology. Waste management. 2015 Jan 1;35:247-54.
43. Tanskanen, P., Butler, E., 2007. Mobile phone take back – learning's from various initiatives. In: Electronics & the Environment, Proceedings of the 2007 IEEE International Symposium, pp. 206–209.
44. Jang YC, Kim M. Management of used & end-of-life mobile phones in Korea: a review. Resources, Conservation and recycling. 2010 Nov 1;55(1):11-9.
45. Rathore P, Kota S, Chakrabarti A. Sustainability through remanufacturing in India: a case study on mobile handsets. Journal of Cleaner Production. 2011 Oct 1;19(15):1709-22.
46. Ongondo FO, Williams ID. Mobile phone collection, reuse and recycling in the UK. Waste management. 2011 Jun 1;31(6):1307-15.

47. Zink T, Maker F, Geyer R, Amirtharajah R, Akella V. Comparative life cycle assessment of smartphone reuse: repurposing vs. refurbishment. The International Journal of Life Cycle Assessment. 2014 May;19(5):1099-109.
48. Diouf B, Pode R, Osei R. Recycling mobile phone batteries for lighting. Renewable energy. 2015 Jun 1;78:509-15.
49. https://straitsresearch.com/report/e-waste-management-market

9

Current E-Waste Management: An Exploratory Study on Managing E-Waste for Environmental Sustainability

Shweta Solanki[1*] and Pramod Singh Rathore[2]

[1]Department of Computer Science, JNVU Jodhpur, Jodhpur, India
[2]Department of Computer and Communication Engineering, Manipal University Jaipur, Jaipur, India

Abstract

The findings of the study indicate that effective management of electronic waste is required in order to achieve environmental sustainability in the present day and age. This is due to the fact that e-waste poses a risk not only to the environment but also to human health. It is also exciting to find that television and social media are the preferred methods of communication for e-waste awareness campaigns among urban respondents. On the other hand, respondents from less populous cities are more likely to favour radio and newspapers for awareness initiatives. The management of the production and disposal of electronic waste, as well as the development of technologies for the environmentally responsible collection and recycling of electronic trash, might be the topic of studies in the future. The execution of rules, training for the unorganised sector, awareness campaigns, an open recycling system, and incentives for e-waste recycling, among other things, are now the most pressing concerns for the government.

In this chapter, the environmental concerns that are linked with discarded electronic equipment, sometimes known as "e-waste," are discussed in detail. In addition, the development of e-waste both now and in the future, as well as any environmental difficulties that may come from its management and disposal approaches, are investigated, and the existing programmes for the management of e-waste are discussed.

Keywords: E-waste management, environmental pollution, recycling, sustainability

---

*Corresponding author: shweta.solanki01212@gmail.com

Abhishek Kumar, Pramod Singh Rathore, Ashutosh Kumar Dubey, Arun Lal Srivastav, T. Ananth Kumar and Vishal Dutt (eds.) Sustainable Management of Electronic Waste, (187–200) © 2024 Scrivener Publishing LLC

9.1 Introduction

The rapid advancement in the electronic industry and the ongoing trend toward computerization and digitalization have presented significant challenges for industrial businesses operating in today's market conditions. This has resulted in a continuing tendency toward shorter product lifecycles and a reduction in the amount of time needed for the development of new products. In addition to this, there is a growing need for customization, which we will refer to as the analogous phase. E-waste is sometimes incorrectly interpreted to refer to obsolete computers or information technology (IT) equipment in general. Waste Electrical and Electronic Equipment (WEEE) is another term that is used in international literature. Even though there has been a fourfold increase in the amount of e-waste collected and processed in India over the past four years, 95% of the country's e-waste is still handled illegally by the informal sector [1][2]. The uncontrolled informal garbage collectors burn objects that cannot be reused or diverted to landfills, which may pose major threats to the environment as well as to people's health.

E-waste has received increased attention as a result of both the General Assembly and the "Waste of Electronic and Electrical Equipment Law" passed by the European Union. The annual volume rise of e-waste is an issue since it is caused by rising consumption and short product lifespans [3][4]. This growth is a result of both factors. E-waste is the type of rubbish that has increased at the quickest rate over the course of the preceding ten years despite the fact that only 15% of it is recycled. In addition, recyclers operating in the informal sector

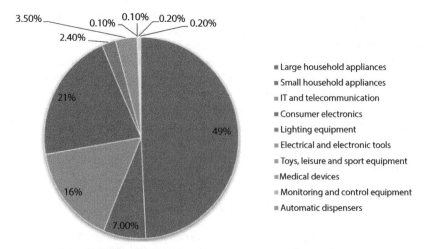

Figure 9.1 Equipment for electronics and electricity waste.

utilise antiquated recycling practises that might result in the emission of compounds that are detrimental to the environment [5]. Several toxic substances that can be found in discarded electronic equipment pose a significant threat to both the health of humans and the environment. The increased usage of electronic devices may be attributed, in part, to the rapid acceleration of technological advancement as well as the growing purchasing power of consumers. The increased consumption has an effect on the environment, whether it is through the extraction of raw materials or the waste created after the product has been used. Equipment for electronics and electricity waste is shown in Figure 9.1. According to the definition of "e-waste," also known as "Waste of Electronic and Electrical Equipment (WEEE)," this word refers to "all components, subassemblies, and consumables, which are part of the product at the time of discarding." There are different definitions supplied for e-waste by a variety of sources, nevertheless, it is essential to keep in mind that the term "e-waste" refers to any electrically powered item, regardless of its size or purpose, that the user has decided they no longer desire [6][7][8][9]. The use of the phrase "no longer desirable" gave the impression that electronic trash may be created for a number of reasons, including those listed above.

9.2 E-Waste Production

Electronic trash may be separated into a variety of distinct classifications according to the physical and chemical components that it is composed of. The references provide a classification system as well as documentation on its properties. It's possible that changes in the composition's physical and chemical make-up occurred as a result of the data being collected over a longer period of time [10][11]. It is possible that the content of e-chemical waste will shift as new technologies emerge. Research on the chemical composition found that the quantity of copper in dynamic RAM would increase by 75% and the quantity of platinum would decrease by 80%, while the quantity of gold and silver would remain the same (DRAM). Changes in technology and form factor will produce differences in the physical makeup of the world's e-waste, despite the fact that there will be an increase in the amount of metals as a result of the decline. Due to the non-metric nature of the assessment and the fact that there is only one dependent variable (level of awareness) and one independent variable, the crosstab is the method of choice for demonstrating the correlation between the various levels of consciousness exhibited by the respondents (cities). The final findings include the essential assertions, a cross tabulation comparison for the opinion status, and comparative bar charts [12][13].

9.3 The Present Predicament

Electronic trash may be classified based on the physical and chemical components that it contains. The classification of its properties is provided in the references, which are listed in Table 9.1. It's possible that the changes in the composition's physical and chemical make-up are the result of data being collected over an extended period of time. It's possible that, as technology advances, the make-up of electronic chemical waste will shift. According to a study that was conducted to investigate the chemical composition of dynamic RAM (DRAM), the quantity of gold and silver would remain the same, the quantity of palladium would decrease by 80%, and the quantity of copper would increase by 75%. This information was gleaned from analysing DRAM. Even though prices for precious metals will go up because of a decrease in the production of DRAM modules, the actual composition of the world's electronic trash will fluctuate because of advancements in technology and variations in the design of modular components [14][15][16].

Figure 9.2 chemical properties of electronics and electricity waste is shown in Figure 9.2. Electronic trash may be separated into a variety of distinct classifications according to the physical and chemical components that it is composed of. The references provide a classification system as well as documentation on its properties. It's possible that changes in the

Table 9.1 The structure of E-waste.

Constituents	Vats and Singh (2014)	Realff et al. (2004)	Kong et al. (2012)	Hossain et al. (2015)
Metals	60.20%	49%	13%	39.50%
- Copper			7%	20.10%
- Iron				8.10%
- Tin				4%
- Nickel				2%
- Lead				2%
- Aluminium				2%
- Zinc				1%
- Silver				0.20%
- Gold				0.10%
- Palladium				0.01%
Plastics	15.20%	33%	21%	30.30%

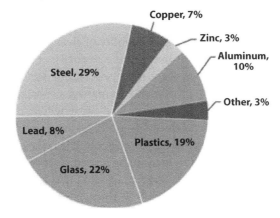

Figure 9.2 Chemical properties of electronics and electricity waste.

composition's physical and chemical make-up were brought about by the collection of data over a lengthier time period. It is possible that the content of e-chemical waste will shift as new technologies emerge. According to the findings of a study on the chemical characteristics, the quantity of copper in dynamic RAM would increase by 75%, the quantity of platinum would fall by 80%, and the number of gold and silver would not change at all (DRAM). Changes in technology will cause variations in the composition of the world's e-waste, and this, in turn, will cause an increase in the proportion of metals in e-waste that has been produced as a direct result of the decrease in the volume of e-waste produced [17][18][19][20].

9.3.1 Prospective Developments

The findings of the study indicate that effective management of electronic waste is required in order to achieve environmental sustainability. This is due to the fact that e-waste poses a risk to both human health and the environment. It is also exciting to find that television and social media are the preferred methods of communication for e-waste awareness campaigns among urban respondents. On the other hand, respondents from less populous cities are more likely to favour radio and newspapers for awareness initiatives. The management of the production and disposal of electronic waste, as well as the development of technologies for the environmentally responsible collection and recycling of electronic trash, might be the topic of studies in the future. Enforcement of regulations, education and training for the informal sector, public awareness campaigns, and an open recycling systems are all necessary components.

9.3.2 Environmental Impacts

We are not going to discuss the exact quantity of electronic trash that is generated by the various countries because the majority of references present varying projections for various years. Because a lack of data is a restriction in the efforts that developing nations make to manage e-waste, the tactics that may be used to anticipate the generation of e-waste will be discussed in this section. As a result, these methods can be useful for anticipating both the benefits and drawbacks of e-waste in developing countries. The easiest technique to calculate the volume of EOL is to make a prediction using data on EEE production and then estimate the percentage of EOL that will be recycled, as well as the percentage of EOL that will be thrown away. In another piece of research, the researchers employed a slightly altered version of a questionnaire that had been developed by the United Nations Environment Program to investigate the possible expansion and circulation of electronic garbage. Because the accuracy of the data is dependent on the roles that the respondents play in the EEE or WEEE life cycle, a mistake in selecting the respondent will nonetheless lead to a misleading evaluation of the flow of e-waste. This is the case because of the nature of the data. When trying to estimate the amount of electronic waste being produced, many people resort to using models such as the Population Balance Model and Material Flow Analysis to develop scenarios based on variations in consumption rate and transfer efficiency among various stakeholders. However, because not all of the relevant facts were easily accessible, it was essential to make some assumptions. These included the amount of stock available on the market and the frequency with which items were reused after being purchased from a second-hand market. Another research that utilised system dynamics modeling came to the conclusion that an increasing population, expanding economies, and shifting consumption habits are all variables that are leading to a rise in the number of EOL electrical and electronic equipment. The results of the simulation suggested that during the last ten years, there would have been an increase in the exports of EOL WEEE products if legislation could have been utilised to accelerate recovery rates. By being knowledgeable about the significant aspects of e-waste creation and recovery, legislators may enhance the quality of decisions they make and the solution to the problem of managing e-waste that engages important stakeholders. End-users (households, institutions, or corporations), collectors (scavengers and service centres), second-hand markets, sub-dealers or recyclers, refurbishers, and dealers who export e-waste that cannot be reused are the four primary players of the opposite e-waste supply chain network in underdeveloped nations. In developed nations, the end-users include households, institutions, or corporations. Impacts of

Figure 9.3 Impacts of recycling of E-waste.

recycling of E-waste is shown in Figure 9.3. The flow of information in the nation under study will determine the number of stakeholders who participate. In several of these investigations, the flow of electronic waste was analysed by using the Markov chain model, Life Cycle Assessment (LCA), or mixed-integer multi-objective linear programming. According to subjective research and literature assessments, up to 90 percent of the electronic trash that is produced in poor nations eventually finds its way into the informal waste sectors. If the government intends to implement the Extended Producer Responsibility (EPR) system, the informal waste sectors need to be given specific consideration in order to improve their participation in the collection and recovery methods. This is necessary in order for the government to achieve its goal of implementing the system.

9.3.2.1 Possible Adverse Effects on the Environment Resulting from Disposal of Electronic Waste

The conditions of the environment have a significant bearing on human health due to the fact that human beings may be harmed in two distinct ways by electronic waste: through direct exposure in recycling locations and through contamination of food chains. Because this is a non-metric style of survey with one dependent variable (consciousness) and one independent variable (cities), a crosstab is used to depict the link between the different degrees of awareness reported by respondents. The final findings include the essential assertions, a crosstabulation comparison for the opinion status, and comparative bar charts.

It is difficult to collect waste from every nook and cranny of the country since it is so extensively spread. In addition, there is a shortage of workers required for the collection and processing of garbage, which is a significant problem. When companies dispose of their garbage in rivers for the sake of a few extra dollars, those profits eventually make their way into people's homes. Urbanization is a factor that directly adds to the production of trash. Handling trash in an unscientific manner results in health risks and contributes to the destruction of the urban environment. Rag collectors are not successfully included into the process of waste management, according to the International Management Journal. Because they are not eligible for social security, rag collectors have no choice but to accept the dangers that come with their line of employment. Additionally, corruption prohibits the correct installation of garbage cans and devices for collecting electronic trash. The following challenges are faced by those who develop policy: inadequate financing for civic organisations, machinery and technology that is decades old, and public apathy. Laws governing trash management are quite outdated and need to be modernised in order to comply with current standards. There is not a separate department within urban local bodies that is responsible for rubbish management. Through education and awareness programmes, both children and people may cultivate a feeling of cleanliness in their own bodies. It is imperative that urban local authorities immediately establish a separate department in order to combat the e-waste catastrophe, and it is strongly recommended that the government conduct a review of the regulations governing waste management. It is probable that an unorganised sector would not be able to fulfill such ambitious targets since the population of the country is so large and the usage of electronic devices is increasing all the time [21][22].

The government may want to explore collaborating with the private sector in order to create official or standard operating procedures and a tiered plan for reaching the aim of reducing the amount of e-waste to the absolute lowest possible. It is strongly recommended that the collection goals for electronic waste be carried out in a phased manner, beginning with the more reasonable and attainable objective limitations. In addition, certain implementation requirements need to be adhered to in order to successfully remove electronic trash that was sourced from the market. In addition, the government may provide recommendations about the procedures that are followed in other countries for the efficient collection and recycling of e-waste. Landfills are used to dispose of any electronic trash that is still present. It is of the utmost importance to place the indiscipline sector under efficient monitoring and oversight in order to ensure that the great majority of the electronic trash that is created may be recycled in a way that is organised. The government may also take into consideration privatising recycling, as was done in the United Kingdom (UK). In that nation, the responsibility of collecting and recycling e-waste has been delegated to a private company that is governed by the public body. The United States government may also take into consideration privatising recycling.

It is essential to involve the workers of the informal sector, educate them on the impacts of wrong electronic waste management, and include them in the process of finding solutions to problems linked to e-waste [23][24].

9.3.2.2 Electronic Trash Disposal and Recycling Both Contribute to Contamination of the Environment

It is possible to recover valuable metals that are utilised in technology by lowering the amount of trash that is produced. The first responders in both places have a good understanding of precious metals and the advantages of recovering them [25]. Despite this, respondents are more confident in its relevance as a new business opportunity than Jalandhar city E-waste management, and they recognise its significance as well. The respondents from both locations are not persuaded that proper management of electronic trash is a prudent choice for a business [26]. This goes opposed to the meaning of the word "urban mining," which depicts the processing of e-waste as the recovery of valuable resources. If you want to recover precious materials from e-waste, then you are "urban mining." The plan for the management of e-waste must take education programmes into consideration [27]. The vast majority of respondents are of the opinion that an educational programme focusing on electronic waste should be a part of the standard curriculum [28][29].

In light of the fact that respondents are aware of how electronic trash is created but do not have confidence, it is important to launch awareness programmes for e-waste. Those from metropolitan areas are more self-assured and aware about the various approaches for managing e-waste than participants from smaller cities. Both municipalities are of the opinion that expanded product responsibility is the most viable solution to the problem of electronic waste. Program of recycling of E-waste is shown in Figure 9.4.

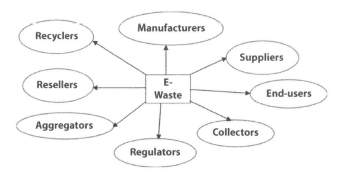

Figure 9.4 Program of recycling of E-waste.

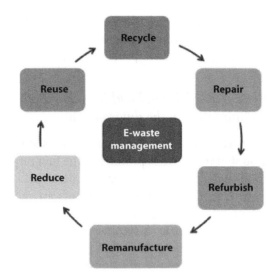

Figure 9.5 Process of E-waste management.

It was shown that although respondents from both cities are aware that e-waste contains potentially harmful components, they are less informed about the diversity of e-waste and its effects on both the environment and human health. However, the responders are aware of the fact that electronic garbage might include precious metals. The handling of electronic trash is a lucrative industry. The respondents who live in less populous places are more positive about the prospect of an e-waste study being included in the curriculum that is currently in place. Process of E-waste management is shown in Figure 9.5. Every department that generates hazardous waste is obligated to organise training sessions on the management of electronic waste under the guidance of the committee that was created at a higher level [30][31].

9.4 Conclusion

Both used electronic equipment and discarded electronic equipment are commonplace in today's world. They stand out from the crowd because of the complexity of the chemicals that make up their makeup and the challenge that comes with accurately quantifying the fluxes that they produce, both locally and worldwide. The improper management of these materials led to significant pollution, which caused severe damage to the environment, particularly in less developed nations that were given these materials for recycling and the recovery of the precious metals they contained.

The effects of electronic waste on ecosystems, human health, and environmental restoration in locations where certain contaminants (such as Li and Sb) burden the environment are not completely addressed in scientific research. This is a significant gap in our knowledge. Numerous technical advancements have been made in recent years in response to mounting concerns over the impact of human electronic waste creation on the natural environment. Manufacturers are competing to portray themselves in a more "green" light in response to pressure from non-governmental organisations and citizen movements to remove potentially harmful components from electronic products. The following is a list of consequences that are suggestive of the pressures listed earlier: The findings of the study indicate that in order to achieve environmental sustainability, effective management of electronic waste is required. This is because e-waste poses a risk not only to the environment but also to human health. It is also exciting to find that television and social media are the preferred methods of communication for e-waste awareness campaigns among urban respondents. On the other hand, respondents from less populous cities are more likely to favour radio and newspapers for awareness initiatives. The management of the production and disposal of electronic waste, as well as the development of technologies for the environmentally responsible collection and recycling of electronic trash, might be the topic of studies in the future.

The development of "halogen-free" appliances, which do not contribute to the generation of PCBs and polychlorinated (though their output is more environment-friendly costly), the substitution of bromide detonation retarders with more ecologically responsible ones based on nutrients, and the enactment of proposed legislation are some of the solutions that have been proposed. This approach (lead, mercury, cadmium, polybrominated biphenyls, and polybrominated diphenyl ethers up to 1000 mg/kg, Directive RoHS - Restriction on Hazardous Substances) is only one example of the steps that may be taken. It is absolutely necessary to separate electronic trash from other forms of solid garbage and recycle it if one wants to save valuable construction resources and key metals.

References

1. Mor, R.S., Singh, S., Bhardwaj, A., & Osama, M. (2017). Exploring the awareness level of biomedical waste management: case of Indian healthcare. Management Science Letters, 7(10), 467-478.
2. Jatindra, P., & Sudhir, K. (2009). E-waste management: a case study of Bangalore, India. Research Journal Environmental and Earth Sciences, 1, 111-115.

3. Sinha, S., Mahesh, P., & Donders, E. (2015). Waste electrical and electronic equipment: the EU and India: sharing best practices. Delhi: Toxic Link, 1-104.
4. Islam, M. T., & Huda, N. (2019). Material flow analysis (MFA) as a strategic tool in E-waste management: Applications, trends and future directions. Journal of Environmental Management, 244, 344- 361.
5. Leclerc, S. H., & Badami, M. G. (2020). Extended producer responsibility for E-waste management: Policy drivers and challenges. Journal of Cleaner Production, 251, 119657.
6. Méndez-Fajardo, S., Böni, H., Vanegas, P., & Sucozhañay, D. (2020). Improving sustainability of E-waste management through the systemic design of solutions: the cases of Colombia and Ecuador. In: Handbook of Electronic Waste Management, 443-478, https://doi.org/10.1016/B978-0-12-817030-4.00012-7.
7. Nnorom, I. C., & Odeyingbo, O. A. (2020). Electronic waste management practices in Nigeria. In: Handbook of Electronic Waste Management, 323-354, https://doi.org/10.1016/B978-0-12- 817030-4.00014-0.
8. Srivastava, R. R., & Pathak, P. (2020). Policy issues for efficient management of E-waste in developing countries. In: Handbook of Electronic Waste Management, 81-99, https://doi.org/10.1016/B978-0-12-817030-4.00002-4.
9. Xu, Y., Yeh, C. H., Yang, S., & Gupta, B. (2020). Risk-based performance evaluation of improvement strategies for sustainable ewaste management. Resources, Conservation and Recycling, 155, 104664.
10. Lara, P., Sánchez, M., Herrera, A., Valdivieso, K., & Villalobos, J. (2019). Modeling reverse logistics networks: a case study for ewaste management policy. In: Int. Conf. on Advanced Information Systems Engineering, 158-169, https://doi.org/10.1007/978-3-030- 21297-1_14.
11. Tran, C. D., & Salhofer, S. P. (2018). Analysis of recycling structures for e-waste in Vietnam. Journal of Material Cycles and Waste Management, 20(1), 110-126.
12. Rahman, M. A. (2017). E-waste management: A study on legal framework and institutional preparedness in Bangladesh. The Cost and Management, 45, 28-35.
13. Awasthi, A. K., & Li, J. (2017). Management of electrical and electronic waste: A comparative evaluation of China and India. Renewable and Sustainable Energy Reviews, 76, 434-447.
14. Islam, A., Ahmed, T., Awual, M. R., Rahman, A., Sultana, M., Abd Aziz, A., & Hasan, M. (2020). Advances in sustainable approaches to recover metals from e-waste- a review. Journal of Cleaner Production, 244, 118815.
15. Gollakota, A. R., Gautam, S., & Shu, C. M. (2020). Inconsistencies of e-waste management in developing nations- facts and plausible solutions. Journal of Environmental Management, 261, 110234.
16. Kumar, A., Dubey, A.K., Ramírez, I.S., Muñoz del Río, A., Márquez, F.P.G. (2022). A Review and Analysis of Forecasting of Photovoltaic Power

Generation Using Machine Learning. In: Xu, J., Altiparmak, F., Hassan, M.H.A., García Márquez, F.P., Hajiyev, A. (eds) Proceedings of the Sixteenth International Conference on Management Science and Engineering Management – Volume 1. ICMSEM 2022. Lecture Notes on Data Engineering and Communications Technologies, vol 144. Springer, Cham. https://doi.org/10.1007/978-3-031-10388-9_36

17. Dubey, A.K., Kumar, A., Ramirez, I.S., Marquez, F.P.G. (2022). A Review of Intelligent Systems for the Prediction of Wind Energy Using Machine Learning. In: Xu, J., Altiparmak, F., Hassan, M.H.A., García Márquez, F.P., Hajiyev, A. (eds) Proceedings of the Sixteenth International Conference on Management Science and Engineering Management – Volume 1. ICMSEM 2022. Lecture Notes on Data Engineering and Communications Technologies, vol 144. Springer, Cham. https://doi.org/10.1007/978-3-031-10388-9_35

18. Rathore, P.S., Chatterjee, J.M., Kumar, A. et al. Energy-efficient cluster head selection through relay approach for WSN. J Supercomput 77, 7649–7675 (2021). https://doi.org/10.1007/s11227-020-03593-4

19. A. Dubey, S. Narang, A. Srivastav, A. Kumar, V. Díaz, Woodhead Publishing, Science Direct, Artificial Intelligence for Renewable Energy Systems. Paperback ISBN: 9780323903967

20. A. Dubey, S. Narang, A. Srivastav, A. Kumar, V. Díaz, Woodhead Publishing, Science Direct, a Visualization Techniques for Climate Change with Machine Learning and Artificial Intelligence. ISBN: 9780323997140

21. Gautam, K., Sharma, A. K., Nandal, A., Dhaka, A., Seervi, G., & Singh, S. (2022). Internet of Things (IoT)-based smart farming system. Internet of Things and Fog Computing-Enabled Solutions for Real-Life Challenges, 39.

22. Sinha, K, D., Kraeuchi, P., & Schwaninger, M. (2005). A comparison of electronic waste recycling in Switzerland and in India. Environmental Impact Assessment Review, 25, 492-504.

23. Kwatra, S., Pandey, S., & Sharma, S. (2014). Understanding public knowledge and awareness on e-waste in an urban setting in India. Management of Environmental Quality: An International Journal, 25(6), 752-765.

24. Weia, L., & Liub, Y. (2012). Present status of e-waste disposal and recycling in China. Procedia Environmental Sciences, 16, 506-514.

25. Schmidt, M. (2005). A production-theory-based framework for analysing recycling systems in the e-waste sector. Environmental Impact Assessment Review, 25(5), 505-524.

26. Wolfram, S., Hans-Jorg, A., Mischa , C., Olivier, J., & Lorenz M, H. (2005). The end of life treatment of second generation mobile phone networks: Strategies to reduce the environmental impact. Environmental Impact Assessment Review, 25, 540-566.

27. Vats, M., and Singh, S. (2014). Status of e-waste in India- a review. International Journal of Innovative Research in Science, Engineering and Technology, 3(10), 16917-16931.

28. Ary, D., Jacobs, L. C., & Razavieh, A. (1996). Introduction to research in education. Fort Worth: Harcourt Brace College Publishers.
29. Needhidasan, S., Samuel, M., & Chidambaram, R. (2014). Electronic waste- an emerging threat to the environment of urban India. Journal of Environmental Health Science and Engineering, 12(1), 36.
30. Roychoudhuri R., Debnath B., De D., Albores P., Banerjee C., & Ghosh S.K. (2019). Estimation of E-waste Generation- A Lifecycle-Based Approach, https://doi.org/10.1007/978-981-10-7290-1_69.
31. Forti V., Baldé C.P., Kuehr R., Bel G. (2020). The Global E-waste Monitor 2020: Quantities, flows and the circular economy potential. United Nations University (UNU)/United Nations Institute for Training and Research (UNITAR) – co-hosted SCYCLE Programme, International Telecommunication Union (ITU) & International Solid Waste Association (ISWA), Bonn/Geneva/Rotterdam.

10
Challenges in E-Waste Management

Himani Bajaj[1], Anjali Sharma[2]*, Deepshi Arora[2], Mayank Yadav[1], Devkant Sharma[3] and Prabhjot Singh Bajwa[3]

[1]*AVIPS Shobhit University Gangoh, Saharanpur Uttar Pradesh, India*
[2]*Guru Gobind Singh College of Pharmacy, Yamunanagar Haryana, India*
[3]*CH. DeviLal College of Pharmacy, Jagadhari Haryana, India*

Abstract

Almost every element of contemporary life has been impacted by information and communications technology (ICT). In this regard, even the most isolated regions of developing countries are seeing a positive impact on human life. The capacity of electronic devices has increased as a result of the swift development of information and communications technology. Electronic devices, such as televisions, smartphones, and refrigerators, have limited useful lives and must therefore be replaced frequently, creating e-waste. E-waste has the fastest rate of growth among municipal solid garbage, producing 20 to 50 million tonnes annually globally. As a result, several nations are currently managing enormous amounts of e-waste. An important issue in handling e-waste is environmental health. Worldwide, nations continue to struggle to increase public awareness and take significant steps to protect the environment from rapid degradation. Because of the aforementioned reasons, effective e-waste management is required constantly. This chapter reveals the issues and difficulties that developing nations face in managing their e-waste.

Keywords: E-waste, electronic devices, environment, developing countries

10.1 Introduction

Electronic and electrical equipment (EEE) has become a necessary part of contemporary life as a result of technological development, scalability in

---

*Corresponding author: sharma.sharmaa.anjali@gmail.com

Abhishek Kumar, Pramod Singh Rathore, Ashutosh Kumar Dubey, Arun Lal Srivastav, T. Ananth Kumar and Vishal Dutt (eds.) Sustainable Management of Electronic Waste, (201–220) © 2024 Scrivener Publishing LLC

Table 10.1 Year-wise production of E-waste in India.

Year	E-waste generation in million tons (MT)
2015	1.97
2016	2.22
2017	2.53
2018	2.86
2019	3.23

electronics, communication and information technologies, and consumer affordability. End-of-life electronics and electric product waste is often referred to as WEEE (waste from EEE) or just e-garbage. E-trash management has thus become a big concern in the current day and has just begun to receive a lot of attention. India is now the third-largest producer of e-waste due to the expanding electronic sector, rising consumerism, and quick technological obsolescence. The amount of e-waste produced in India in the past few years is represented in Table 10.1.

10.2 E-Waste: Meaning and Definition

Electronics that have been discarded, are obsolete, or are on the verge of the end of their "useful life" are considered to be "e-waste." Electronic items that are used every day include computers, televisions, VCRs, stereos, copiers, and fax machines. The never-ending debate about the best way to get rid of old and unwanted gadgets is nothing new; it's been going on at least since the 1970s. But, much has changed since then, especially the quantity of devices being thrown out today. After numerous titles were suggested, including "Digital trash," the term "e-waste" came to be accepted. E-waste is any dumped electrical or electronic equipment. This covers both functional and damaged things that are discarded. If the product remains unsold in the store, it is frequently thrown away. The hazardous compounds that naturally leak from the metals in e-waste when it is buried make it extremely risky. The term "e-waste" typically refers to obsolete consumer and commercial electronics that have components that make them dangerous when disposed of in landfills [1, 2]. The various sources of e-waste are shown in Table 10.2.

Metal, polymers, cathode ray tubes, printed circuit boards, wires, and other materials are frequently found in e-waste. Metals including copper, silver, gold, and platinum may be recovered chemically from e-waste.

Table 10.2 Various E-waste sources, their constituents, and health impacts [4-10].

E-waste sources	Constituents	Health benefits
Solder in printed circuit boards, glass panels, and gaskets in computer monitors	Lead	Damage to the central and peripheral nervous system, blood systems, and kidney damage. Adverse effect on the brain development of children causes damage to the circulatory system and kidneys
Chip resistors and semiconductors	Cadmium	Toxic irreversible effects on human health. Accumulates in the kidney and liver. Causes neural damage
Relays, switches, and printed circuit boards	Mercury	Chronic damage to the brain. Respiratory and skin disorders due to accumulation in fishes
Galvanized steel plate decorator or hardener for steel housing	Chromium	Causes bronchitis
Cabling and computer housing	Plastic and PVC	Burning produces dioxin that causes reproductive and developmental problems
Electronic equipment and circuit boards	Brominated flame retardants	Disrupts endocrine functions
Front panels of CRTs	Barium, phosphorus, and heavy metals	Causes muscle weakness and damage to the heart, liver, and spleen
Copper wire, printed circuit board tracks	Copper	Stomach cramps, nausea, liver damage or Wilson's disease
Nickel-cadmium rechargeable batteries	Nickel	Allergy to the skin results in dermatitis while allergy to the lung results in asthma

(Continued)

Table 10.2 Various E-waste sources, their constituents, and health impacts [4-10]. (*Continued*)

E-waste sources	Constituents	Health benefits
Lithium-ion battery	Lithium	Lithium can pass into breast milk and may harm a nursing baby Inhalation of the substance causes lung oedema
Motherboard	Beryllium	Carcinogenic (lung cancer) Inhalation of dust and fumes causes chronic beryllium

If e-waste is disassembled and processed using elementary processes, it is enormously harmful due to the presence of poisonous compounds such liquid crystal, lithium, mercury, nickel, selenium, arsenic, barium, brominated flame retardants, cadmium, chrome, cobalt, and lead. E-waste poses a serious threat to people, animals, and the environment [3].

10.3 Environmental Sustainability in E-Waste Management

It is crucial to realise that e-waste is a special kind of waste stream that needs to be managed, treated, and disposed of properly. Understanding the notion and the terminology used to describe it accurately can help you decide how to manage it best.

EEE was defined by Luciano Morselli in 2009 as "everything whose proper operation depends on electric currents or electromagnetic fields for functional reasons. When its owner discard it, it becomes WEEE " [11].

Electrical and electronic equipment (EEE) was described by the Step Initiative in 2014 Step Initiative (2014), *One Global Definition of E-waste* [12] as "anything that has electronic components and a power source or battery." A. Step further explained that "all sorts of electrical equipment and parts 'discarded' by its owner as waste without the purpose of re-use" are included in the definition of "e-waste." To distinguish between waste and an item or something that can be reused, Step stresses the word "discarded."

EEE products are defined as "those that work on electric currents and are intended for use with a power not above 1000 V alternating current (AC) and 1500 V direct current (DC)" by the European Directive 2012/19/EU [13].

Following the end of its useful life, the features of each of these groups will indicate the optimum method of treatment and disposal for the corresponding equipment (EoL). This equipment should be managed, collected, and recycled logistically in accordance with its unique properties. Electronic garbage can be generated by domestic, commercial, industrial, institutional, or other uses. Its production is influenced by a number of variables such as the equipment's useful life (computers, televisions, etc.), the need for users to upgrade their equipment (mobile phones, for example), and significant technological advancements (such as the switch from the global system for mobile communications (GSM) to the universal mobile telecommunications system (UMTS) mobile telephony). For instance, the Massachusetts Institute of Technology (MIT) and Step Initiative predict that a mobile phone has a three to five-year lifespan [14].

This application needs to be considered because it has a significant impact on the world's rapidly increasing e-waste.

10.4 Sustainable Management of E-Waste

Depending on the country's capacity for recycling, the majority of e-waste is recycled via formal or informal programs. Businesses can be established to address the need for refurbishing equipment and recovering raw materials if e-waste is properly managed. E-waste requires specialized techniques for collection, transportation, segregation, treatment, and disposal since it is a complex mixture of hazardous and non-hazardous components. Understanding the EEE life cycle is vital to understand any probable environmental effects. Figure 10.1 describes the life cycle of EEE [15].

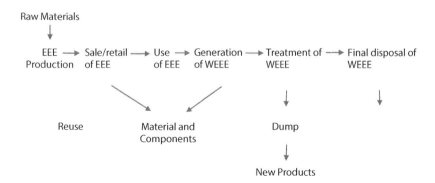

Figure 10.1 Life cycle of EEE.

10.5 Life Cycle of E-Waste

The Toolkit on Environmental Sustainability for the ICT Sector, issued by ITU in 2013, discusses the significance of managing EEE at EoL. E-waste is a notable part of ICT equipment. The research suggests that it shouldn't be presumed that an appliance is broken or outdated if it doesn't suit the ICT demands of its initial user; it can still be beneficial to someone else. According to the EoL concept, it might be feasible to increase its lifespan by having additional users or by using it for different things. The idea also emphasises the possibility of recovering or recycling the materials and component parts at the conclusion of a product's useful life [16].

10.6 Terminology of E-Waste

a) **Reuse**: Expanding the end of life of equipment or component parts so that they can continue to be used for the original intent; this may or may not involve changing the equipment's ownership. This approach tries to encourage the best use of the resources that are already available, but social or environmental hazards linked to inadequate management should be kept in mind.

b) **Dismantling and Segregation**: This pertains to manually separating each part and component of a piece of equipment that is no longer in use.

c) **Recycling and Recovery**: Devices, parts, and material recovery are all part of this process. Both semi-manual and manual dismantling are possible. Part of the WEEE recycling process involves material recovery, particularly metal recovery, which necessitates specialised facilities and financial expenditure [17].

d) **Refurbishment**: Any procedure that enables the reuse of WEEE that was previously WEEE falls under this category. Changes to both hardware and software are included.

e) **Final Disposal**: Non-recoverable materials can be disposed of by incineration during the process of final disposal of garbage or materials [18].

By utilizing cutting-edge and effective methods and technologies, it is possible to minimize the formation of trash throughout an object's life cycle by adhering to the principles of reduce, reuse, and recycle (3Rs).

The effective implementation of a collection and renovation process should be part of the reuse alternative. There are restrictions to reusing products and it can only momentarily extend the equipment's life. On the extension of the lifespan of ICT equipment recycling and material, recovery depends. Utilizing technologies and management practices that are environmentally sustainable is important for these recycling operations. For instance, certain ICT equipment requires particular approaches and specialized recovery techniques, which are frequently only possible with the use of specialized equipment by recycling businesses and expert operators [19].

10.7 Key Stakeholders in the E-Waste Management System

Because so many different parties are involved along the entire value chain, the e-waste management system contrasts greatly with other waste streams. Due to their overlapping duties, the complexity of the stakeholder base is also a significant challenge. The three major stages of the e-waste value chain are e-waste generation; collection, treatment, and recovery; and final disposal. A list of major stakeholders of the value chain with their responsibilities are as shown in Table 10.3.

10.8 Status of E-Waste Management in India

Despite the country's stringent environmental laws, there was no specific legislation or guidance regarding electronic or computer waste in India in 2004. The Hazardous Waste Rules (1989) do not classify e-waste as hazardous waste unless a higher concentration is established for a few different compounds [20]. There are a few grey areas that need to be handled, even though PCBs and CRTs would always exceed these limits [21].

Electronic trash is included by List-A and List-B of Schedule-3 of the Hazardous Wastes (Management & Handling) Rules [22]. Therefore, the import of this trash requires specific approval from the Ministry of Environment and Forests. Because the nation's informal sector currently collects and recycles electronic waste, the government has taken the following actions to improve and raise awareness about environmentally sound treatment of electronic trash. Several workshops on handling electronic trash were organised by the Central Pollution Control Board (CPCB) [23]. The CPCB has taken steps to quickly analyse the amount of e-waste

Table 10.3 Stakeholders & their responsibilities.

Stakeholders	Responsibilities
Government agencies (central and state)	The regulatory framework for e-waste management is established by the government. Governmental organisations are also in charge of implementing these rules.
Producers, manufacturers	They coordinate, fund, and manage the e-waste takeback system individually or collectively through PROs in accordance with EPR-based regulations.
Producer responsibility organisations (PROs)	On behalf of the manufacturers, they manage the e-waste collection system and make sure the garbage is taken to the right facilities for treatment.
Retailers	In e-waste collecting retailers rarely participate. There may be a take-back programme in place for some EOL products.
Consumers	The fate and course of e-waste disposal are mostly determined by individual and bulk users. Disposable behaviour is heavily influenced by awareness levels and accessibility to systems and infrastructure.
Waste collectors and aggregators (Kabariwala, scrap dealers)	The market for e-waste in India is dominated by informal collectors. They are in charge of rubbish collection (door to door, municipal dumpsite).
Recyclers	There are both formal and informal recyclers here. They are in charge of a fraction recovery and recycling. Industrial recyclers that operate large-scale material recovery operations or use mechanical shredding and sorting are known as formal recyclers. Informal recyclers perform material pre-processing and basic recycling. They employ ineffective methods that have a negative impact on the environment and human health.
Refurbishers	Refurbishers work in the used goods industry. They prepare the recyclable materials for secondary use, contributing significantly to effective waste management.
Civil societies and development organizations	They are crucial in raising awareness of the e-waste problem. To address the broader e-waste problem, they aid in knowledge transfer and stakeholder involvement, among other things.

produced in the major cities. The National Working Group was established to develop and e-waste management strategy.

- ✓ Environmental Management for the Information Technology Industry in India is a comprehensive technical reference that has been created and widely disseminated by the Department of Information Technology (DIT) and Ministry of Communication and Information Technology [24]. To recover copper from printed circuit boards, the DIT has also built demonstration projects at Indian Telephone Industries. The absence of trustworthy data makes it difficult for policymakers to develop an e-waste management strategy and for the business community to make wise investment choices.
- ✓ Due to the lack of a successful consumer take-back programme, only a small portion of electronic waste makes it to recyclers.
- ✓ The environment and human health are seriously endangered by the formal sector's reliance on the informal sector's capabilities due to the absence of a secure e-waste recycling infrastructure.
- ✓ The current e-waste recycling systems are entirely business-driven and developed without any help from the government.

10.9 Challenges in E-Waste Management

The ecosystem for managing e-waste is currently thriving in India, but there are many obstacles to overcome, including those related to policy and regulation, environmental and health protections, operational issues, the dominance of the informal sector, financial limitations, supply chain logistics, and others. Numerous gaps and discrepancies between the operational elements of the regulatory framework and the ground realities require quick attention in order to be properly filled [25-26].

Some significant challenges that require prompt response are listed below.

10.9.1 Discrepancies in Estimate of E-Waste

Compared to developed nations, India produces less electronic trash per capita. Because of its vast population and illegal trash dumps, our nation may soon face a problem as economic expansion will raise purchasing power. There are many different sources of electronic waste, including individual families, large customers, commercial buildings, manufacturers, and retailers. All electrical and electronic equipment (EEE), in whole or in part, disposed of as trash by users (individually or in bulk), as well as rejects from manufacturing, refurbishment, and repair processes, are classified as e-waste under the 2016 legislation for its management. Under Schedule I of the 2016 E-waste Management Standards, WEEE has been divided into two main categories: waste from information, technology, and communication (ITEW) and waste from consumer electronics and electrical equipment (WEEE) (CEEW). Despite the fact that there are many different kinds of EEE trash, only 21 of them are now covered by the rule. The worldwide e-waste monitor considers 54 different forms of electronic waste. In addition to classification problems, e-waste data management and data gathering concerns also exist. The fact that the accountabilities of producers under the ERP scheme are not fully enforced is largely attributable to this disparity. In addition, Indian estimates are based on a far smaller number of producers than the real figures, those who hold extended producer responsibility authorization (EPRA) registrations with the CPCB. Additionally, import data is also not represented. As a result, the actual figures for e-waste generation are probably far higher than the estimates made at this time [27-28].

10.9.2 Lack of Awareness

The majority of those involved in processing e-waste are unaware of any potential concerns, acceptable disposal techniques, or the applicable regulatory framework. The majority of consumers are either completely unaware of the dangerous nature of e-waste components or are only vaguely aware of the consequences of improper disposal. Although the EPR requires producers to spread awareness about e-waste, most of them are hesitant to do so. In the current policy, the function of urban local bodies is also not very clear, which causes ULBs in urban communities to shirk their obligations. There are very few ULB-established collection depots or drop-off locations in cities [28].

Bulk consumers typically pursue the auction route without doing enough checks on the final disposal of the waste being disposed of. Institutional consumers frequently do not keep track of the obsolete equipment that is disposed of. The informal industry uses extremely ineffective techniques to recover valuables from e-waste while ignoring environmental and health safety concerns. To strengthen the entire ecosystem of e-trash management, it is necessary to examine the general lack of awareness of health and environmental risks, the responsibilities of the producers, the provisions of the e-waste legislation, and the role of each player in the ecosystem.

10.9.3 The Dominance of the Informal Sector

It is widely recognised that the informal sector now handles the majority of the collection and handling of e-waste as well as other waste streams. The issue is made more complicated by factors including inadequate knowledge of the hazardous nature of e-waste, a lack of appropriate formal collecting methods, customary waste stream separation practices, etc. As a result, the informal sector manages more than 95 percent of this garbage in the unorganised markets. In the informal recycling industry, recycling facilities are often found outside of industrial zones and are set up like a scrap yard, where employees are exposed to danger while handling harmful e-waste components [29].

Their lack of investment in safety precautions or infrastructure enables informal employers to operate at large profit margins. The main driver of a booming informal industry is low-cost operations.

Although the informal sector has exceptional reach and access to garbage, it is not recognised in the existing legislative system. They can afford to pay more affordable rates for e-waste collection because they have little money invested in the proper disposal of e-waste. Making the legal e-waste recycling sector viable and profitable is challenging because there aren't many EEE waste goods available and a lot of waste gets dumped into the black markets.

10.9.4 Inadequate Formal Recycling Sector

India's recycling facilities are substantially underutilised, according to a number of studies and publications, because rubbish is not effectively sent to the official sector. Leakage is a problem with the current take-back and collection systems. It can be expensive to set up a recycling factory at first and it is challenging to sustain a business by investing extensively in cutting-edge equipment. Because of this, with a few exceptions, qualified recycling businesses in India can only pre-process e-waste; crushed e-waste containing valuable metals is sent to smelting facilities abroad. Many of the officially recognized recyclers are actually dismantlers [30].

The absence of affordable recycling technology is one of the main issues facing formal businesses. Finding enough e-waste and a lack of technological knowledge are further problems. The development of technologies for

WEEE management, environmental monitoring, and disposal techniques should get government funding [31].

10.9.5 Lack of Technological Advancement

Due to their relatively low technological levels, many nations lack the infrastructure and facilities needed to treat WEEE properly. As a result, large amounts of toxic waste are typically managed using outdated techniques including manual demolition, open burning, and acid leaching [32].

These archaic technologies cause a variety of hazardous, organic, and inorganic contaminants to be emitted into the environment. Polycyclic aromatic hydrocarbons, halogenated substances like flame retardants, polychlorinated biphenyls (PCBs), and other combustion by-products are also discharged into the environment during the thermal decomposition of the non-metal components of e-trash. Additionally, heavy metals, including lead, cadmium, mercury, copper, and chromium, leak into the environment and seriously pollute the air, water, and land at the same time. According to numerous studies, open site burning is the principal method of producing and dispersing pollutants [33-34].

10.9.6 Involvement of Child Labour

In India, it is estimated that 4.5 lakh children between the ages of 10 and 14 are working in various e-waste-related activities without the proper protections and safeguards in different yards and recycling facilities. Therefore, it is vital to enact strong regulations to stop child labour from entering the market for the collection, sorting, and distribution of e-waste [35].

10.9.7 Ineffective Legislation

On the majority of SPCB/PCC websites, there is no publicly accessible information. On their websites, which serve as their primary point of contact with the public, 15 of the 35 PCBs/PCC have no information about e-waste. Even the simplest e-waste rules and guidelines are not yet available online. Citizens and institutional generators of E-waste are completely at a loss as to how to deal with their junk and do not know how to fulfill their responsibilities because there is no information on their website, particularly on specifics of recyclers and collectors of e-waste. So, the 2012 E-waste Management and Handling Rules have not been successfully implemented [36].

10.9.8 Health Hazards

Over 1,000 hazardous substances found in e-waste damage land and groundwater. Headache, irritability, nausea, vomiting, and eye pain can all be symptoms of exposure. Liver, kidney, and brain diseases can affect recyclers. They are endangering both their health and the environment as a result of their ignorance [37].

10.9.9 Lack of Incentive Schemes

The unorganised sector lacks clear procedures for handling e-waste. Additionally, no incentives are specified to persuade those involved in treating e-waste to follow a formal path. Only somewhat worse than in the formal sector are the working conditions in the informal recycling industry. There are no programmes to reward producers for handling e-waste [38].

10.9.10 Reluctance of Authorities Involved

There is a lack of coordination between the numerous agencies in charge of managing and discarding e-waste, including municipalities' exclusion.

10.9.11 Security Implications

Computers that are nearing the end of their life frequently contain sensitive personal data and bank account information that, if not erased, could be used for fraud.

10.9.12 Lack of Research

The government must support research into environmental monitoring, hazardous waste legislation, and the creation of standards for managing hazardous waste [39-40].

10.10 E-Waste Policy and Regulation

From manufacturing and trade to final disposal, the policy will address all issues, including technology transfers for recycling electronic trash. In order to ensure the handling of e-waste in an environmentally sound manner and to control both legal and illicit exports and imports, clear regulatory tools must be in place. To prevent the arrival of e-waste from industrialised nations for disposal in the country, it is also necessary to close the gaps in the current regulatory framework. The Port and Customs authorities must keep an eye on these factors. The rules should forbid the disposal of e-waste in public landfills and promote proper recycling among owners and producers of e-waste. It is necessary that manufacturers must be responsible financially, physically, and legally for their products [41-42].

10.11 E-Waste Recycling

Many abandoned machines have salvageable components that might be merged with other used machinery to build a functional unit. Removing, testing, and inspecting individual components before putting them back together to form fully functional machines takes a lot of work. For the environmentally sound management of e-wastes, institutional infrastructures including e-waste collection, transportation, treatment, storage, recovery, and disposal must be built at national and/or regional levels. The regulatory authorities should authorise these facilities and, if necessary, offer suitable incentives. Establishing e-waste collection, exchange, and recycling facilities in collaboration with governments, NGOs, and manufacturers should be encouraged. Environmentally responsible e-waste recycling necessitates the use of sophisticated equipment and processes that are not only very expensive but also complicated to execute. The correct recycling of complex materials requires the ability to recognise or determine the existence of hazardous or potentially hazardous constituents as well as desirable constituents (i.e., those with recoverable value). Then, both of these streams must be correctly recycled using the company's skills and process systems. Appropriate air pollution control methods are required for transient and point source emissions. Guidelines for recycling that are environmentally friendly will be created.

10.12 Life Cycle Assessment (LCA) Analysis of E-Waste

To analyse the environmental effects of treating electronic trash (e-waste), a life cycle evaluation was conducted. E-waste reuse with the last disposal scene of life is environmentally friendly due to the minimal environmental load caused by human toxicity, earth ecotoxicity, marine ecotoxicity, and freshwater ecotoxicity categories. When it comes to end-of-life drainage, recycling e-waste has a smaller and more significant environmental impact than using landfills or engineers, respectively. Reducing the amount of solid waste effluent and wastewater, increasing the timing of e-waste treatment, maximising energy efficiency, avoiding the disposal of e-waste in landfills and incinerators, and clearly allocating waste management responsibilities to all stakeholders are crucial factors in reducing the negative environmental effects of e-waste recycling (recycling industry, retailers, manufacturer, and consumers).

10.13 Existing Laws Relating to E-Waste

- ✓ Trans-boundary movement of e-waste is covered under the Basel Convention
- ✓ India ratified the convention in 1992
- ✓ Waste importers exploit gaps as listed in the convention
- ✓ Allowed to import against a license
- ✓ Covered under the hazardous waste amended rules 2003 in List A and B of Schedule 3
- ✓ The rule is inadequate to handle generation, transportation, and disposal of this complex waste
- ✓ Regulations unable to monitor and regulate the informal sector.

10.14 Management Options

10.14.1 Responsibilities of the Government

In order to coordinate and consolidate the regulatory responsibilities of the numerous government bodies addressing hazardous substances and to provide for harsh penalties, governments should establish regulatory agencies in each district. Promoting research into environmental monitoring, hazardous e-waste legislation, and the improvement and standardisation of hazardous waste management is important. This is more encouraging given that the Swachh Bharat Abhiyan was launched in India in October 2014 and the revenue generated from recycling e-waste and recovering metals [43][44][45].

10.14.2 Responsibility and Role of Industries

Waste producers must be accountable for identifying the output characteristics of their wastes and, if hazardous, provide management choices.

By informing and paying the consumers financially, producers, distributors, and retailers should take on the duty of recycling and disposing of their own products [46][47].

10.14.3 Responsibilities of the Citizen

Perhaps more favoured than any other waste management strategy, including recycling, is waste prevention. Donated electronics have a longer shelf life and are kept out of the waste management system for a longer period of time. Care should be taken to make sure the donated items are in useable shape, though. Reuse is not only advantageous to society, but it is also the most environmentally friendly option. By donating used equipment, non-profit organisations, schools, and low-income families can use technology that might otherwise be out of reach. E-waste should never be thrown out with regular trash or home rubbish. This ought to be separated on the premises, sold, or given to other charities. NGOs should use a participatory management strategy for e-waste.

10.15 Conclusion

The management of e-waste is a significant challenge to the governments of many developing nations, including India. This is growing tremendously and is turning into a major public health problem. Integration of the informal and formal sectors is crucial for the separate collection, efficient treatment, and disposal of e-waste, as well as its diversion from traditional landfills and open burning. Procedures for the safe and responsible handling and treatment of e-waste must be established by the accountable authorities in developing and transitioning countries.

Expanding informational efforts, improving capacity, and raising awareness are required to promote ecologically friendly e-waste management programs. Further efforts are urgently needed to enhance current processes, including collection plans and management approaches, in order to reduce the illegal traffic in e-waste. The specific e-waste streams can be handled more easily by minimizing the number of hazardous substances in e-products because it will help with prevention.

One of the very few businesses that appear to have made real efforts in this approach since 2008 is the mobile phone manufacturer Nokia. In accordance with an EPR Authorization plan in India that was approved by the Central Pollution Control Board (CPCB), companies were given the responsibility of developing channels for the ethical collection and disposal of e-waste. Some major corporations recently had their import permits suspended for breaking the e-waste standards. Such actions significantly influence how efficiently e-waste management is implemented in India. Any project that is done needs to have certain incentives to draw in stakeholders. To ensure compliance throughout the electronics industry, the government must announce incentives in the area of e-waste management. These might take the shape of tax breaks or refunds. To maintain compliance with the collection of e-waste throughout India, the e-waste collection targets also need to be routinely evaluated and renewed.

References

1. Shekdar AV. Sustainable solid waste management: An integrated approach for Asian countries. Waste management. 2009 Apr 1;29(4):1438-48.
2. Williams E. International activities on E-waste and guidelines for future work. In Proceedings of the Third Workshop on Material Cycles and Waste Management in Asia, National Institute of Environmental Sciences: Tsukuba, Japan 2005.
3. Iles A. Mapping environmental justice in technology flows: Computer waste impacts in Asia. Global Environmental Politics. 2004 Nov 1;4(4):76-107.
4. Iqbal M, Breivik K, Syed JH, Malik RN, Li J, Zhang G, Jones KC. Emerging issue of e-waste in Pakistan: a review of status, research needs and data gaps. Environmental pollution. 2015 Dec 1;207:308-18.
5. Jugal K. E-waste management: as a challenge to public health in India. Indian Journal of Community Medicine. 2010;35(3):382-5.
6. Herat S. Environmental impacts and use of brominated flame retardants in electrical and electronic equipment. The Environmentalist. 2008 Dec;28:348-57.
7. Pinto VN. E-waste hazard: The impending challenge. Indian journal of occupational and environmental medicine. 2008 Aug;12(2):65.
8. Keirsten S, Michael P. A Report on Poison PCs and Toxic TVs. Silicon Valley Toxic Coalition. 1999.
9. India TL. Scrapping the Hi-tech Myth: Computer Waste in India. New Delhi, India. 2003.
10. Streicher-Porte M, Widmer R, Jain A, Bader HP, Scheidegger R, Kytzia S. Key drivers of the e-waste recycling system: Assessing and modelling e-waste processing in the informal sector in Delhi. Environmental impact assessment review. 2005 Jul 1;25(5):472-91.
11. Morselli L, Passarini F, Vassura I. Waste recovery: strategies, techniques and applications in Europe. Waste recovery. 2009:0
12. http://www.stepinitiative.org/files/step/_documents/StEP_WP_One%20Global%20Definition%20of%20Ewaste_20140603_amended.pdf

13. Medina AM. Directiva 2012/19/UE del Parlamento Europeo y del Consejo de 4 de julio de 2012 sobre residuos de aparatos eléctricos y electrónicos (RAEE). Refundición.(DOUE L 197/38, de 24 de Julio de 2012). Actualidad Jurídica Ambiental. 2012(16):19-20.
14. http://www.gsma.com/latinamerica/wp-content/uploads/2014/05/eWaste-Latam-EspResEje.pdf
15. Babyrani Devi S, Shobha SV, Kamble RK. E-Waste: The Hidden Harm of Technology Revoluation. Journal of Indian Association for Environmental Management. 2004;31:196-205.
16. Chatterjee S, Kumar K. Effective electronic waste management and recycling process involving formal and non-formal sectors. International Journal of Physical Sciences. 2009 Dec;4(13):893-905.
17. Sum EY. The recovery of metals from electronic scrap. Jom. 1991 Apr;43:53-61.
18. Joseph K. Electronic waste management in India–issues and strategies. InEleventh international waste management and landfill symposium, Sardinia 2007 Oct 1.
19. Forti V, Balde CP, Kuehr R, Bel G. The Global E-waste Monitor 2020: Quantities, flows and the circular economy potential.
20. Nivedha R, Sutha DA. The challenges of electronic waste (e-waste) management in India. European Journal of Molecular & Clinical Medicine. 2020 Dec 19;7(3):4583-8.
21. Gupta V, Kumar A. E-waste status and management in India. Journal of Information Engineering and Applications. 2014;4(9):41-8.
22. Borthakur A, Singh P. Electronic waste in India: Problems and policies. International Journal of Environmental Sciences. 2012;3(1):353-62.
23. Kishore J. E-waste management: as a challenge to public health in India. Indian journal of community medicine: official publication of Indian Association of Preventive & Social Medicine. 2010 Jul;35(3):382.
24. Kishore J. E-waste management: as a challenge to public health in India. Indian journal of community medicine: official publication of Indian Association of Preventive & Social Medicine. 2010 Jul;35(3):382.
25. http://www.indianenviornmentportal.org.in/files/file/e-waste-management-NGT-CPCB-report-pdf
26. Shalini S, Joseph K, Yan B, Karthikeyan OP, Palanivelu K, Ramachandran A. Solid Waste Management Practices in India and China: Sustainability Issues and Opportunities. InWaste Management Policies and Practices in BRICS Nations 2021 Aug 2 (pp. 73-114). CRC Press.
27. Herat S, Agamuthu P. E-waste: a problem or an opportunity? Review of issues, challenges and solutions in Asian countries. Waste Management & Research. 2012 Nov;30(11):1113-29.
28. Sinha S. The informal sector in e-waste management. VIKALPA. 2019;44(3):133-5.

29. Priti Banthia, Mahesh and Manjusha Mukherjee, 2019. Informal e waste recycling in Delhi. Retrieved from http://www.toxiclink.org/docs/informal20E-waste.pdf.
30. Khan SS, Lodhi SA, Akhtar F, Khokar I. Challenges of waste of electric and electronic equipment (WEEE): Toward a better management in a global scenario. Management of Environmental Quality: An International Journal. 2014 Mar 4;25(2):166-85.
31. Ongondo FO, Williams ID, Cherrett TJ. How are WEEE doing? A global review of the management of electrical and electronic wastes. Waste management. 2011 Apr 1;31(4):714-30.
32. Duarte AT, Dessuy MB, Silva MM, Vale MGR, Welz B, Microchem J. 2010,96,102-107.
33. Wu Q, Du Y, Huang Z, Gu J, Leung JY, Mai B, Xiao T, Liu W, Fu J. Vertical profile of soil/sediment pollution and microbial community change by e-waste recycling operation. Science of the Total Environment. 2019 Jun 15;669:1001-10.
34. Leung AO, Luksemburg WJ, Wong AS, Wong MH. Spatial distribution of polybrominated diphenyl ethers and polychlorinated dibenzo-p-dioxins and dibenzofurans in soil and combusted residue at Guiyu, an electronic waste recycling site in southeast China. Environmental science & technology. 2007 Apr 15;41(8):2730-7.
35. Kumar, A., Dubey, A.K., Ramírez, I.S., Muñoz del Río, A., Márquez, F.P.G. (2022). A Review and Analysis of Forecasting of Photovoltaic Power Generation Using Machine Learning. In: Xu, J., Altiparmak, F., Hassan, M.H.A., García Márquez, F.P., Hajiyev, A. (eds) Proceedings of the Sixteenth International Conference on Management Science and Engineering Management – Volume 1. ICMSEM 2022. Lecture Notes on Data Engineering and Communications Technologies, vol 144. Springer, Cham. https://doi.org/10.1007/978-3-031-10388-9_36
36. Dubey, A.K., Kumar, A., Ramirez, I.S., Marquez, F.P.G. (2022). A Review of Intelligent Systems for the Prediction of Wind Energy Using Machine Learning. In: Xu, J., Altiparmak, F., Hassan, M.H.A., García Márquez, F.P., Hajiyev, A. (eds) Proceedings of the Sixteenth International Conference on Management Science and Engineering Management – Volume 1. ICMSEM 2022. Lecture Notes on Data Engineering and Communications Technologies, vol 144. Springer, Cham. https://doi.org/10.1007/978-3-031-10388-9_35
37. Rathore, P.S., Chatterjee, J.M., Kumar, A. *et al.* Energy-efficient cluster head selection through relay approach for WSN. J Supercomput 77, 7649–7675 (2021). https://doi.org/10.1007/s11227-020-03593-4
38. A. Dubey, S. Narang, A. Srivastav, A. Kumar, V. Díaz, Woodhead Publishing, Science Direct, Artificial Intelligence for Renewable Energy Systems. Paperback ISBN: 9780323903967

39. A. Dubey, S. Narang, A. Srivastav, A. Kumar, V. Díaz, Woodhead Publishing, Science Direct, a Visualization Techniques for Climate Change with Machine Learning and Artificial Intelligence. ISBN: 9780323997140
40. Singh R, Chari KR. Socio-economic issues in waste management by informal sector in India. InProceedings of the International Expert Conference on the Project TransWaste, Going Green–Care Innovation 2010, Vienna, Austria 2010.
41. CPCB retrieved from: http://www.indianenviornmentportal.org.in/files/file/e-waste-management-NGT-CPCB-report-pdf
42. Babyrani Devi S, Shobha SV, Kamble RK. E-Waste: The Hidden Harm of Technology Revoluation. Journal of Indian Association for Environmental Management. 2004;31:196-205.
43. Manish A, Chakraborty P. E-waste management in India: challenges and opportunities. TerraGreen. 2019;12:22-8.
44. Biswas A, Singh SG, Singh SG. E-waste Management in India: Challenges and Agenda. Centre for Science and Environment. 2020.
45. Chatterjee S. India's readiness on ROHS directives: a strategic analysis. Global Journal of Science Frontier Research. 2012;10(1):14-26.
46. Widmer R, Oswald-Krapf H, Sinha-Khetriwal D, Schnellmann M, Böni H. Global perspectives on e-waste. Environmental impact assessment review. 2005 Jul 1;25(5):436-58.
47. Sadala S, Dutta S, Raghava R, Jyothsna TS, Chakradhar B, Ghosh SK. Resource recovery as alternative fuel and raw material from hazardous waste. Waste Management & Research. 2019 Nov;37(11):1063-76.

11

Recycling of Electronic Wastes: Practices, Recycling Methodologies, and Statistics

Suresh Chinnathampy M.[1,2*], Ancy Marzla A.[1,3], Aruna T.[1,4], Dhivya Priya E. L.[1,5], Rindhiya S.[1,2] and Varshini P.[1,2]

[1]Department Electronics and Communication Engineering, Francis Xavier Engineering College, Tirunelveli, India
[2]Francis Xavier Engineering College, Tirunelveli, India
[3]Ponjesly College of Engineering, Nagercoil, India
[4]Paavai Engineering College, Namakkal, India
[5]Erode Sengunthar Engineering College, Erode, India

Abstract

In general, up to 60 different elements can be found in modern electronics, many of which are beneficial, some of which are dangerous, and some of which are both. In a more colloquial sense, the term electronic waste, or "e-waste", refers to obsolete electronic devices such as computers, televisions, videocassette recorders (VCRs), stereos, photocopiers, and fax machines. These devices are all considered to be nearing the end of their "useful life." Increasing transboundary secondary resource movement and Asia's rapid economic growth will require both 3R initiatives (reduce, reuse, recycle) in every nation and effective management of the global material cycle. Management of electrical and electronic garbage, or "E-waste," has gained significant attention. For material cycles from the national and international environmental and resource preservation perspectives, the use of digital goods is at an all-time high in the developing digital world and it is difficult to envision our everyday lives without these. Although they are most significant throughout their lifetime, they could endanger the environment if they are burned or disposed of in landfills. It is a fast-growing waste globally and is hazardous to the environment. Globally, e-waste production totals more than 50 million metric tons per year, or about seven pounds per person. When this waste

---

Corresponding author: sureshchinnathampy@gmail.com

Abhishek Kumar, Pramod Singh Rathore, Ashutosh Kumar Dubey, Arun Lal Srivastav, T. Ananth Kumar and Vishal Dutt (eds.) Sustainable Management of Electronic Waste, (221–236) © 2024 Scrivener Publishing LLC

is burned, poisons can also be released into the atmosphere. Most electronic gadgets include a variety of harmful compounds that can seep into the ground or soil.

Keywords: E-waste, digital goods, recycle, environment, electronic garbage

11.1 Introduction

E-waste is one form of waste that is growing quickest in the world. We already produce 50 million tons of it yearly and that number will only increase as more people gain access to technology. Electronic waste (or "e-waste") is the term for used, outdated, or damaged electrical and electronic equipment. This covers everything from old refrigerators to iPhones. In other words, anything that you have opted to get rid of that uses energy. It is shocking and sad that just 10% of the world's electronic garbage is recycled. 90% of the rubbish we produce is not recycled and this waste is eventually buried, burned, or illegally sold. E-waste is full of toxins that are detrimental for both people and the environment, as well as mercury, lead, beryllium, brominated flame retardants, and cadmium—stuff that sounds as nasty as it is [6]. These chemicals end up in our land, water, and air when devices are improperly disposed of. To make problems worse, electronic garbage is occasionally illegally transported to nations with no regulations on how to handle and dispose of it. It is then dumped there. Sometimes valuable materials are retrieved, although frequently under risky working circumstances. While disposing of these goods in landfills is no longer permitted in many nations, it is nevertheless done in certain others. Burning is a popular procedure for electronic waste that is good for nothing, but it can also be used to remove copper and other valuable alloys from gadgets. Because flaming waste releases tiny bits that can travel thousands of miles, imperiling both human and animal health, it increases the threat of developing chronic diseases and cancer. When electronic waste is burned, toxins like lead, cadmium, and mercury are released into the air and water [14]. These chemicals have the potential to bioaccumulate within the food chain, which has an adverse effect not just on wildlife but also on human health. PVC polymers and brominated flame retardants are also common components of electronics and, when they burn, they generate hazardous dioxins and furans. Even though many people in the developed world now recycle their outdated electronics, up to 50% to 80% of them are still exported to developing countries (often illegally) where they are "recycled," frequently without the same safety and environmental safeguards as in the developed world, including those pertaining to child labor. The disposal of e-waste is an increasing issue as a result. Environmental contamination is caused by the mounting piles of used gadgets and parts [1]. More significantly, when unprotected, components like lead, beryllium, cadmium, and others constitute a health risk [3]. These carcinogenic substances may emit radiation, which could put workers at the disposal site, residents in the surrounding areas, and even locals at risk.

11.2 Recycling E-Waste

The term "e-waste" refers to waste that is generated by electronic devices. This refers to the waste that is generated by broken, surplus, and obsolete electronic equipment, as shown in Figure 11.1. The term "e-scraps" might also be used in some contexts.

These electronic devices frequently contain potentially hazardous substances and chemical compounds. In addition, the haphazard disposal of these electronic devices may result in the release of toxic substances into the surrounding environment. E-waste, or waste produced by broken, surplus, and old electronic equipment, is referred to as electronic rubbish. The phrase "e-scraps" may also be used. These electronics frequently contain toxic and deadly elements. Furthermore, the incorrect disposal of these electronics could lead to the release of hazardous substances into the environment. So, in this section, we will talk about the sustainable alternatives

Figure 11.1 Damaged, surplus, and outdated electronic equipment.

to recycling them. We will observe the procedures these computer devices and electronic gadgets go through to become various valuable things. E-waste is subjected to several procedures, including sorting, separating, primary shredder, hammermill, and others. Televisions with cathode ray tubes (CRTs) are completely disassembled. Leaded glass, which is present in CRT televisions, can be dangerous if broken. Leaded and non-leaded glasses, as well as the steel and masks inside, are separated as the CRTs are fed into the CRT separation container. Hazardous materials and items are eliminated once the materials are pre-sorted. The raw material is prepared for the primary shedder. Here, material will be shredded into substantial pieces that can be manually removed. After leaving the primary shedder, the material is moved via an over band magnet to where materials like steel and magnetic objects are gathered and sent to a picking line. Staff members pick up contaminants and partially separated items during picking, resulting in a very clean steel product at the end of that line. The material that was removed then continues to flow through until it is appropriately separated. The secondary shredder and hammermill get the leftover material after that where it is still smaller. At this stage, any aluminum fragments left adhered to the plastic will be separated. Any remaining ferrous metals that were imbedded in the plastic before its liberation are subsequently removed as the material is transferred to a drum magnet [2]. From there, the material enters an optical sorter where we may further purge our plastic stream of impurities like circuit boards and copper wire. Finally, the procedure concludes. Both the hazardous elements that are also found in the e-waste and the precious materials that are contained inside have been successfully retrieved. The valuable materials are sold off and used in manufacturing processes to create new products while the other materials we have recovered undergo further refinement.

11.2.1 Process of E-Waste Recycling

Recycling electronics can be challenging since they are intricate machines consisting of different proportions of glass, alloys, and polymers. The recycling procedure can vary depending on the matter being recycled and the techniques employed, but here is a general outline.

11.2.2 Collection and Transportation

Two of the first steps in reprocessing, especially for e-waste, are group and transport. Recycling agencies set up group bins or booths for returning unwanted electronics to certain places, then they transport the e-waste they have gathered to processing plants and provision [11].

11.2.3 Shredding, Classify, and Uncoupling

After being collected, the materials that make up the electronic waste stream need to be processed and separated into clean material before they can be used to manufacture new products or sent to recycling resources. Effective material separation is the foundation of the electronic waste recycling process. The first step in sorting and separating plastics from metals and internal electronics is shredding the e-waste runnels. Therefore, to prepare for further classification, electronic waste is broken down into bits as small as 100mm (about 3.94 in). Iron and steel are separated from the conveyor's waste stream by a strong raised magnetite. The steel components are then processed for sale as recycled steel after being separated. Circuit boards, copper, and aluminum are separated from the material stream, which is now primarily plastic, by means of further mechanical processing. Then, glass and plastic are separated using a water separation technique. The quality of retrieved materials is improved by visual inspection and manual sorting. Circuit boards, copper, and other segregated streams are gathered and made ready for sale as recycled commodity materials. The procedure makes use of cutting-edge separation technology. To further clean the stream, the final step in the separation procedure finds and removes any leftover metal particle from the polymers. After the phases of shredding, classifying, and uncoupling have been completed, the materials that have been separated are then able to be offered for sale as valuable raw materials that can be used in the production of new electronics or other products.

11.2.4 Most Effective Method to Remove Metals

The hydrometallurgical process, in which we utilize chemical reagents to leach out our necessary metals, is an even more successful way to remove the valuable metals. The hydrometallurgical method entails several phases, including the breakdown and collecting of life's final products, water washing and cleaning to eliminate any unwanted components, alkali/alkaline salts, extraction, heat treatment, and filtration. Generally, hydrometallurgical methods were used to extract precious metals from mineral ores, but they can also be used to recycle waste electrical and electronic equipment (WEE) by selectively dissolving the desired metals using acid or caustic leaching. The solution that contains the desired metallic species and any potential contaminants is then further filtered, with the metallic compounds being concentrated using techniques such as solvent extraction, adsorption, or ion exchange. Since it can regulate the contaminants and is also one of the most economical and environmentally benign recovery procedures, the hydrometallurgical technique is the most effective and efficient way to recover metallic species. For example, recovery efficiencies from leaching through thermal transformation to extract metallic materials from PCBs are possible up to 93%, 65%, 91%, and 95%, respectively, for Cu, Zn, Pb, and Sn. The efficiency of hydrometallurgical e-waste recovery is reported to be very high, ranging between 81% and 99.9%. The other example given is the recovery of Sn from PCBs at room temperature using a hydrometallurgical method followed by electrodeposition, which leads to a 100% recovery rate after only two hours of processing.

11.3 Smart Phones at the End of Their Life

Every year, approximately 1.6 billion smart phones are sold. Additionally, 233 million non-smartphones are produced, bringing the total number of phones produced in a year to roughly 1.7 billion. Thus, there have been approximately 22.6 billion mobile phones produced in the past decade. 4.5 billion phones, on average, are discarded annually. Inadvertent damage to a cell phone occurs in about 45% of cases. During the projection period, it is expected that the global market for smartphones will grow by 5%. (2022-2028). Asia's smartphone market is growing because of things like a rise in disposable income, a rise in the number of new products coming out, an expansion of the telecom infrastructure, the release of more affordable smartphones, and so on [8]. Both the number of models on the market and the size of the market as a whole have been growing steadily [15]. Ericsson thinks that over the next few years, several hundred million more people will sign up for smartphones, bringing the total number of smartphone subscriptions around the world to more than six billion. The most people use smartphones in China, India, and the United States.

The various stages in the process of mobile phone recycling is shown in Figure 11.2.

Stage 1:
Mobile phones are frequently separated by recyclers. They might distribute the ones that can still be used, most likely to certain underdeveloped countries. Others decide to keep them and start recycling. They remove the batteries in this case and distribute them. Battery recycling departments

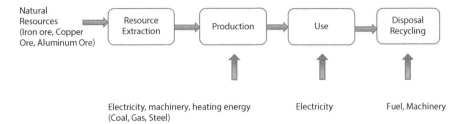

Figure 11.2 Drivers and barriers to return and recycling of mobile phones.

frequently get the batteries. Recycling companies then start shredding the cell phones. They then heat the shredded phones at temperatures that could reach 1100C.

Stage 2:
Before using powerful chemicals to treat the shredded cell phones, recycling organizations heat them. Mobile phones are frequently reduced to dust because of this. They are then transported to the smelter. Here, the recyclers extract significant metal products using various recycling techniques. These metal components are used to make other items. The recyclers burn plastic materials as a source of energy. These plastic substances are ground up and transformed into different substances [12].

Stage 3:
Manufacturers first remove these various materials, after which they deliver them to their various divisions. Here, they are utilized to create various reusable items. They must undergo detoxification before being used to create anything that will come into touch with food. No one could possibly be harmed in this way.

11.4 Recycling of Printed Circuit Boards (PCB)

Any electrical product would be incomplete without printed circuit boards (PCBs). However, recycling PCBs effectively and sustainably can be quite difficult. 40% of PCBs are made up of metals, 30% of them organic, and 30% of them ceramic. They have organic components, chemical byproducts, and heavy metal components that can be extremely harmful to the wellbeing of people and the planet [10]. They also include premium precious metals including palladium, silver, copper, and gold. Recycling waste PCBs successfully is consequently a challenging procedure (up to 7% of the world's gold may already be contained in e-waste). Recycling a PCB recovers the components and precious metals, turning trash into treasure. The first step in the physical process of recycling PCBs entails either drilling or sorting the boards in order to remove components, followed by cutting the boards into smaller pieces. A Printed Circuit Board (PCB) can have its capacitors, motors, batteries, plugs, semiconductors, LEDs, and other components removed using this method. In addition, magnetic separators are used to separate non-ferrous metals and trash from ferromagnetic materials and ferromagnetic materials from non-ferrous metals. Under the Waste Electrical and Electronic Equipment Directive, different recycling methods are required for these kinds of components. Thermal or chemical recycling techniques, in contrast, can provide pure metals and a significantly better level of efficiency and financial return. However, these procedures frequently call for high pressures or temperatures during processing, which might result in pollution from dangerous gases. Small portions make it considerably simpler to separate metals from ceramics, fiberglass, and other materials from a management standpoint. Additionally, sorting serves the crucial purpose of eliminating any unwanted items from the recycling process. Dismantled printed circuit board assemblies have a serious impact on the environment since they comprise of heavy metals as well as halogen-containing residues like brominates, mercury, lead (soldering tin), and a mixture of plastics which can leak into the surroundings if not managed appropriately. Battery cells may burn up or start to emit potentially harmful organic gases if subjected to high heat or fire. An explosion could occur if strong currents and heat are applied to a capacitor. Therefore, large, or PCP-containing battery cells and capacitors should be removed manually from this procedure and discarded separately in a proper manner. Delivering circuit boards to a facility will allow for further dismantling tasks. Hg switches are no longer being produced. The main sources of Mercury and Cadmium in printed circuit board fabrication are accumulators [9].

11.5 Solar Panel Recycling

End-of-life solar panel waste gives chances for recycling that can recover valuable materials and generate employment. According to the International Renewable Energy Agency, the value of recyclable raw materials from these panels would be approximately $452 million over the course of a decade. This amount is comparable to the costs of the raw materials that are required to produce approximately 65 million new panels.

The value of the material can be determined through recycling, which also helps to preserve landfill space. Crystalline silicon technology currently accounts for the vast majority of the market share for solar panels. Polymer layers, glass, silicon solar cells, a back sheet, copper wire, and a junction box made of plastic are the components that go into the construction of this type of panel. The polymer layers shield the panel from the outside, however, they make the dismantling and recycling process difficult because it frequently requires high temperatures to dissolve the adhesives. Many of these components can be recycled. The percentage of glass in the solar panel is about 70% and recycling of glass is a thriving industry. Additional easily recyclable components are the aluminum frame, plastic junction box, and copper cable. It might be more difficult to recycle some of the other components included inside solar cells. Even though these minerals are valuable components, silver and internal copper are typically present in relatively small concentrations in panels. Lead (pb) and cadmium (cd) are other potential hazardous materials that might be discovered in solar panels. In addition to aluminum, antimony, tellurium, and tin solar panels may also contain the elements gallium and indium. Included on the list of potential add-on components for a solar energy system are battery back-ups, racking systems, and inverter systems. These components may all be reused or recycled. Racking and inverters are two examples of items that could be recycled together with other types of scrap metal and electronic waste. Existing programs for recycling batteries could potentially manage energy storage systems that are based on batteries connected to the grid. The ideal recycling plan would recover as much solar panel material as is feasible. Numerous methods exist for recycling solar panels, some of which may require all three of the stages listed below: removing the junction box and frame; glass and silicon wafer separation using thermal, mechanical, or chemical processes; or silicon cell and specialty metal separation via chemical and electrical processes. Researchers are exploring ways to commercialize recycling so that the majority of a solar panel's components can be profitably recovered in this young and expanding sector. Although there are parts of this recycling process in the United States, it is not currently taking place on a significant scale. Glass, metals, and electronics recycling are well-established sectors that may recycle solar panels and other parts of solar power systems. The frame and junction box are often removed before engaging in these procedures, which commonly comprised of crushing, shredding, and grinding. These procedures use copper, aluminum, and glass. Recycling of c-Si modules currently results in a net cost activity since the real environmental costs and externalities for landfills are avoided. However, these procedures can shorten the energy payback time (EPBT) for the entire PV industry, boost the recovery of embedded materials and energy, and guarantee the long-term sustainability of the supply chain.

11.6 How Has E-Waste Management in India Evolved Through the Years?

11.6.1 Promoting Formal E-Waste Recycling

E-waste Recycling Credits are a point-based payment system that the Indian government developed to encourage businesses to transport e-waste through government-authorized reprocessing facilities. This was done through the creation of E-waste Recycling Credits (ERCs). The E-Waste Guidelines have in the past defined the various categories of electronic waste and the labelling that should be applied to it for items such as cell phones and laptops. Various ERC award levels must be assigned to these categories. Organizations must get the relevant ERCs, that can be used to cut energy usefulness charges based on the kind of e-waste provided. A program like this will also provide a vast inducement for unauthorized e-waste businesses to formalize their actions and establish supply chain relationships with approved reprocessing amenities for possible piloting of the ERCs [7].

By jointly supporting infrastructure improvements and processing systems at current government-approved recycling facilities, the Indian administration can similarly increase official e-waste recycling volume. Through public-private partnerships with significant e-waste corporations, it can offer co-funding incentives to governments for establishing new recycling facilities. State governments might also create incentive programs to encourage small-scale, improvised e-waste recycling facilities to modernize their infrastructure and conform to environmental and occupational health and safety requirements. States are eligible to apply for national financial programs for urban development, which can be utilized to connect the extensive informal sector network of dice.

11.6.2 Training and Upskilling Informal Sector Players

Most of the untrained e-waste recovery industry needs an upgrade, specifically for managing and dismantling dangerous goods. In addition to providing a connection between informal and formal sector manufacturers, it is essential that it protect the health and safety of workers as well as the surrounding environment. The National Skill Development Mission, which is part of the Indian government, is working to advance it. When the understanding of the Electronics Sector Skill Council, the Green Jobs Sector Skill Council, and regulatory bodies such as the Central and State Pollution Control Boards are combined, it is possible to develop innovative short training and educational programs that are especially suited for e-waste collectors, handlers, and dismantlers [17]. These programs can be especially useful because they are specifically tailored to meet their needs. Both training to raise public awareness of the dangers associated with e-waste, as well as the launch of a coordinated, nationwide program to upskill employees in the informal sector [4] should be the responsibility of the central and state governments. The significance of the unofficial sector in the disposal of e-waste, as well as the locations of official e-waste collection facilities that have been authorized by the government to operate in the area. The national initiative to upskill workers in the informal sector should be launched by the federal and state governments in coordination with education to increase public understanding of the dangers associated with e-waste [5] and the significance of the black market in the management of electronic waste, in addition to the locations of official e-waste collection facilities that have been approved by the government [18][19][20][21].

11.6.3 International Statistics of E-Waste

The amount of electronic waste that is discarded each year across the globe ranges from 20 million to 50 million metric tons, which is greater than the total weight of all the commercial airplanes that have ever been constructed. It is anticipated that the annual production of electronic waste will rise to 120 squillion tons by the year 2050. Only 21% of electronic waste is recycled in an official capacity, while the remaining 79% is either thrown away in landfills or recycled informally. Asia is responsible for the generation of the greatest quantity of electronic waste in absolute terms [16], followed by Europe and then North America. However, Europe generates the most electronic waste, with 16 kilograms per person, and Africa generates the least electronic waste, with 1.7 kilos per person. This disparity will continue to be a problem if there is an increase in the global population. According to the United Nations, the amount of e-waste generated globally increased by 21% between 2014 and 2019. At this rate, e-waste will double in size in just 16 years. In 2015, 62 counties or so had a policy on electronic trash; by 2020, 79 counties had one. In 2020, 72% of people on the planet were subject to a national e-waste policy, legislation, or regulation. In 2018, Europe (42%), Asia (11%), the USA (9.5%), Africa (0.8%), and Oceania (8.9%) had the greatest rates of e-waste recycling and collecting.

2019 saw the reduction of 16 Mt of carbon dioxide emissions due to recycled copper, iron, and aluminum e-waste components, which is the same number of emissions avoided by recycling secondary materials in lieu of new materials. E-waste production reached 57.4 Mt (Million Metric Tons) in 2021. An average of 2 Mt is added to the amount each year. By 2022, the amount of electronic waste that has not been recycled will be above 347 Mt. The biggest producers of e-waste are China, the US, and India. The amount of known collected and correctly recycled e-waste is only 17.4% [13]. The maximum rates of e-waste reuse are found in Estonia, Norway, and Iceland. The market for recycling e-waste was estimated to be worth $48,880 million in 2020. Every year, approximately 1.5 billion smartphones are sold. Additionally, 233 million non-smart phones are produced, bringing the total number of phones produced in a year to roughly 1.8 billion. Thus, there have been approximately 22.4 billion mobile phones produced in the past decade. 5 billion phones, on average, are discarded annually. Inadvertent damage to a cell phone occurs in about 45% of cases [6].

11.6.4 National Statistics of E-Waste

Electronic garbage in India has become a major issue for the surroundings and general public's health. After China and the United States, India is the 3rd largest contributor to electronic trash, creating about 2 million tons annually and importing an undocumented amount from other nations. It produces approximately 3 million tons (MT) of e-waste yearly. In 2020–2021, the nation managed 3.6 lakh tons of e-waste. Garbage production is expected to reach 7.3 lakh tons in 2017-18 and 10.15 lakh tons in 2019–20, according to the (Central Pollution Control Board) CPCB. Plastic waste material production is growing yearly by 3% and e-waste generation will be even higher. There is an annual growth of 31%. Only 468 authorized recyclers and 2,808 pickup locations are spread throughout 22 states.

The 13 lakh tons of capacity that 468 recyclers have is not enough to handle the e-waste production in India. Only ten states and 65 cities account for 60% of the country's 70% e-waste production. The report further identified the leading trash producers in India as being computers and cell phones.

States generating E-waste is shown in Figure 11.3. The telecommunications sector accounts for 12 percent of the total of all e-waste created annually, followed by the medical equipment business with 8%, computer devices with 7%, and electrical equipment with 8%. Only 16% of electronic waste is created by individual households, with the majority coming from the government, nonprofit organizations, and for-profit companies. Generation of E-waste every year (in metric tons) is shown in Figure 11.4.

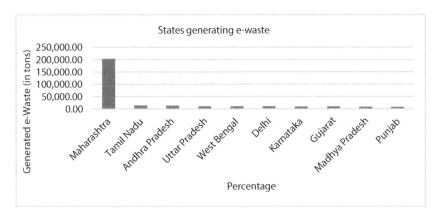

Figure 11.3 States generating E-waste.

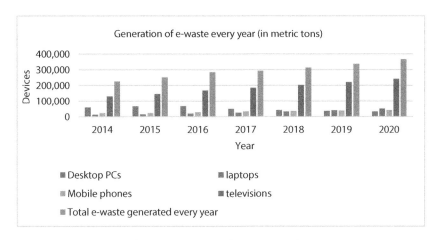

Figure 11.4 Generation of E-waste every year (in metric tons).

11.7 Conclusion

E-waste is a comparatively recent problem in the global garbage problem. The situation is escalating as a result of an increase in e-waste that results from a lack of awareness and expertise. In terms of discarded garbage, it also continues to expand at the fastest rate in the globe. Most people ignore or misunderstand this issue on a worldwide scale. Many people do not understand how it works or how it impacts themselves, the community, or the environment. Users in wealthy nations routinely replace digital products because technology is constantly evolving. Such updates leave an excess of proposed product or service idle. The problem regarding e-waste is what happens to old cellphones and computers. Although few individuals are aware of how crucial it is to get rid of this old equipment correctly, many others still throw it away in incinerators or dumpsters. Sadly, there is still one more e-waste cleanup technique in use currently. E-waste generated in underdeveloped nations is largely transferred to other developing nations. There are many technology options for handling e-waste, but even before they can be used in the management system, the required legal framework, manpower, collection system, and logistics, must be developed. Operational research and evaluation studies can be necessary for this. E-waste is often shipped, although many industrialized countries have implemented laws to prevent it. Most of the electronic waste produced worldwide is shipped to nations like Nigeria, Pakistan, and India. Although it may appear odd that a country would choose to import another country's waste, it occasionally happens unlawfully. This strategy produces work and beneficial byproducts. E-waste provides beneficial metals like iron, copper, silicon, nickel, and gold.

References

1. T. Kiyokawa, H. Katayama, Y. Tatsuta, J. Takamatsu and T. Ogasawara, "Robotic Waste Sorter with Agile Manipulation and Quickly Trainable Detector," *in IEEE Access*, vol. 9, pp. 124616-124631, 2021, doi: 10.1109/ACCESS.2021.3110795.
2. A. U. R. Khan and R. W. Ahmad, "A Blockchain-Based IoT-Enabled E-Waste Tracking and Tracing System for Smart Cities," *in IEEE Access*, vol. 10, pp. 86256-86269, 2022, doi: 10.1109/ACCESS.2022.3198973.
3. Kumar, K. Suresh, AS Radha Mani, S. Sundaresan, T. Ananth Kumar, and Y. Harold Robinson. "Blockchain-based energy-efficient smart green city in IoT environments." In Blockchain for Smart Cities, pp. 81-103. Elsevier, 2021.
4. Bruce A. Fowler (2017). "Current E-Waste Data Gaps and Future Research Directions" Electronic Waste Toxicology and Public Health Issues, (Pages 77-81).
5. N. Othman, L. M. Sidek, N. E. A. Basri, M. N. M. Yunus and N. A. Othman, "Electronic plastic waste management in malaysia: the potential of waste to energy conversion," *3rd International Conference on Energy and Environment (ICEE)*, 2009, pp. 337-342, doi: 10.1109/ICEENVIRON.2009.5398623.
6. V. Agarwal, S. Goyal and S. Goel, "Artificial Intelligence in Waste Electronic and Electrical Equipment Treatment: Opportunities and Challenges," International Conference on Intelligent Engineering and Management (ICIEM), 2020, pp. 526-529, doi: 10.1109/ICIEM48762.2020.9160065.

7. Kumar, A., Dubey, A.K., Ramírez, I.S., Muñoz del Río, A., Márquez, F.P.G. (2022). A Review and Analysis of Forecasting of Photovoltaic Power Generation Using Machine Learning. In: Xu, J., Altiparmak, F., Hassan, M.H.A., García Márquez, F.P., Hajiyev, A. (eds) Proceedings of the Sixteenth International Conference on Management Science and Engineering Management – Volume 1. ICMSEM 2022. Lecture Notes on Data Engineering and Communications Technologies, vol 144. Springer, Cham. https://doi.org/10.1007/978-3-031-10388-9_36
8. Dubey, A.K., Kumar, A., Ramirez, I.S., Marquez, F.P.G. (2022). A Review of Intelligent Systems for the Prediction of Wind Energy Using Machine Learning. In: Xu, J., Altiparmak, F., Hassan, M.H.A., García Márquez, F.P., Hajiyev, A. (eds) Proceedings of the Sixteenth International Conference on Management Science and Engineering Management – Volume 1. ICMSEM 2022. Lecture Notes on Data Engineering and Communications Technologies, vol 144. Springer, Cham. https://doi.org/10.1007/978-3-031-10388-9_35
9. Rathore, P.S., Chatterjee, J.M., Kumar, A. *et al.* Energy-efficient cluster head selection through relay approach for WSN. J Supercomput 77, 7649–7675 (2021). https://doi.org/10.1007/s11227-020-03593-4
10. A. Dubey, S. Narang, A. Srivastav, A. Kumar, V. Díaz, Woodhead Publishing, Science Direct,Artificial Intelligence for Renewable Energy Systems. Paperback ISBN: 9780323903967
11. A. Dubey, S. Narang, A. Srivastav, A. Kumar, V. Díaz, Woodhead Publishing, Science Direct, a Visualization Techniques for Climate Change with Machine Learning and Artificial Intelligence. ISBN: 9780323997140
12. S. Parveen, S. Yunfei, J. P. li, J. Khan, A. U. Haq and S. Ruinan, "E-waste Generation and Awareness on Managing Disposal Practices at Delhi National Capital Region in India,"*16th International Computer Conference on Wavelet Active Media Technology and Information Processing*, 2019, pp. 109-113, doi: 10.1109/ICCWAMTIP47768.2019.9067630.
13. D. A. Hazelwood and M. G. Pecht, "Life Extension of Electronic Products: A Case Study of Smartphones," in *IEEE Access*, vol. 9, pp. 144726-144739, 2021, doi: 10.1109/ACCESS.2021.3121733.
14. P. Mou, D. Xiang and G. Duan, "Products made from nonmetallic materials reclaimed from waste printed circuit boards," in *Tsinghua Science and Technology*, vol. 12, no. 3, pp. 276-283, June 2007, doi: 10.1016/S1007-0214(07)70041-X.
15. E. V. Korchagina and O. A. Shvetsova, "Analysis of Environmental Consequences of Tourism Activity in Baikal Lake Area: Regional Practice of Solid Waste Management,"*IEEE International Conference* "Management of Municipal Waste as an Important Factor of Sustainable Urban Development" (WASTE), 2018, pp. 19-21, doi: 10.1109/WASTE.2018.8554177.
16. R. Dassi, S. Kamal and B. S. Babu, "E-Waste Detection and Collection assistance system using YOLOv5," *IEEE International Conference on Computation System and Information Technology for Sustainable Solutions (CSITSS)*, 2021, pp. 1-6, doi: 10.1109/CSITSS54238.2021.9682806.

17. B. Debnath, A. Das, S. Das and A. Das, "Studies on Security Threats in Waste Mobile Phone Recycling Supply Chain in India," *IEEE Calcutta Conference (CALCON)*, 2020, pp. 431-434, doi: 10.1109/CALCON49167.2020.9106531.
18. Jinglei Yu, Meiting Ju and E. Williams, "Waste electrical and electronic equipment recycling in China: Practices and strategies," *IEEE International Symposium on Sustainable Systems and Technology*, 2009, pp. 1-1, doi: 10.1109/ISSST.2009.5156728.
19. R. I. Moletsane and C. Venter, "Electronic Waste and its Negative Impact on Human Health and the Environment," *International Conference on Advances in Big Data, Computing and Data Communication Systems (icABCD)*, 2018, pp. 1-7, doi: 10.1109/ICABCD.2018.8465473.
20. Devi, L. Renuka, N. Arumugam, J. E. Jayanthi, TS Arun Samuel, and T. Ananth Kumar. "Investigation of High-K Gate Dielectrics and Chirality on the Performance of Nanoscale CNTFET." Journal of Nano-and Electronic Physics 13, no. 2 (2021).
21. H. Nixon, O. A. Ogunseitan, J. -D. Saphores and A. A. Shapiro, "Electronic Waste Recycling Preferences in California: The Role of Environmental Attitudes and Behaviors," *Proceedings of the 2007 IEEE International Symposium on Electronics and the Environment*, pp. 251-256, doi: 10.1109/ISEE.2007.369403.

12

Sustainable Development Through the Life Cycle of Electronic Waste Management

D. Magdalin Mary[1], S. Jaisiva[2]*, C. Kumar[3] and P. Praveen Kumar[4]

[1]*Department of EEE, Sri Krishna College of Technology, Coimbatore, Tamil Nadu, India*
[2]*Department of Electrical and Electronics Engineering, Sri Krishna College of Technology, Coimbatore, Tamil Nadu, India*
[3]*Department of Electrical and Electronics Engineering, Karpagam College of Engineering, Coimbatore, Tamil Nadu, India*
[4]*Department of IT, Sri Manakula Vinayagar Engineering College, Puducherry, India*

Abstract

Electronic waste (E-waste) is one of the solid waste sources that is growing fastest and its intelligent management has grown to be a main issue that calls for a deep draw near the central guiding principle of prolonged manufacturer accountability (EPR). Over the previous 20 years, China, one of the world's biggest manufacturers of electrical and virtual tools (EEE), has made great efforts to enhance e-waste management along with the massive era of e-waste. The usage of electronics has expanded significantly over the past 20 years, transforming the way people exchange information, connect, and enjoy entertainment. Simple actions like reusing old gadgets may extend the lifespan of priceless goods. Recycling electronics also protects priceless components from the waste stream. The overall amount of trash managed locally and worldwide may be decreased by improving the lifecycle management of electronics by reducing the source of materials consumed, enhancing reuse, restoration, extending product life, and recycling electronics. The waste management hierarchy of the EPA is in line with the life cycle approach. The hierarchy underlines that reducing, reusing, and recycling are essential components of sustainable materials management and grades the different management options from greenest to least green.

Keywords: EPA, E-waste, E-production, recycling, ISO, electronic garbage, RoHS, E-trash

Corresponding author: jaisiva1990@gmail.com

Abhishek Kumar, Pramod Singh Rathore, Ashutosh Kumar Dubey, Arun Lal Srivastav, T. Ananth Kumar and Vishal Dutt (eds.) Sustainable Management of Electronic Waste, (237–254) © 2024 Scrivener Publishing LLC

12.1 Introduction

This section presents a general concept of E-waste, its basic technical definitions, and the primary sources of its generation. In addition, this section describes technical and environmentally sustainable alternatives that can be used to manage e-waste globally, properly. It also describes some examples of policies and strategies adopted by many countries to promote sustainability in handling e-waste. The introduction includes examples of regulatory frameworks in other regions, such as the WEEE directive in Europe.

12.1.1 Electronics Production, Waste, and Impacts

The electronic reformation wholly altered how people connected and led their lives and shrunk the planet's size. There is a generation obsessed with technology because of a period when there was a great deal of technological advancement (Murthy *et al.* 2022) and wide use of technology, such as home automation, personal computers, and laptops and constantly developing cellular devices. Almost as regularly as the updation of wardrobes to upgrade our various forms of innovation, technology becomes outdated and new tech is purchased each time an upgrade is released.

Uses of E-Production
"E-waste" refers to electronic devices that have reached or are nearing the end of their useful life, which includes old, discarded electronics, cellphones, laptops, game consoles, and their parts. The following information is both intriguing and alarming:

- Over forty thousand tons of electronic waste are created annually around the world. That equates to 757 handbooks and laptops are being produced per second (Kumar *et al.* 2017).
- Every 18 months, the typical cellphone user replaces their device.
- Of all of our harmful waste, 70% is made up of e-waste.
- Recyclability for e-waste is only 12.5%.
- About 90 percent of electronic garbage is taken to landfills, where it is primarily burned and contributes to the emission of harmful contaminants.

- Lead, which is present in electronics, can harm our kidneys and central nervous system.
- Several factors can influence a child's mental development.

12.1.2 E-Production: Recycle

Our electronic garbage is often disposed of in dumps located in either Asia or Africa. The practice of "composting" in such dumps is not ideal. Electronic garbage is hazardous not just to the local habitat but also to humans and, most importantly, to the individuals who work in landfills. When electronic waste is deposited or incinerated at dumpsters, the air, water, and land become contaminated with the chemicals that are released from the garbage. The vast majority of the time, laborers in impoverished countries are the ones to collect them as well as extract rare minerals from discarded electronic gadgets. The hourly compensation for workers at facilities that process electronic waste is typically around $2.75. On the job site, they are not covered when managing potentially dangerous items (Cucchiella *et al.*, 2015). When home appliances are burned, they are the first to inhale aerosol toxins, some of which are produced as a byproduct of the process.

12.1.3 Uses of Recycling

Precious metals like gold and silver are found in phones and other electronic devices. Annually, the United States of America throws out telephones with a combined value of $75 billion worth of gold and silver-recycling and 1 million laptops, resulting in annual energy savings equal to 3600 US houses. Most e-waste components are made up of electronic equipment, which may be recycled and utilized to recover resources (Dhir *et al.* 2021, Kumar *et al.*, 2022). When adequately removing your electronic equipment, you have a few options:

- Others may use outdated cell phones in a manner analogous to an old TV or a computer (Kang *et al.* 2005). Multinationals, community mobilization projects, or even a complete option is acceptable. This is a significantly better alternative than having the possibly hazardous e-waste goods wind up in a landfill. Consequently, you are the source of delight for another person. It is a wise decision to go that route.

- A licensed electronic waste processing facility in a particular area. Anyone not possessing the necessary accreditation will likely just transfer it to a location that may be harmful. Contribute to the alternative and help us reduce the amount of junk caused by electronic devices. To prevent the issues, it must be replaced with gadgets as frequently as possible.

12.2 Impact on the Environment

The global production of electronic garbage, sometimes known as "e-waste," which includes old computers, phones, televisions, and other electronic hardware, is expected to be 40 million metric tons annually. (Tabelin *et al.* 2021, Dubey *et al.*, 2022). Even though it only makes up 2% of the garbage in landfills, electronic trash is responsible for 70% of the dangerous heavy metals. In the Philippines, Pakistan, China, India, and Vietnam, poorly regulated informal recycling markets handle about 65 to 90% of electronic waste being burned in the outdated devices. Along with harming human health, these improper e-waste disposal methods also seriously harm the ecosystem. E-waste reproduction unit is shown in Figure 12.1.

Burning of electronic trash results in the production of hazardous particles that pollute the environment. Eczema, as well as other breathing problems linked to contamination of materials employed at the smoldering locations is

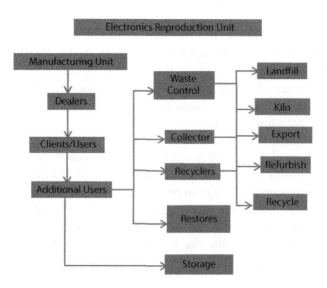

Figure 12.1 E-waste reproduction unit.

Figure 12.2 Waste electrical and electronic equipment.

found to be more likely to affect those who live nearby (Pramila *et al.* 2012). Waste electrical and electronic equipment is shown in Figure 12.2.

Lithium, mercury, and lead heavy metals that are included in computer and mobile phone batteries are discharged into the earth and groundwater streams when they are inappropriately disposed of. Whenever these contaminants make their way directly into the water stream, wildlife and individuals who consume the toxic water risk experiencing negative repercussions, including contamination (Rathore *et al.*, 2021).

Especially alarming amounts of toxins were found in water samples close to some of the uncontrolled informal marketplaces that shred and burn e-waste in underdeveloped nations. For instance, lead levels in water samples from Mandoli, India were 11 times higher than the exposure limit set by the Indian government. In addition, the mercury concentrations in these samples were roughly 710 times greater than the exposure limit set by the Indian government. MAYER alloys corporation is shown in Figure 12.3. However, the aforementioned detrimental impacts on the environment and public health can be lessened when e-waste is recycled and properly disposed of (K. Suresh Kumar *et al.* 2021).

Figure 12.3 MAYER alloys corporation.

12.3 Environmental Impact of Electronics Manufacturing

According to the European Environment Agency, about 10,000 metric tons of discarded electrical and electronic equipment, often known as WEEE or e-waste, is created every decade across the nation. Figure 12.3 does not include garbage released in other parts of the world (Huisman et al., 2017). Only around 40% of it is being collected for recycling. There are comparable statistics throughout the world. Approximately 32% of the e-waste that has accumulated from the $95.6 million in consumer technology sales that were recorded in the United States of the Americas in 2017 is reused. A growing number of everyday home items are added to the list each year (Chang et al. 2021). The list goes on and on. People discard these devices when they break, perform poorly, or simply when a newer model becomes available.

12.3.1 Built-In Obsolescence is Also to Blame

Based on the findings of a survey conducted by the organization of Europe, the percentage, including all worldwide sales to upgrade defective devices, climbed from 5.2% in 2005 to 9.2% in 2009 as a direct result of constructed deterioration. The percentage of large home equipment that needed to be changed in the initial two decades rose from 8 percent of all restorations in 2005 to 19 percent in 2017; this represents a significant increase in the frequency of such upgrades. According to a study conducted by Vanguard in 2018, 39% of people in the general public operated VCRs at the same age and 89% of people in the millennial population (18 to 29) possessed mobile phones.

The Atlantic said it is not a new concept to intentionally shorten product lifespans to entice consumers to acquire new things. The Phoebus cartel, made up of General Electric, Osram, Phillips, and Tungsram, promised that light bulbs would not survive for more than 1,000 hours in 1924. The cartel was broken up in 1939 when producers in Eastern Europe started making inexpensive bulbs (Zhang et al. 2022). The effects of built-in obsolescence have grown and become more severe. Electronics manufacturing has an ecological influence on the world that extends beyond trash production. Concerns about infringements, exploitation of employees, degradation of the environment, groundwater, and land, and toxicity of all three have been mentioned as potential challenges. The air might be contaminated when e-waste is burned rather than recycled (Deva Kumar *et al.* 2021).

Although many things continue to end up in landfills, several nations worldwide have legislation dictating how e-waste should be treated. The remaining was just thrown away. Who can recall their first (or second, third, or fifth) iPhone's location?

12.3.2 Usage of ISO 14001 Helps to Lessen How Much of an Impact Producing Electronics has on the Environment

In order to lessen the environmental effect of their operations, several contract electronics manufacturers are examining every step of the manufacturing cycle, from purchasing and storage through manufactured goods creation and sharing. The most extensively used standard for evaluating an organization's environmental effect is undoubtedly ISO14001: 2015. Despite being a voluntary standard, ISO 14001 is crucial for any business trying to acquire a competitive edge. Each EMS may be customized to match the unique demands of the firm while sticking to its own business procedures because no two businesses are the same. As David points out, manufacturers are learning that there are many places where they can find efficiencies, such as reviewing their internal use of single-use plastics, lowering the quantity of manufactured goods wrapping, researching and developing better sustainable power conservation programs, coming up with innovative storage technologies, and putting in action sustainable transportation strategies are some examples of things that could be done (Aishwarya *et al.* 2021). Businesses may gain from real, reliable data to help with decision-making and tracking since an EMS is evidence-based.

12.4 E-Waste Management Initiative

The "polluter pays principle" was implemented as part of the Environmental (Protection) Act of 1986. This principle states that the entity that is responsible for producing pollution is obligated to compensate for the harm caused to the local habitat and is considered a part of international environmental law. Product stewardship accountability is another name for the "polluter pays" system (EPR). The Environment (Protection) Act of 1986 gives the federal authority and each public authority the authority to make laws that preserve individuals and the natural environment from being exposed to the dangerous and poisonous properties of pollution. Any

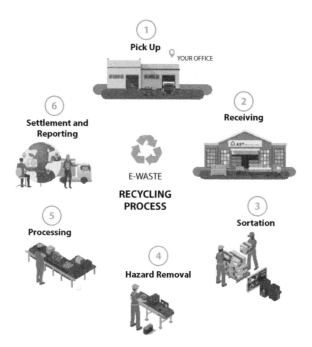

Figure 12.4 E-waste recycling process.

person who violates the terms of this act or any of the stated rules is subject to disciplinary action. E-waste recycling process is shown in Figure 12.4.

The person who violates the rules and regulations about e-waste could be liable to repercussions if they are found to be in violation. Currently, India is in the process of establishing a collection of regulations and has just set some instructions for the correct and environmentally acceptable burial of E-waste. The assistance of non-governmental organizations (NGOs) is currently reviewing the guidelines that computer component producers drafted. These rules were published in the year 2000. When a component of biological trash reaches the point where it can no longer be used, the producer is "directly" liable for the secure and proper disposing of the goods. A complicated and extensive handbook titled "Eco Monitoring for the Software Development Industry in India" was established, a component of the Ministry of Communication and Information Science that forms sustainable initiatives to reduce electronic waste.

Several firms in the electronics industry have launched a variety of green initiatives in an attempt to increase clients' awareness of the importance of recycling their electronic trash. The "materials recovery facility" for the Indian region was something that Nokia India launched. The program urged people who used mobile phones to recycle their old devices and peripherals that were placed throughout primary vendors and daycares. This was done regardless of the manufacturer of the cellular phone that was currently being used. In addition to this, Nokia intends to initiate a program for the treatment of electronic trash.

In addition, the Department of Environment and the government of Delhi have concluded that ragpickers should be included in the overall trash strategic plan in the nation's capital. These individuals would be employed to clean up and receive training, clothes, and identification cards as part of the job. It is those eco-clubs that would be connecting with techniques for a specific region and the administration aims to incorporate eco-clubs in this program. Currently, eco-clubs are operating in more than 2100 schools across the Capital Area.

12.4.1   Present Barriers in Recycling of E-Waste

According to Statistics, the amount of electronic waste recycled in 2019 was just 21.6 percent. It is due, in particular, to the fact that many electronic products nowadays need to be constructed to be repurposed. The reduction in size, weight, and thickness of today's mobile phones, combined with elimination of detachable battery packs, makes upgrading these devices far more challenging and labor-intensive.

12.4.2   Global E-Waste Problem

The transfer of toxic substances across countries is an issue being addressed by treaty obligations like the Basel Convention, which aims to reduce and regulate this transportation. Even with the Convention in existence, there is still some shipping and disposal of electronic trash done illegally. It is projected that fifty metric tons of electronic garbage will be produced worldwide in 2018. Portable gadgets, including laptops, displays, cellphones, iPads, and televisions, constitute a portion of that total, while the other side consists of significant home appliances. About 20% of the country's e-waste gets transferred annually, even though that regulation covers 66% of the earth's population. In the United States, about 10 Lakhs of computer peripherals are discarded each year and only about 15 percent is reused appropriately. China throws out 160 million pieces annually. China was once thought of as the country with the most extensive electronic trash dumps in the globe. Many hundreds of thousands of people are skilled in the process of disassembling old electrical equipment.

The quantity of electronic garbage is expanding at a rate that is anywhere between 5 and 10 percent each year around the world. The amount of electronic garbage produced annually in India equals 146,000 tons. Nevertheless, current statistics consider the e-waste produced extensively; they need not consider garbage importation, which is significant in developing nations. The rationale is that India receives significant garbage from nations outside its borders. The European nations have set an objective of 4 kilograms of electronic trash per capita. However, Switzerland has already recovered 11 kilograms of electronic waste per capita, making India the first nation across the globe to develop and maintain a comprehensive recycling system.

The Waste Electrical and Electronic Equipment, specifically goals for the collection, recovery, and recycling of recyclable materials are mandatory in the participating nations of the EU. As a result, it mandates a baseline acquisition objective of about 3 Kilograms per capita yearly for every sovereign nation in the organization. These reusing objectives, premised on compendium and mass rather than volume, are intended to cut down on the number of toxic materials dumped into garbage dumps and boost the accessibility of recyclables, which will, in turn, promote minimal usage of substances in newly manufactured goods.

According to reports, the e-waste produced is successfully separated, gathered, and processed. The significant move taken in South Korea was the adoption of the EPR system in 2003. As a result, around 85% of e-waste was composed of its consumers. During the same period, the number of recyclable materials repurposed was 12, and the amount composted was 69 percent. The remaining, which accounts for 19%, was distributed throughout several cemeteries and combustion units.

The handling of electronic waste in developing countries is made worse by the slack or nonexistent implementation of current regulatory frameworks, a lack of understanding and sensitivity on the part of individuals participating in those processes, and insufficient professional protective measures. As a result of this, it is essential for developing nations to implement efficient strategies to promote the reusability, refurbishment, or composting of recyclable materials in infrastructure that are specifically designed for this purpose. This will help reduce the risk of contaminating the atmosphere and endangering people's lives.

Most e-waste comes from industrialized countries like Britain, Mexico, Europe, Italy, France, and India. The mission of the Basel Action Network (BAN) is to promote environmentally responsible practices for the disposal of electronic trash. It protects the environment from the trafficking of hazardous substances. A system of green advocacy non-governmental organizations (NGOs) in the United States includes the Silicon Valley Toxic Coalition (SVTC), Technology Take-Back Coalition (ETBC), and the Breast Cancer Action Network (BAN). The promotion of alternatives at the federal level for the management of hazardous materials is the goal that all three organizations have in agreement. E-Stewards is a relatively new organization that audits and certifies recyclers and carry schemes. The organization's objective is to let ethical customers know which of these initiatives and recycling plants satisfy high-quality requirements.

12.5 Issues with E-Waste in India

As per the survey that was presented at the Global Economic Forum 2020, India is listed as number 175 out of 190 nations and made up the bottom four on the Environmental Performance Index for 2020. It was determined that its low effectiveness was concerning to public health and environmental affairs and mortality was due to smog classifications. In addition, our nation has ranked seventh in the globe, which is remarkable as the best nation creating electronic garbage, behind the United States of America, Britain, France, Denmark, and Russia, officially recycling just under 2% of the entire amount of electronic waste that India creates each year. Since the year 2017, Asian countries have produced annual production of more than 2 thousand metric tons of e-waste consumed all over the world.

Treating ravage in open dumping sites is a behavior that is all too common and leads to a range of issues, including poisoning of waterways, poor health, and several others. The private economy is responsible for most e-waste activities, including collecting, shipping, treatment, and recovery. The industry is highly interconnected yet is not controlled in any way. In many cases, not all of the materials and values that can be retrieved are bounced back. In addition, there are significant concerns over the health and safety of employees, as well as spillages of toxic substances into the natural atmosphere.

Seelampur in Delhi is home to Asia's most extensive collection of e-waste disposal facilities. Children and adults devote eight to ten hours every day to dismantling electronic equipment to save valuable materials such as copper, gold, and other valuable gems. Recycling plants of electronic waste utilize recyclable and working equipment. This scenario could be addressed by promoting consciousness, strengthening the infrastructure of recycling facilities, and revising the regulations that are currently in place. The unorganized Indian industry manages the vast bulk of the electronic garbage that is gathered there.

A significant chunk of electrical and electronic waste is collected by indirect mechanisms of reusing and composting electronic waste, including businesses like secondhand equipment merchants, service centers, and distributors operating on e-commerce portals. These streams collect the computers to reuse them for their components and parts.

12.6 Impact of E-Waste Recycling in Developing Nations

The vast majority of electronic wastes include a few recyclables, such as plastic, glass, and metals; however, because of incorrect burial procedures and processes, these resources cannot be salvaged for use in other contexts. The harmful elements of e-waste that can wreak havoc on a person's health are treated and disassembled in an imprecise approach. When disposing of garbage, the components are disassembled and either burned or subjected to aqueous chemical treatment to get rid of the waste. These operations lead to immediate exposure to dangerous substances and the consumption of those compounds. Professionals often need the competence and knowledge needed to do their tasks appropriately and there is no prohibition on wearing protective gear like gloves and face masks when there should be.

In contrast, manually extracting harmful elements results in the introduction of potential hazards into the bloodstream of the person performing the process. There is a wide range of possible health risks, from injury to the kidneys and liver to spinal stenosis. Reclaiming debris from electronic waste contaminates the air, groundwater, and environment. Heating lines and connections to collect metal have resulted in the release of brominated and chlorinated dioxins, as well as other chemicals, that pollute the air and, consequently, lead to a risky life for livelihoods.

All through the reprocessing, harmful substances that are worthless from an economic standpoint are thrown away as waste. Because of the leaching of these harmful toxins into the subterranean watershed, the integrity of the groundwater in the area is degraded and the liquid is no longer suitable for agriculture or human use. Toxic substances, namely palladium, iridium, arsenic, and PCBs, are introduced into the subsurface layers of the disposal of electronic trash in dumpsters. This renders the ground unsuitable for commercial use. Research carried out not too long ago on reusing electronic waste is digested by an unorganized sector. These urban areas are Visakhapatnam, Kochi, Bangalore, and Trivandrum. According to those research findings, the areas that are actively involved in the operations of

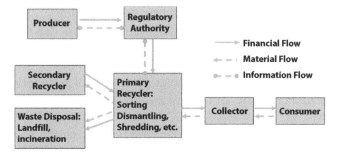

Figure 12.5 Modern structure of e-waste recycling network.

resource recovery are the best locations for the presence of such persistent hazardous chemicals. According to additional research carried out by the same team, the moderate nature of the environmental contaminants that are created allows them to disperse into the surrounding air. The Figure 12.5 shows the modern structure of the e-waste recycling network. Modern structure of e-waste recycling network is shown in Figure 12.5.

The tremendous rise in the use of computer gadgets can be attributed to the expansion of the information data and management industries. Because electronic devices are being upgraded quicker, consumers are forced to dispose of their old electronic goods more rapidly than ever before. This, in turn, contributes more e-waste to the cycle of substantial trash. The ever-expanding challenge of electronic trash necessitates a heavy focus on the disposal of recyclable materials and improved e-waste handling.

12.7 Opportunities and Challenges in E-Waste Management in India

Hazardous equipment, sometimes known as e-waste, is produced if the electronic equipment has either outlived its usefulness for the purpose it was designed for or had beyond its expiration. Because of the rapid technological improvements and the development of new electronic equipment, this older electronic equipment is quickly rendered obsolete and supplanted with versions of more recent vintage. Because of this, the quantity of electronic waste has grown in recent years. The livelihood in the nation has the propensity to promote newer versions and the average lifespan of products has also shrunk in recent years.

Common characteristics of electronic wastes were subjected to a technological processing method. It would be possible to extract precious metals.

If electronic waste is disassembled and analyzed in a clumsy circumstance using fundamental procedures, it poses a significant risk due to toxic substances. Electronic garbage presents a significant hazard to people, wildlife, and the environment. Sometimes, in extremely minute quantities, the buildup of toxic substances and potentially corrosive substances such as mercury, lead, beryllium, and cadmium poses a significant threat to the local habitat.

The subscribers are an essential factor in improving the management of electronic trash. Patented equipment for Connecting the Industry to Facilitate a Green Wealth are some examples of initiatives that encourage clients to marshal electronic waste properly and implement green consumption behavior. In industrialized economies, the governance of electronic waste is regarded as a top priority. In contrast, in underdeveloped nations, the dilemma is made worse by the comprehensive adoption or replication of the e-waste planning of developed nations, as well as a variety of issues about this, such as shortage of the venture. In connection with this, there is a deficiency in equipment and a lack of moral laws that are expressly geared toward handling e-waste.

12.8 Recent Investigations on Electronic Waste Management

The present scenario states that electronic waste disposal in India needs to be evaluated, as well as the number of units of electronic waste and the accurate intensity of the issue in Indian cities. This will necessitate a significant number of additional ecologic observational studies. The results of these investigations will provide researchers with helpful information that can be incorporated into formulating a strategy for treating electronic trash. India needs to initiate a monitoring program for the ailments and health effects associated with e-waste. Enhancing recovery and transportation methods is required to guarantee the long-term viability of existing electronic waste disposal systems. Establishing governmental collaborations to establish acquire or knock locations might be a preferable course of action. The imposition of advanced recycling fees is another different strategy for ensuring the continued viability of trash handling. Finding the most effective methods for managing electronic trash worldwide and implementing them could be the solution to achieving healthy advancement in the future. This initiative aims to eliminate potentially harmful components found in electronic and electrical equipment and the widespread adoption of safer alternatives to those components. In producing these goods, the Restriction of Hazardous Substances (RoHS) Regulations has been enacted by several nations around the world. It is essential to locate an increasing number of alternatives that pose less risk and are suitable for use in electronic devices.

12.9 Conclusion

Among the most pressing environmental issues facing the globe today is the potentially dangerous characteristic of electronic trash. The issue is made worse by the growing volume of e-waste produced and by a need for more understanding and the requisite skills to deal with it. As a result of the significant amount of employees in India who make their living from the rudimentary disassembly of electronic products while putting their own life in jeopardy, there is an immediate must to devise a preventative plan that targets the risks associated with the care of electronic waste by these employees. These technicians must be supplied with the necessary details about the safe management of e-waste and personal safety measures. Many engineering solutions are currently critical in managing e-waste; however, for these technologies to be integrated into the processing system, the necessary circumstances, which include regulation, a collecting system, infrastructure, and trained personnel, must first be developed. This may call for specific investigations involving study and assessment.

References

Murthy, Venkatesha, and Seeram Ramakrishna. "A review on global E-waste management: urban mining towards a sustainable future and circular economy." Sustainability 14, no. 2 (2022): 647.

Cucchiella, Federica, Idiano D'Adamo, SC Lenny Koh, and Paolo Rosa. "Recycling of WEEEs: An economic assessment of present and future e-waste streams." Renewable and sustainable energy reviews 51 (2015): 263-272.

Kumar, Amit, Maria Holuszko, and Denise Crocce Romano Espinosa. "E-waste: An overview on generation, collection, legislation and recycling practices." Resources, Conservation and Recycling 122 (2017): 32-42.

Kang, Hai-Yong, and Julie M. Schoenung. "Electronic waste recycling: A review of US infrastructure and technology options." Resources, Conservation and Recycling 45, no. 4 (2005): 368-400.

Tabelin, Carlito Baltazar, Ilhwan Park, Theerayut Phengsaart, Sanghee Jeon, Mylah Villacorte-Tabelin, Dennis Alonzo, Kyoungkeun Yoo, Mayumi Ito, and Naoki Hiroyoshi. "Copper and critical metals production from porphyry ores and E-wastes: A review of resource availability, processing/recycling challenges, socio-environmental aspects, and sustainability issues." Resources, Conservation and Recycling 170 (2021): 105610.

Pramila, Sharma, M. Fulekar, and Pathak Bhawana. "E-waste-A challenge for tomorrow." Research Journal of Recent Sciences ISSN 2277 (2012): 2502.

Chang, Tai-Wu, Chun-Jui Pai, Huai-Wei Lo, and Shu-Kung Hu. "A hybrid decision-making model for sustainable supplier evaluation in electronics manufacturing." Computers & Industrial Engineering 156 (2021): 107283.

Dhir, Amandeep, Suresh Malodia, Usama Awan, Mototaka Sakashita, and Puneet Kaur. "Extended valence theory perspective on consumers'e-waste recycling intentions in Japan." Journal of Cleaner Production 312 (2021): 127443.

Zhang, Zhen, Muhammad Zeeshan Malik, Adnan Khan, Nisar Ali, Sumeet Malik, and Muhammad Bilal. "Environmental impacts of hazardous waste, and management strategies to reconcile circular economy and eco-sustainability." Science of The Total Environment 807 (2022): 150856.

Kumar, A., Dubey, A.K., Ramírez, I.S., Muñoz del Río, A., Márquez, F.P.G. (2022). A Review and Analysis of Forecasting of Photovoltaic Power Generation Using Machine Learning. In: Xu, J., Altiparmak, F., Hassan, M.H.A., García Márquez, F.P., Hajiyev, A. (eds) Proceedings of the Sixteenth International Conference on Management Science and Engineering Management – Volume 1. ICMSEM 2022. Lecture Notes on Data Engineering and Communications Technologies, vol 144. Springer, Cham. https://doi.org/10.1007/978-3-031-10388-9_36

Dubey, A.K., Kumar, A., Ramirez, I.S., Marquez, F.P.G. (2022). A Review of Intelligent Systems for the Prediction of Wind Energy Using Machine Learning. In: Xu, J., Altiparmak, F., Hassan, M.H.A., García Márquez, F.P., Hajiyev, A. (eds) Proceedings of the Sixteenth International Conference on Management Science and Engineering Management – Volume 1. ICMSEM 2022. Lecture Notes on Data Engineering and Communications Technologies, vol 144. Springer, Cham. https://doi.org/10.1007/978-3-031-10388-9_35

Rathore, P.S., Chatterjee, J.M., Kumar, A. et al. Energy-efficient cluster head selection through relay approach for WSN. J Supercomput 77, 7649–7675 (2021). https://doi.org/10.1007/s11227-020-03593-4

A. Dubey, S. Narang, A. Srivastav, A. Kumar, V. Díaz, Woodhead Publishing, Science Direct,Artificial Intelligence for Renewable Energy Systems. Paperback ISBN: 9780323903967

A. Dubey, S. Narang, A. Srivastav, A. Kumar, V. Díaz, Woodhead Publishing, Science Direct, a Visualization Techniques for Climate Change with Machine Learning and Artificial Intelligence. ISBN: 9780323997140

Aishwarya, M., Rajesh Gopinath, L. R. Phanindra, K. Clarina, Rashmi R. Kagawad, and S. G. Ananya. "Reduction of Significant Aspects and Enhancement of Non-Significant Aspects for Hazardous Wastes in a Medical Electronics Manufacturing Firm." In Integrated Approaches Towards Solid Waste Management, pp. 203-213. Springer, Cham, 2021.

Kumar, T. Deva, TS Arun Samuel, and T. Ananth Kumar. "Transforming Green Cities with IoT: A Design Perspective." In Handbook of Green Engineering Technologies for Sustainable Smart Cities, pp. 17-35. CRC Press, 2021.

Kumar, K. Suresh, T. Ananth Kumar, S. Sundaresan, and V. Kishore Kumar. "Green IoT for Sustainable Growth and Energy Management in Smart Cities." In Handbook of green engineering technologies for sustainable smart cities, pp. 155-172. CRC Press, 2021.

13
E-Waste Challenges & Solutions

K. Dhivya* and G. Premalatha

IFET College of Engineering, Villupuram, India

Abstract

E-waste is a significant problem in the technological world. To put it simply, e-waste refers to any unwanted or obsolete electronic equipment. Obsolete technology is inevitably phased out, increasing amounts of WEEE garbage. Large appliances and electronics such as refrigerators, air conditioners, computers, and mobile phones are all broken. This trash has the potential to kill humans. Humans and ecosystems alike are negatively affected by improper waste management. Polluting the environment by incinerating, burying, or dumping electronic waste is unacceptable. Rare earth elements used in reusable electronics are programmable. The World Health Organization warns that prolonged exposure to e-waste can harm human health. Metals like lead and cadmium and chemicals like polychlorinated biphenyls, chromium, cadmium, chromium, and PCBs can cause both land and marine animals to perish. The effects of electronic waste are magnified in countries with limited resources. According to studies, e-waste negatively impacts both global labor and local communities. To help cut down on electrical and electronic waste, the European Union passed the WEEE Directive and the RoHS Directive (WEEE). As a result, many countries have enacted WEEE regulations to deal with these problems. Life-Cycle-Analysis (LCA) and tunable design are topics of conversation in the electrical and hardware industries in light of the E-Squander age report, Structure, and its impacts on the global climate, the environment, human health, and social cohesion. Regarding recycling electronic devices, India uses the same guidelines as the rest of the world.

Keywords: E-waste, green, management, nature

---

*Corresponding author: k.dhivya91@gmail.com

Abhishek Kumar, Pramod Singh Rathore, Ashutosh Kumar Dubey, Arun Lal Srivastav, T. Ananth Kumar and Vishal Dutt (eds.) Sustainable Management of Electronic Waste, (255–276) © 2024 Scrivener Publishing LLC

13.1 Introduction

E-waste, also known as the management of electronic trash, is one of the most frequently discussed problems that have arisen as a direct result of the advancement of technology. E-waste is produced whenever an electronic device is no longer being used for its intended purpose or has reached the end of its usable lifespan. Old devices are rendered useless as new ones are developed, as technological progress is ephemeral. Because of this, there is a reduction in waste material classified as WEEE (waste from electrical and electronic equipment) [1]. It is comprised of garbage produced when significant home appliances, such as air conditioners, refrigerators, computers, mobile phones, and so on, have reached the end of their useful lives. This type of garbage is made up of a convoluted assortment of components, some of which are highly hazardous to human health. Problems can arise for the environment and human health if they are not disposed of appropriately. The land, air, and water all suffer when pollutants from electronic waste are released into the environment. If they are kept under the appropriate levels of control, the rare earth materials that have accumulated in today's electronics can be recycled and used for other purposes.

According to the World Health Organization (WHO) [5], the harmful elements that seep from e-waste can cause a variety of health problems. Minerals like lead, cadmium, chromium, or polychlorinated biphenyls are among them (PCBs). This may also be life threating to animals both on land and in the water. When opposed to developed countries, underdeveloped nations are more at risk from e-waste since they are less equipped to treat these pollutants. The results of the studies indicate that workers and the community at large will suffer because of global e-waste. To protect both the present and future generations, it is crucial to correctly implement these E-waste recycling recommendations. The efficient use of policies such as collecting, processing, and recycling Electrical and Electronic Equipment (EEE) [2][3] at the point of expiration in order to enhance current production and consumption, increase resource effectiveness, and support the circular economy is needed [4].

To address the growing problem of discarded electrical and electronic equipment, the European Union enacted the WEEE Directive and the Restriction of Hazardous Substances Directive (WEEE). Many countries have recently enacted numerous regulations to guarantee the proper management of WEEE after realizing these issues exist. The 2016 E-Waste Management Rules have been published on the official website of India's Ministry of Environment, Forests, and Climate Change (MoEF&CC).

The E-Waste (Management and Handling) Rules of 2011 have been replaced by the newer regulations that have been issued. The 2018 E-waste (Management) revision rules, which were stringent standards and a part of the public authority's expanding responsibilities for natural administration, saw very few revisions throughout their revision process. For example, in India, which concentrates on the management of e-waste, Brilliant Urban Communities and the Swachh Bharat Mission are some of the initiatives the government of India has recently launched. This chapter provides information regarding the report of the E-Squander age, Structure, their implications for the climate, their effects on the environment, its detrimental effects on health and society, Life-Cycle-Analysis (LCA), and controllable designing from the points of view of the electrical and hardware industries. The management of electronic waste in India follows the same international standards as in other countries. A few specific examples will be used to illustrate the challenges associated with the management of e-waste. Following this, there will be a brief summary of the challenges that are connected with the management of e-waste in India and smart devices [7][8][9][10].

13.2 Related Works

Prior to the year 2004, there were not many written works on the subject of the management of e-waste and, according to the statistics, there have been approximately 6563 authors who have researched e-waste. The only thing that was lacking in quantity was the number of times their findings were published. The Chinese Academy of Sciences was the most successful academic institution and China was recognized as the nation that has made the most scientific discoveries (370 publications). The journal with the most publications was Waste Management, which had 225 total, and the journal with the most co-citations was Environment Science and Technology (9704 co-citations). As a consequence of this, it maintains control over e-waste. There are many different approaches to the management of electronic waste, each of which has been reported for a specific problem that researchers have uncovered. The following are some of the recommendations that they made for the effective management of WEEE. Within the context of the overall procedure for waste management, the terms "reverse logistics" (RL) and "closed-loop supply chain" (CLSC) refer to two essential components. E-waste, also known as wasted electrical and electronic equipment, is one of the critical end-of-life (EOL) commodities that the RL/CLSC takes into consideration. A great number of studies have been published in both the RL and CLSC sectors that independently focus on

WEEE. However, there was not a single review article that could be located that dealt with the problems that are specific to the device. In order to fill in this void, 157 publications released between May 1999 and 1999 were selected, placed into appropriate categories, and given a content analysis. The four steps that comprise the methodology are namely material collection, descriptive analysis, category selection, and material evaluation. Following the completion of the procedures, the conceptual framework, reverse distribution design and planning, decision-making and performance evaluation, and qualitative studies were selected as the four primary areas of study in the field of RL and CLSC of e-waste, and each of these was reviewed individually. Research voids in the existing body of knowledge were singled out to hone in on prospective topics for further investigation. This review is the first of its kind, so academics, researchers, and businesspeople may find it useful as a resource to understand better the activities and research of RL/CLSCs that are focused on WEEE. RL/CLSCs conducted this review [14][15][16].

There was a significant increase in the number of papers published in the field of e-waste research after the year 2004, with 2800 new papers appearing. 7.001% of the 6,573 writers who participated in the e-waste research project did not publish more than a single paper. In this particular area of research, China was the most productive nation and the Chinese Academy of Sciences was the most productive organisation (370 publications). Most publications were found in the field of Waste Management, while Environment Science and Technology had the most co-citations (225) (9704 co-citations). The key hot topics in the field of e-waste management were:

- Management and recycling of electronic waste in developing nations
- Health risk assessment following exposure to organic contaminants
- Degradation and recovery of waste metal components
- Effects of heavy metals on the health of children

There is currently no information that can be considered reliable regarding the importation or generation of electronic waste in India. The WEEE deposit is typically located in a variety of underdeveloped countries around the world. As a direct consequence of this, the economies of these nations have expanded at a more glacial pace. It is unclear what the actual numbers are for India's production of electronic waste.

To date, a number of studies have been done to determine the real numbers. In 2005, the Central Pollution Control Board (CPCB) carried out a survey. In 2005, the nation produced an estimated 1.347 lakh MT of electronic waste and by 2012, that number is predicted to rise to over 8.0 lakh MT. However, an inventory of e-waste from three products, computers, mobile phones, and televisions, was done in 2007 by the GTZ of India and the Manufacturers' Association for Information Technology (MAIT) of India. 3, 320, 979 metric tons (MT) of electronic garbage were created in India in 2007.

13.3 E-Waste: A Preamble

Any abandoned, outdated, or obsolete electrical or electronic equipment, such as lighting fixtures, home and office appliances, commercial computers, and consumer electronics, is referred to as electronic trash or "e-waste." It also includes toys, leisure, sports, and recreation gear that is powered by electricity. It holds a pseudonym of 'Digital Rubbish' [6]. It has resulted in numerous issues that will cause greater threat to lives of living beings and the environment. Pertaining to India, the E-waste growth rate is 10%.

13.4 Six Categories of E-Waste

There are six major categories of e-waste that are widely considered. They are:

Category	Equipment
Temperature Transfer Devices	Freezers, Refrigerators, Air conditioners, Heat pumps, etc.
Display Devices	LED and LCD displays, laptops, notebooks, tablets, televisions, and monitors
Types of Lamps	LED lamps, high-intensity discharge lamps, and fluorescent lamps, among others
Major Apparatus	Washing machines, photocopy equipment, printing equipment, electronic stoves, dryers, dishwashing equipment, and so on

Minor Apparatus	Video cameras, digital toys, electric kettles, electric shavers, radio sets, calculators, small medical devices, and so on
Telecommunications and IT Hardware	Smartphones, routers, laptops, printers, phones, GPS units, etc.

The operation, size, weight, and chemical composition of the aforementioned equipment varies. Also, life expectancy is another essential factor for consideration. On further discussion, the e-waste produced also varies in quality and quantity, which may cause different economic, environmental, and health issues if not disposed of well. Hence, the process of recycling is to be introduced in this case. The proper recycling technique is to be identified according to the EEE.

13.5 Composition of Materials Found in Equipment

The following materials are found to be available in the composition of the EEE. They are equally dangerous if they are not treated properly and might cause various issues to the environment. E-waste frequently contains valuable and potentially harmful materials. The type of electronic equipment, the model, the manufacturer, the year of manufacturing, and the age of the waste all have a significant impact on its composition. More valuable metals can be found in scrap from IT and telecom systems than in scrap from home appliances [18]. A cell phone, for instance, has over 40 different elements, including expensive metals like silver, gold, and palladium, as well as more common ones like copper and tin and specialized ones like lithium, cobalt, indium, and antimony. E-waste must go through a specific process to prevent losing rare minerals and precious components. It is possible to mine materials like palladium. On the other hand, it has been discovered that electronic waste includes flame retardants called PBDEs. These PBDEs are added to plastics and other components to make them more fire-resistant. There is a possibility of hazardous materials, including arsenic, cadmium, chromium, lead, and mercury, on the majority of electrical circuit boards [22], [23]. The printed circuit board of an electronic device typically contains pre-treated lead solder in an amount equal to 50 grams of tin-lead solder for every square meter of circuit board [19]. In air conditioners, freezers, and refrigeration systems that are more than ten years old, chlorofluorocarbons, which are harmful to the ozone layer, can be found (CFCs). End-of-life (EOL) cathode ray tubes (CRTs), which are

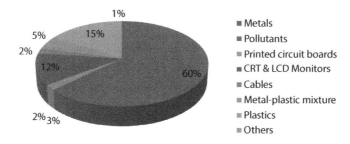

Figure 13.1 Composition of WEEE [11].

used in computers, televisions, and other electronic equipment, contain the common metals barium, lead, cadmium, copper, zinc, and other rare earth metals. EOL cathode ray tubes are referred to as "CRTs." CRTs also include other rare earth metals in their construction. By using leaded glass, for instance, users are shielded from the X-rays generated during the picture projection process.

CRT monitors typically have between 1.6 and 3.2 kg of lead inside of them. Because cathode ray tubes emit toxic radiation, their disposal in landfills is prohibited in the United States, other developed countries in the European Union, and Japan. E-waste management solutions are difficult to develop and implement [17] because the composition of the various parts is changing due to technological innovation. The typical components that makeup WEEE are illustrated in Figure 13.1. This is especially true for electronic components. Therefore, recycling and disposal practices for e-waste should develop in order to accommodate the ever-changing characteristics of this type of waste. The composition of electronic waste can be affected by several factors, including the availability of a market for reuse, the infrastructure of the recycling industry, trash segregation programs, and the implementation of regulations.

13.6 Recycling of WEEE

Promoting Recycling in the Formal E-Waste Sector

To encourage formal businesses to send their e-waste through government-approved recycling facilities, the Indian government has developed a point-based incentive program called E-waste Recycling Credits (ERCs). The E-Waste Rules currently provide classification and coding for e-waste products such as laptops, PCs, and cell phones. Different ERC reward levels must be associated to different categories. Depending on the type of e-waste given, organizations are obliged to secure the necessary ERCs, which can be used to lower energy utility bills. Unlicensed e-waste businesses will be strongly encouraged by such a program to formalize their activities and establish supply chains with authorized recycling facilities. The ERCs might be piloted for a duration of three to five years in order to assess effectiveness and make modifications.

Unregulated E-waste
The amount of garbage actually processed in the formal sector of dismantling and recycling, despite its expansion, is still quite small. The bulk of these formal facilities are operating at or below their permissible limitations since there aren't enough waste sources. Households and institutional customers are hesitant to recycle their waste in the formal economy because they are unaware of e-waste and the costs involved. In contrast to the formal sector, which has not yet made significant investments in reliable collection and processing facilities, the informal sector plays a crucial role in encouraging customers to return their waste by making household vehicles simple and offering financial incentives. This contrasts the formal sector, which has not yet made significant investments in reliable collection and processing facilities. The public sector is also a significant source of motivation for consumer waste recycling. The unregulated electronic waste industry has employed millions of people, a large percentage of whom are residents of economically depressed regions.

Extended Producer Responsibility (EPR)
Extended producer responsibility renders manufacturers financially liable for the costs of maintaining their products after their useful lives, which encourages them to design environmentally friendly products. By anticipating manufacturers to include the cost of reuse within the item cost, this plan seeks to relieve local states from the costs of managing a few necessary things. This strategic approach differs from item stewardship, which distributes responsibility throughout an item's chain of care, in that it [12] aims to relieve local governments of the costs associated with managing a few essential objects. EPR is predicated on the idea that manufacturers, or brand owners, have the greatest influence over how goods are designed and promoted as well as the greatest capacity and willingness to reduce harm and waste. [13] Examples of EPR include reuse, buyback, or reusing initiatives.

The Organization for Economic Co-operation and Development, also known as the OECD, is a global organization that works to develop better systems in order to improve people's lives. They want to craft policies that will facilitate communication, opportunities, flourishing, and prosperity for everyone. They work together to address a wide range of social, economic, and environmental problems and to establish general norms that are founded on facts. They provide a unique environment for conversation and the sharing of information, as well as opportunities for networking, the exchange of methods, and the promotion of open methods and international standard-setting. Their objectives range from enhancing their current financial performance to better positioning themselves in the market, as well as creating and addressing global tax avoidance hotspots.

Numerous governments have examined their choices for carrying out agreements in light of rising trash levels and have come to the conclusion that holding manufacturers of specific products responsible for their products' post-buyer lifecycles may be a good answer. Extended Producer Responsibility (EPR) is a contracting tactic that places a significant financial or actual responsibility on producers to handle or remove post-buyer goods. In theory, assigning such responsibilities might aid in waste prevention at the source, progress the ecosystem item plan, and aid in achieving the board's goals for cultural reuse and material reuse. The OECD is extending EPR to cover additional goods, product categories, and waste streams, including electrical and electronic equipment.

13.7 Procedures in the E-Waste Management

Methods need to be in place for the proper collection, sorting, and disposal of electronic waste. Regulations, guidelines, and legislation have been enacted in the vast majority of industrialized countries to encourage the proper collection, management, and recycling of e-waste, as well as the safe disposal of non-recyclable components [21]. By taking these steps, we hope to lessen the damage done to the environment when discarding materials that can't be recycled. The 3Rs program, Extended Producer Responsibility, product stewardship, and the Advance Recycling Fund are all examples of the many programs that fall under this umbrella (ARF). Two new regulations have been adopted in the European Union (EU) that mandate the free collection and recycling of electronic devices at the end of their useful lives (EOL) [22]. You can read the rules in detail here and here. One of the main goals of these regulations is to reduce the amount of trash that ends up in landfills.

In developing and transitional nations, however, e-waste is typically disposed of by burning it in the open air, filtering it with cyanide, and using simple smelters to recover valuable materials, primarily copper, gold, and

Table 13.1 Process comparison for treating E-waste.

Agricultural nations	Created nations
Informal setting	Formal room
Manual division	Quasi partition
Dismantling manually	Dismantling manually
To recover metals, e-waste scrap is heated, consumed, and filtered in small workshops.	The recovery of metals in smelter and treatment plants uses cutting-edge techniques

silver, with low returns. On the other hand, the vast majority of electronic waste is disposed of in open dumps, surface water bodies, and unlined, unmonitored landfills [20], all of which are highly detrimental to the health of humans and the environment. Quite a few countries have developed efficient methods for dealing with electronic waste. Process comparison for treating E-waste is shown in Table 13.1.

13.7.1 Disposal Systems

Landfilling is the main technique of e-waste disposal. Discarded electronic equipment frequently burns outdoors or ends up in landfills with other municipal waste, where it releases toxic and cancer-causing substances into the air. The informal disposal of e-waste in developing and transitioning nations, which employs often risky processes and practices, leads to low material recovery [24].

Developed countries handle e-waste differently from developing and transitory states. In developing and emerging nations, there are no regulations or educational initiatives about the fate of e-waste. Notably, less advanced disposal methods like open burning, dumping, and uncontrolled landfill sites are employed, seriously endangering the environment and exposing personnel to chemicals created by e-waste [26]. In developing nations like Brazil [25], China [26], and India, significant obstacles to the proper disposal of e-waste have been analysed. These obstacles incorporate the trouble of implementing/enforcing current guidelines and clean innovations, which are upheld by an absence of capacity building and awareness.

On the other hand, it has created expensive infrastructure and complex disposal plans that make handling garbage less dangerous. The availability of data, however, prevents a thorough analysis of the situation. Therefore, the restrictions on managing e-waste in developing and transitional nations, as well as the problems associated with doing so, are caused by variations in the financial and legal settings between typical developing and created country scenarios. The removal and disposal procedures used by agricultural nations for e-waste are typically ineffective and require oversight unlike industrialized countries where laws are stringent and powerful policing is in place.

13.7.2 E-Waste Management Challenges

In India, the unorganized sector is mostly responsible for recycling e-waste. Thousands of low-income households scavenge stuff from rubbish dumps to make a living. For example, middle-class urban households frequently sell their recyclable trash to unregulated, low-end purchasers known as "kabadiwalas," who then sort it and sell it as a raw material to artisanal or industrial processors. This is particularly true of recyclable goods including plastic, paper, textiles, and scrap metal. Similar regulations apply to how e-waste is handled in India. Many urban households collect, identify, repair, refurbish, and disassemble outmoded electrical and electronic equipment as part of the unregulated e-waste recycling sector.

These significant troubles are brought about by the considerable dependence on an unregulated area for recycling e-waste, as listed below:

- First, it is ineffective to try and impose financial penalties for disregarding or breaking the rules for handling and processing e-waste.
- Second, since laborers who reuse e-waste are saved money and come up short on legitimate preparation, the overall population realizes less about market costs and wellbeing and security costs.
- Third, regardless of the massive development in how much e-waste is created yearly, very little is being spent by huge scope ventures on recuperation.

13.7.3 Market Mismanagement for End-of-Life Products

Confidential administrators' ability to lay out e-waste management frameworks in a conventional industry is constrained by their failure to depend on solid hotspots for a lot of e-waste to produce economies of scale. For example, to carry out productive reusing advancements for e-waste the board in India might require significant forthright capital expenses, which private firms can't legitimize in that frame of mind of certainty in regard to the accessibility of adequate amounts of e-waste. Enlightening requirements additionally influence these business sectors adversely.

Most importantly, since the reusing of e-waste is a moderately youthful industry, there might be a market boundary because of a potential absence of information about reusing innovations that are reasonable. Second, the activity of business sectors is hampered by unfortunate information among shoppers, mainly caused by the absence of accurate information on e-waste management.

13.8 E-Waste (Management) Rules

On March 23, 2016, the Ministry of Environment, Forest, and Climate Change released the E-Waste Management Rules, 2016, which replaced the E-waste (Management & Handling) Rules from 2011.

13.8.1 2016 E-Waste (Management) Rules

1. In addition to the manufacturer, the dealer, and the refurbisher, the regulations now include the Producer Responsibility Organization (PRO) as a new stakeholder.
2. The restrictions now apply to the equipment listed in Schedule I, as well as the parts, consumables, spare parts, and EEE. In addition, the restrictions do not exclude the equipment.
3. The regulation of light bulbs that contain mercury, such as compact fluorescent lamps (CFLs).
4. In order to facilitate the collection of electronic waste by producers who are subject to extended producer responsibility, a strategy that is predicated on collection mechanisms has been put into place. Take-back systems, collection stations, and collection centers are all included in this category.
5. The possibility of constructing a PRO, an e-waste exchange, an online retailer, and a deposit return program has been made available in order to facilitate the correct routing of e-waste.
6. The provision that allowed EPRs to be authorized on a state-by-state basis has been replaced with a provision that allows the CPCB to authorize EPRs throughout India.

7. In order to comply with the requirements of an Extended Producer Responsibility Authorization, those in charge of the collection and distribution of e-waste must adhere to the goals outlined in Schedule III of the Rules. During the first two years of the rule's implementation, the phase-wise collection goal for e-waste is 30 percent of the waste generation quantity specified in the EPR Plan. This is followed by 40 percent of the waste generation quantity in the third and fourth years, 50 percent of the waste generation quantity in the fifth and sixth years, and 70 percent of the waste generation quantity in the seventh year and beyond.
8. Beginning today, manufacturers of electrical and electronic equipment have the legal right to collect a deposit from customers in exchange for the right to return the deposit to the customer, along with interest, after the product has reached the end of its useful life. This innovative new financial product has been given the name the Deposit Refund Scheme.
9. The regulations make it possible for agencies or organizations that have been granted permission to operate in accordance with the regulations to trade electronic waste. This waste is generated by electrical and electronic equipment that has outlived its usefulness and has become obsolete. The electronic waste exchange is a suitable replacement for this. It is either an independent electronic system that offers support for the market or an independent market tool that offers support for the market.
10. Starting today, any company that makes electrical or electronic equipment is required to collect any e-waste that is produced during the manufacturing process, direct it toward recycling or disposal, and obtain SPCB certification.
11. If the dealer has been given the responsibility of collecting electronic waste for the producer, they are obligated to provide each customer with a box for them to place their unwanted electronics in before sending it back to the producer.
12. The dealer, retailer, or online retailer is obligated to make a payment to the e-waste depositor in accordance with the take-back program or deposit reimbursement plan implemented by the producer.
13. The electronic waste that is created during the refurbishing process must be collected by the refurbisher, transported to

a licensed dismantler or recycler through the refurbisher's collection center, and the refurbisher must receive a single SPCB approval for the entire process.
14. The Rules provide additional specificity regarding the responsibilities of the State Government with regard to ensuring the safety, well-being, and development of the individuals who are engaged in the activities of dismantling and recycling.
15. The State Department of Industry or any other government agency designated in this regard by the State Government shall ensure that industrial space or sheds are designated or allocated for the dismantling and recycling of electronic waste in the current and future industrial park, estate, and industrial clusters. This requirement applies to both existing and future industrial parks, estates, and clusters.
16. The State Department of Labor or any other government agency designated in this regard by the State Government is responsible for ensuring the identification and registration of workers involved in dismantling and recycling, helping to organize these workers into groups to make it easier to establish dismantling facilities, carrying out industrial skill development activities for these workers, conducting annual monitoring, and ensuring the safety and health of these workers. In addition, the State Department of Labor or any other government agency designated in this regard by the State Government is responsible for ensuring the safety and health of citizens.
17. The State Government is obligated to develop a comprehensive plan for the efficient implementation of these regulations and to provide a yearly report to the Ministry of Environment, Forestry, and Climate Change.
18. The party that is sending the item must provide a document that contains the specifics of the transporter who will be carrying it. When transporting electronic waste, you are required to use this manifest method.
19. The concept of liability for environmental or third-party harm brought on by incorrect e-waste disposal has also been established, along with a provision for financial penalties for rule violations.

20. Orphan products must be sent to authorized recyclers or dismantlers, and the Municipal Committee, Council, and Corporation are in charge of gathering these products.

13.8.2 Central Issues for Rules 2016 of E-Waste Management

- Strong waste management facilities must be built in two years or less by local government bodies with a population of at least one lakh.
- Towns with fewer than a lakh residents would be given three years to build effective waste management facilities.
- In five years or less, abandoned dump sites will need to be closed down or biocured.
- After 16 years, the guidelines for effective waste management must be updated.
- The management of waste is the responsibility of the metropolitan authorities, who reserve the right to charge fines and client costs immediately for littering and non-isolation.
- After establishing a transition period of two to five years, sanctions would be applied in accordance with the nation's existing clergyman situation.

13.9 Report from the Central Pollution Control Board

- As of 2013–2014, metropolitan specialists only had 553 compost and vermin-fertilizer plants, 56 bio-methanation plants, and 22 rejected estimated fuel plants remaining after distributing 12 waste-to-energy plants.
- If current trends continue, city solid waste production should reach 16,500,000 tons by 2031, necessitating the need for 1240 hectares of land for disposal.
- Roughly 6,20,00,000 tons of waste are produced every year in India, of which just 11,90,00,000 are dealt with and around half, i.e., 31,00,00,000 tons are unloaded in landfill destinations.

13.10 An Integrated Waste Management Systems Web Application

The Service of Climate, Forest, and Environmental Change submitted the proposal in May 2016. The internet application was created to make dealing with rubbish more convenient. The program can be used to keep an eye on the development of hazardous waste and will help to ensure its management.

13.10.1 Intent Behind the Application

1. Citizens can quickly and simply request and acquire online permission for the import and result of certain waste classes for reuse, reusing, recovery, co-taking care of, and finally, sharing of primary resources.
2. It will make it easier to conduct commerce with electronic care and to provide different forms of affirmations and agreements to the company's visionaries and endeavors.
3. Online application provisions have been supplied, together with any necessary annexures or supporting documents.
4. The application will check how many endorsements or announcements are allowed under consent to spread out or agreement to work.
5. The competitors can review the web application they created and can also in-person examine the situation with their application.
6. By combining the data on Consent to Spread out, Consent to Work and Endorsement, and import/exchange assent, informational collection on squander making/taking care of current units, a work in progress, or movement is created.
7. Creates informational collections on various types of garbage produced, such as biomedical, hazardous, electronic, urban, and plastic waste.

13.11 E-Waste Management Rules 2016 Amendments

According to the GSR 261(E) notification, the E-Waste Management Rules 2016 have been revised as of March 22, 2018. By sending the nation's manufactured e-waste to approved recyclers and dismantlers, amended regulations seek to formally establish the e-waste reuse area. The assortment priority under the Extended Producer Responsibility (EPR) arrangements in the Rules have been altered and new manufacturers who have recently started their sales obligations have objectives to work toward. Some of the main components of the 2018 Amendment to the E-waste (Management) Rules are listed below:

1. The EPR's improved e-waste assortment targets will take effect on October 1st, 2017. The weight-based stage-by-stage assortment targets for e-waste in 2017–18 will be 10% of waste age, with an annual increase of 10% through 2023, in accordance with the EPR Plan.
2. In accordance with the EPR Plan, the goal is now to minimize waste by 70% by the end of 2023. The updated EPR targets will take into account the amount of e-waste collected by producers between October 1, 2016, and September 30, 2017, up until March 2018.
3. Separate e-waste assortment targets have been created for new manufacturers, such as those whose number of prolonged sales activities are not exactly the ordinary existence of their items. The normal existence of the products will be in accordance with the guidelines periodically provided by CPCB.
4. In order to participate in the activities that are permitted by the regulations, Producer Responsibility Organizations (PROs) must apply for membership with the Central Pollution Control Board (CPCB).
5. The public authority responsible for overseeing Reduction of Hazardous Substances (RoHS) arrangements will bear the expense of inspecting and testing. If the goods don't adhere to RoHS requirements, makers will pay for the test.

13.12 Management of Battery Waste Rules, 2022

As more research is presented about renewable energy, the utilization of power will build up forward movement. This would thusly prompt an age of colossal battery waste. This present circumstance requests a proactive strategy to manage what is happening in terms of flooding battery waste. Remembering this, the public authority introduced the Management of Battery Waste Rules, 2022.

Features:

- The 2001 new battery (handling and management) recommendations, which are currently insufficient to solve the escalating issue of battery waste
- The updated regulations take into account a variety of battery types, including modern, adaptable, and electric vehicle batteries
- The Expanded Maker Obligation concept underlies the introduction of the new regulations
- This standard requires battery manufacturers to create plans for collecting, removing, and reusing the battery
- The new regulations also forbid the removal of battery waste from landfills and burning.
- In accordance with this specification, the manufacturer may assign responsibility for trash collection and disposal to a third party
- It also lays out plans for establishing a focused online platform for the exchange of EPR declarations between e-waste producers and recyclers
- Used batteries served as the basis for the idea of a "base recovery level of the material"
- This strengthens the structure and enhances how fresh advancements are received
- The reused battery would essentially be present in the new batteries, which must be assembled with particular components. By doing this, the reliance on the novel, unrefined chemical for battery production will be reduced.
- A penalty known as the "Polluter Pay Standard" would be imposed on anyone who disregarded the concept of EPR.
- The resource obtained through natural remuneration will be used to repurpose used batteries that haven't been recovered or reused. The significance of the new guidelines lies in the fact that they are "outcome driven" in nature. Regarding the collection and reuse of garbage within a consistent time frame, these targets are quantifiable.
- Beginning in 2022–23, two-wheeler makers must begin collecting 70% of the batteries they sell on the market. This requirement will be in effect starting in 2026–27.

13.13 Conclusion

This chapter has provided a tremendous amount of information that emphasizes the challenges and issues of electronic waste management, which are the hazards in today's generation of technology. These challenges and issues are highlighted in this chapter. As was previously mentioned, the most significant problems stem from a general lack of knowledge, skills, and awareness among citizens and manufacturers' improper handling of materials. They could make the next generation more susceptible to a variety of natural catastrophes by the way they raise their children. As a result, more emphasis is being placed on the correct disposal of metals, which in turn helps to make the earth a better place by reducing the amount of radiation present.

References

1. M. Shahabuddin, M. Nur Uddin, J. I. Chowdhury, S. F. Ahmed, M. N. Uddin, M. Mofjur, M. A. Uddin, "A review of the recent development, challenges, and opportunities of electronic waste (ewaste)",International Journal of Environmental Science and Technology, 2022.
2. Gao, Ya, Long Ge, Shuzhen Shi, Yue Sun, Ming Liu, Bo Wang, Yi Shang, Jiarui Wu, and Jinhui Tian. "Global trends and future prospects of e-waste research: a bibliometric analysis." Environmental Science and Pollution Research 26 (2019): 17809-17820.
3. Doan, Linh Thi Truc, Yousef Amer, Sang-Heon Lee, Phan Nguyen Ky Phuc, and Luu Quoc Dat. "E-waste reverse supply chain: A review and future perspectives." Applied Sciences 9, no. 23 (2019): 5195.
4. Md Tasbirul Islam, Nazmul Huda, "Reverse logistics and closed-loop supply chain of Waste Electrical and Electronic Equipment (WEEE)/E-waste: A comprehensive literature review", Resources, Conservation and Recycling, 2018.
5. Y Krishnamoorthy, M Sakthivel, G Sarveswaran,"Emerging public health threat of e-waste management: global and Indian perspective", Rev Environ Health, 2018.
6. Kishore, Jugal. "E-waste management: as a challenge to public health in India." Indian journal of community medicine: official publication of Indian Association of Preventive & Social Medicine 35, no. 3 (2010): 382.
7. Ongondo, Francis O., Ian D. Williams, and Tom J. Cherrett. "How are WEEE doing? A global review of the management of electrical and electronic wastes." Waste management 31, no. 4 (2011): 714-730.

8. Li, Yan, Xijin Xu, Junxiao Liu, Kusheng Wu, Chengwu Gu, Guo Shao, Songjian Chen, Gangjian Chen, and Xia Huo. "The hazard of chromium exposure to neonates in Guiyu of China." Science of the total environment 403, no. 1-3 (2008): 99-104.
9. Neira, Joaquin, Leigh Favret, Mihoyo Fuji, R. Miller, S. Mahdavi, and V. Doctori Blass. "End-of-life management of cell phones in the United States." Master's Thesis, University of California, Santa Barbara, Santa Barbara (2006).
10. Liu, Qiang, Ke Qiu Li, Hui Zhao, Guang Li, and Fei Yue Fan. "The global challenge of electronic waste management." Environmental Science and Pollution Research 16 (2009): 248-249.
11. Robinson, Brett H. "E-waste: an assessment of global production and environmental impacts." Science of the total environment 408, no. 2 (2009): 183-191.
12. Chancerel, Perrine, Christina EM Meskers, Christian Hagelüken, and Vera Susanne Rotter. "Assessment of precious metal flows during preprocessing of waste electrical and electronic equipment." Journal of industrial ecology 13, no. 5 (2009): 791-810.
13. Lepawsky, Josh. "The changing geography of global trade in electronic discards: time to rethink the e-waste problem." The Geographical Journal 181, no. 2 (2015): 147-159.
14. Rolf, Widmer. "Global perspectives on e-waste." Environment Impact Assessment Review 25 (2005): 436-458.
15. BK Gullett, WP Linak, A Touati, SJ Wasson, S Gatica, CJ King, "Characterization of air emissions and residual ash from open burning of electronic wastes during simulated rudimentary recycling operations", Journal of Material Cycles and Waste Management 2007.
16. Bo, Bi, and Kayoko Yamamoto. "Characteristics of e-waste recycling systems in Japan and China." International Journal of Environmental and Ecological Engineering 4, no. 2 (2010): 89-95.
17. Kumar, A., Dubey, A.K., Ramírez, I.S., Muñoz del Río, A., Márquez, F.P.G. (2022). A Review and Analysis of Forecasting of Photovoltaic Power Generation Using Machine Learning. In: Xu, J., Altiparmak, F., Hassan, M.H.A., García Márquez, F.P., Hajiyev, A. (eds) Proceedings of the Sixteenth International Conference on Management Science and Engineering Management – Volume 1. ICMSEM 2022. Lecture Notes on Data Engineering and Communications Technologies, vol 144. Springer, Cham. https://doi.org/10.1007/978-3-031-10388-9_36
18. Dubey, A.K., Kumar, A., Ramirez, I.S., Marquez, F.P.G. (2022). A Review of Intelligent Systems for the Prediction of Wind Energy Using Machine Learning. In: Xu, J., Altiparmak, F., Hassan, M.H.A., García Márquez, F.P., Hajiyev, A. (eds) Proceedings of the Sixteenth International Conference on Management Science and Engineering Management – Volume 1. ICMSEM

2022. Lecture Notes on Data Engineering and Communications Technologies, vol 144. Springer, Cham. https://doi.org/10.1007/978-3-031-10388-9_35
19. Rathore, P.S., Chatterjee, J.M., Kumar, A. *et al.* Energy-efficient cluster head selection through relay approach for WSN. J Supercomput 77, 7649–7675 (2021). https://doi.org/10.1007/s11227-020-03593-4
20. A. Dubey, S. Narang, A. Srivastav, A. Kumar, V. Díaz, Woodhead Publishing, Science Direct,Artificial Intelligence for Renewable Energy Systems. Paperback ISBN: 9780323903967
21. A. Dubey, S. Narang, A. Srivastav, A. Kumar, V. Díaz, Woodhead Publishing, Science Direct, a Visualization Techniques for Climate Change with Machine Learning and Artificial Intelligence. ISBN: 9780323997140
22. Nnorom, Innocent C., and Oladele Osibanjo. "Overview of electronic waste (e-waste) management practices and legislations, and their poor applications in the developing countries." Resources, conservation and recycling 52, no. 6 (2008): 843-858.
23. Yu, Jinglei, Eric Williams, Meiting Ju, and Chaofeng Shao. "Managing e-waste in China: Policies, pilot projects and alternative approaches." Resources, Conservation and Recycling 54, no. 11 (2010): 991-999.
24. de Oliveira, Camila Reis, Andréa Moura Bernardes, and Annelise Engel Gerbase. "Collection and recycling of electronic scrap: A worldwide overview and comparison with the Brazilian situation." Waste management 32, no. 8 (2012): 1592-1610.Babu, Balakrishnan Ramesh, Anand Kuber Parande, and Chiya Ahmed Basha. "Electrical and electronic waste: a global environmental problem." Waste Management & Research 25, no. 4 (2007): 307-318.
25. He, Wenzhi, Guangming Li, Xingfa Ma, Hua Wang, Juwen Huang, Min Xu, and Chunjie Huang. "WEEE recovery strategies and the WEEE treatment status in China." *Journal of hazardous materials* 136, no. 3 (2006): 502-512.
26. Dwivedy, Maheshwar, and R. K. Mittal. "An investigation into e-waste flows in India." Journal of Cleaner Production 37 (2012): 229-242.

14

Global Challenges of E-Waste: Its Management and Future Scenarios

Pranay Das[1] and Swati Singh[2*]

[1]*Central Institute of Petrochemicals Engineering and Technology, Lucknow, India*
[2]*CSIR-National Botanical Research Institute, Lucknow, India*

Abstract

The increasing global population and rapid urbanization have led to a significant increase in the production of e-waste in recent times. In addition, its improper disposal and poor management is a global challenge that threatens human health and the environmental system. Electronic waste or e-waste is a popular, informal name for electronic products that are nearing the end of their usefulness. Every year, thousands of consumers replace old electronics, TVs, fridges, computers, printers, phones, and other portable devices, mainly because of innovation in technological advances or substantial utility. The expected life span of electronics is a maximum of three to five years due to increasing consumption rate, advancement, and population expansion. E-waste contains several harmful substances such as heavy metals, radioactive substances, halogenated compounds, and nano-sized dust particles, all of which require proper management during the collection, storage, recycling, and disposal phases. This chapter highlights modern developments on e-waste production and streams, current recycling technologies, human health, and environmental impacts of recycled materials and processes. The background, challenges, and problems of e-waste disposal and its proper management are also discussed along with the implications of the analysis.

Keywords: Environment, electronic waste or E-waste, global impact, management, research need

---

*Corresponding author: swati.sikarwar12@gmail.com

Abhishek Kumar, Pramod Singh Rathore, Ashutosh Kumar Dubey, Arun Lal Srivastav, T. Ananth Kumar and Vishal Dutt (eds.) Sustainable Management of Electronic Waste, (277–292) © 2024 Scrivener Publishing LLC

14.1 Introduction

Discarded electronics products, popularly known as electronic waste (e-waste), range from equipment used in computers, information and communication technology (ICT), home appliances, audio and video products, and all their appurtenant (Osibanjo & Nnorom, 2008). However, there is no standard or generally accepted definition of e-waste in the world. In several cases, e-waste includes relatively expensive and basically robust products used for data processing, broadcastings, or entertainment in reserved homes and companies (Basiye, 2008; Khayanje, 2008; Patil et al., 2021). E-waste is not very hazardous if it is retained in safe storage or systematically recycled by experts or transported in parts or as a whole in the formal sector (Ahirwar & Tripathi, 2021). However, e-waste can be deliberated hazardous if it is recycled by primitive techniques. E-waste contains many substances like plastics, heavy metals, glass, etc., which, if not handled in an eco-friendly manner, can be contaminated and very dangerous for human health and the environment. In the non-formal sector, recycling e-waste in primitive ways can harm the environment. The side effects of e-waste can be on the soil through the leaching of hazardous materials from landfills; in water due to river contamination, and other sources; or due to the emission of gases into the air and burning of e-waste (Patil & Ramakrishna, 2020). The e-waste recycling procedure, if not done appropriately, can cause harm to humans through inhalation of gases during recycling, skin contact of workers with hazardous substances, and contact during acid treatments used in the recovery process (Pinto, V. N. 2008). Various hazardous and toxic elements are found in e-waste, such as Cadmium (Cd) and Lead (Pb) in Printed Circuit Boards. Lead (Pb) is mainly found in all electronic product assemblies, cathode ray tubes, etc. Cadmium is found in screen CRTS, while switches and flat screen monitors may contain mercury (Kaya, M. 2016). Mercury is also present in CFLs, transmitters, and some other specialty products (Pahari & Dubey, 2019). In addition to cadmium in computer batteries, cadmium is also used for plating the metal parts of metal enclosures in sub-assemblies. 'Polychlorinated biphenyls' are found in transformers, capacitors, and as a brominated flame retardant on 'PBD/PBDE' for isolation on printed circuit boards, plastic casings, cables and polyvinyl chloride (PVC) cable sheathing, and in plastic fragments of integrated circuit technology (Chmielewski & Szurgott, 2015). No specific study has been done so far to know the impact of e-waste in the environmental system. However, some NGOs have found that recycling of e-waste in the non-formal sector is dangerous. These components use primitive, non-scientific and non-eco-friendly methods. As these units are working in the well-organized sector, there is no data available to confirm the fact that there are heretical laws prevailing for labor, environment protection, and industry (Needhidasan *et al.*, 2020). Greenpeace surveyed environmental pollution during the manufacturing of electronic products in several countries like China, the Philippines, Thailand, and Mexico (Iles, A. 2004). The study is based on assessment of pollution caused by the use of certain hazardous chemicals in the manufacturing of several electronic products in these countries (Terazono *et al.*, 2020). These industries are involved manufacturing units of semiconductor chip and printed circuit boards and various assembly units of computers, television, monitors, etc. (Chatterjee & Kumar, 2009).

This article presents a brief overview of e-waste and its global contribution. Initially, a comprehensive analysis is provided with a brief discussion on the state-of-the-art for its global production and management.

14.2 Worldwide Production of E-Waste

Since 2010, the amount of e-waste generated globally has gradually increased. As of 2019, about 53.6 million metric tons had been generated (Andeobu *et al.*, 2021). This was an increase of 44.4 million metric tons in a span of just five years, of which only 17.4 percent were recognized to be systematically stored and recycled (Stevens, A. 2020).

14.3 Global Availability of E-Waste: Additional Information

Technological advances and rising consumer demand have defined a period in which computer electronics have become a major part of the waste creek (Osibanjo & Nnorom, 2007). The global volume of electronic waste in 2014 mainly consisted of 12.8 million metric tons of minor apparatus, 11.8 million metric tons of huge equipment, and 7 million metric tons of temperature conversation equipment (Shrivastava & Shrivastava, 2017; You *et al.*, 2020; Kaur *et al.*, 2020). The volume of e-waste is estimated to grow to around 50 million metric tons by 2018 with a year-on-year growth rate of 4 to 5 percent (Ari, V. 2016). In the year 2014, the majority of e-waste worldwide was generated in Asia, a country predicted to experience the robust growth in the electronics business from 2014-2016, manufacturing

Table 14.1 Worldwide amount of electronic waste generated from 2010 to 2019 (Akram *et al.*, 2019).

S. no.	Year	Waste (Million Metric Tons)
1	2010	33.8
2	2011	35.8
3	2012	37.8
4	2013	39.8
5	2014	44.4
6	2015	46.4
7	2016	48.2
8	2017	50
9	2018	51.8
10	2019	53.6

16 million metric tons of e-waste (Forti *et al.*, 2020; Boubellouta & Kusch-Brandt, 2021). In contrast, both the US and Europe produced about 11.6 million metric tons (Ilankoon *et al.*, 2018). However, the highest amount of e-waste per capita was the highest in Europe, with 15.6 kg per capita. This was followed by 15.2 kg per capita in Oceania and 12.2 kg per capita in the Americas (Magalini, F. 2016). Table 14.1 shows the worldwide amount of electronic waste generated from 2010 to 2019.

14.4 Environmental Impact of E-Waste

When e-waste is exposed to heat, toxic substances are released into the air that cause extreme damage to the environment; it is one of the biggest environmental effects of e-waste (Robinson, B. H. 2009). Those toxic substances can then leach into groundwater, distressing both land and marine faunas (Issac & Kandasubramanian, 2021). Electronic waste can also dangerously contribute to air pollution. As stated, electronic waste encompasses toxic constituents that are hazardous to human wellbeing, such as mercury, polybrominated flame retardants, lead, barium, cadmium, and lithium (Ryneal, M. 2016). The adverse health effects of these pollutants on humans embrace impairment to the brain, heart, liver, kidneys, and skeletal structure (Pal *et al.*, 2015). E-waste is frequently measured and ignored widespread, as the long-term influence of this waste is still indistinct (Davis & Garb, 2015). Nevertheless, several e-waste recycling centers have been built in recent years in an effort to safeguard human beings and the planet. The disturbing effects of e-waste are discussed in greater depth as follows.

14.4.1 E-Waste Impacts on Soil Ecosytem

Majorly, e-waste can have a detrimental consequence on the soil system of an area. As e-waste breakdowns, it discharges deadly heavy metals (AMOLO, 2017; Ahmed *et al.*, 2019). These heavy metals include cadmium, lead, and arsenic. When these pollutants get into the soil ecosystem, they disturb the vegetation and foliage that is cultivated from this soil (Okereafor *et al.*, 2020). Therefore, these pollutants can enter the food supply and cause birth defects as well as many other health complications.

14.4.2 Impacts of E-Waste in the Water

Toxins also enter groundwater from improperly disposed e-waste by residents or businesses. This groundwater is what is at the bottom of many surface streams, ponds, and lakes. Many animals depend on these channels of water for nutrition. Thus, these contaminants can make wildlife sick and create inequities in the terrestrial ecosystem. E-waste can also affect humans who depend on this water ecosystem. Toxins such as barium, lead, mercury, and lithium are also known to be hazardous.

14.4.3 E-Waste Impacts on Air Pollution

When e-waste is disposed of in a landfill sites, it is generally burned by furnaces on that particular site (Ari, V. 2016). This procedure can discharge hydrocarbons into the air, which contaminate the air. In addition, these hydrocarbons may contribute to the greenhouse gas influence, which several researchers think is a major contributor to global warming or climate change (Rao, C. S. 2007; Grachev, V. A. 2019). In certain parts of the biosphere, concerned individuals examine through landfills to retrieve e-waste for money. Nevertheless, some of these individuals sting undesirable parts such as wires to remove the copper, which can also cause air pollution.

14.4.4 Final Considerations on the Effect of E-Waste on the Environment

The environmental impact of e-waste can be demoralizing (Rahman, M. S., & Alam, J. 2020). Even although the continuing possessions of e-waste are still unidentified, it undoubtedly has certain undesirable effects on soil, water, and air quality. All these are essential parts of a healthy planet.

14.5 Management of E-Waste

Global E-Waste Day, celebrated on 13 October, was first celebrated when WEEE Forum organizers requested the public reflect on some serious facts about how waste management and e-waste is affecting our lives (Murthy, V., & Ramakrishna, S. 2022; Islam, & Huda, 2018). This new annual day was designed to not only educate on awareness of e-waste, but to encourage consumers to store all their used electronics and recycle them properly. The evidence is actually an eye opener. Around 50 million tons of e-waste is expected to be generated globally every year (Premalatha *et al.*, 2014). Some of these will be personal products such as smartphones, tablets, and computers. Huge domestic machines and heating and cooling apparatuses will make up the rest. If anything, this digit will upsurge rapidly over the next few years. This is challenging because, as noted by the WEEE Forum, only 20% of worldwide e-waste is being recycled each year (Méndez-Fajardo *et al.*, 2020). This means, 40 million tons of e-waste are being dumped in landfills, incinerated ,or fading (Singh, N., Li, J., & Zeng, X. 2016). It is a global environmental and health apprehension. This is disappointing, as most of the world's population, up to 66%, is covered by e-waste legislation (Edge, J. L. 2011). A new study by the International Telecommunications Union is warning that by 2100, the amount of global waste will exceed 4 billion tons annually (Valone, T. F. 2021).

As noted by the ITU, e-waste is a rapidly growing aspect of the global waste management heave (Osibanjo, O., & Nnorom, I. C. 2007). The level of e-waste rose to 44 million tons in 2016 and will exceed 52.2 million tons by 2021 (Rajput, R., & Nigam, N. A. 2021). In fact, as electronic devices are produced and customers urgency to exchange their older devices grows, the ITU warns that landfill sites will rapidly convert to an "e-waste memorial park". Still, organizations are trying their best to raise awareness about the upcoming problem (Kubásek, M. (2013) and they are anticipated to inspire and assist advanced reprocessing charges. Low recovering degrees for e-waste generate several difficulties, including heavy loss of limited and hazardous raw resources inside e-waste, which can be rectified and recovered for production of new products; dangerous environmental hazards regarding e-waste in landfill sites where contaminated material within them can pollute soil and water ecosystems; determination of health hazards from exposure to those pollutants; challenges for emerging countries to tackle restricted consignments of e-waste (Kaya, M. 2016). A good start is to fully understand the magnitude of this challenge, but addressing this will require a tougher and more synchronized strength to raise purchaser awareness about e-waste. This involves properly disposing of undesirable devices by taking them to be recycled. That is the ITU memorandum – unless the equipment used is recycled properly, responsibly, and proficiently, the environmental dangers from e-waste will only increase (Amuzu, D. 2015). While only 20% of e-waste is unruffled and recycled appropriately, the ITU said many industries may not be aware that electronic equipment, including office devices, is used by a reprocessing firm rather than as scrap (Isimekhai, K. A. 2017). Another concern is that companies don't even realize is that e-waste still has a lot of worth, even if the device itself isn't working. This creates a serious waste of valuable resources only to be discarded in landfills.

Most prominently, there is no need to change attitudes about recycling. Numerous studies show that many individuals appreciate the value and well-being of recycling and embrace the concept. As a result, the issue is awake. A lot of customers and companies may not realize what can be recycled, which includes electronics. More education will be a key factor for businesses and consumers alike in the years to come (Kumar *et al.*, 2022). Our challenge nowadays is that most of the e-waste is not actually recycled. For example, in Europe, e-waste is delimited by the 'Waste Electrical and Electronic Equipment Ordinance'. That law was premeditated to expand the collection, handling, and recycling of electronics items at the end of their lives (Lazzarin, L., & Kusch, S. 2015). The European Union has some of the sternest rules in the world. Their goal is to collect and treat 65 percent of its e-waste properly every year. But Europe still has a long way to go. Presently, the reprocessing rate in the European Union is 35 percent. At least Europe is employed to the objective of a higher reprocessing percentage. In the U.S., there is no centralized rule leading e-waste. Part of the countries have passed their own legislation to report the issue (Schumacher, K. A., & Agbemabiese, L. 2019). Most states make computer electronics producers accountable for a creation through its end-of-life treatment. We recognize that all electronics comprise of a lot of treasured metals. This is one of the many reasons why we should be recycling them. Similarly, we know that as more and more of the ecosphere becomes advanced, our main problem will be e-waste growth and we know that a melodramatically growing volume of old headsets, processers, and additional electronics will end up in landfills. So, let's preserve the efforts of various environmentalists, administration, and healthcare supporters to increase consciousness around leftover organization subjects (Dubey et al., 2022). As an outcome, reprocessing is the best method to inhibit e-waste from damaging our environment and wellbeing. This is a message we need to keep echoing.

14.6 Concerns and Challenges

The Basel Action Network (BAN), which works to inhibit the globalization of toxic substances, said in a report that 50-80% of the e-waste composed by the US is disseminated to other countries like India, China, Pakistan, Taiwan, and Africa (Premalatha et al., 2014; Sasaki, S. 2018). This transpired because inexpensive employment is accessible for recycling in these nations. In the US, disseminating e-waste is lawful (Makhale, N. O. 2016). The reprocessing and removal of e-waste is extremely contaminating in various ecosystems in China, India, and Pakistan. Lately, China has debarred the import of e-waste (Mahmud et al., 2008).

Export of e-waste by the US is seen as a lack of responsibility on the part of the Federal Government, electronics industry, consumers, recyclers, and local governments towards viable and sustainable options for disposal of e-waste (Annamalai, J. 2015, Rathore et al., 2021). In India, recycling of e-waste is almost entirely left to the informal sector, which does not have adequate means to handle either increasing quantities or certain processes, leading to intolerable risk for human health and the environment (Vats, M. C., & Singh, S. K. 2014).

The export of e-waste by the US is perceived as a lack of accountability on the part of federal management, electronics manufacturers, customers, recyclers, and local governments toward viable and supportive alternatives to e-waste disposal (Namias, J. 2013). In India, the reprocessing of e-waste is almost entirely left to the informal sector, which does not have the cumulative capacity or satisfactory means to handle assured processes, posing unbearable risks to human health and the environment.

Most of the SPCB/PCC websites lack any public information. 15 out of 35 PCBs/PCCs do not have any info associated to e-waste on their websites or in their major public edge facts (Chaudhary, K., & Vrat, P. 2017). Even the basic e-waste guidelines and strategies have not been uploaded. In the absence of any information, especially on the details of e-waste, recyclers and collectors, citizens, and institutional generators of e-waste are at a complete loss on how to deal with their waste and do not know how to fulfill their responsibility. Consequently, the successful implementation of the E-Waste Management and Handling Rules, 2012 is a disaster (Jeyaraj, P. 2021; Dhas *et al.*, 2021).

14.6.1 Absence of Infrastructure

There is an enormous gap amongst existing reprocessing and assortment conveniences and the quantity of e-waste that is produced. There is no collection and removal system. Recycling facilities are lacking.

14.6.2 Hazardous Impact on Human Health

E-waste comprises of more than 1,000 deadly ingredients, which pollute soil and groundwater (Gupta, S. 2011). Direct contact can be a major source of several diseases. Recyclers can have several diseases related to liver, kidney, and neurological disorders. Due to an absence of consciousness, they are placing their own health and atmosphere at risk (Kumar, A., & Gupta, V. 2021).

14.6.3 Lack of Enticement Schemes

No guidelines exist for e-waste treatment for the unorganized sector. Similarly, no inducement has been reported to motivate complex people to adopt documented methods for the treatment of e-waste (Bandyopadhyay, A. 2008). The employment situation in the Conversational Recycling sector is lower than in the designated sector. There is almost no temptation for fabricators to handle e-waste.

14.6.4 Poor Awareness and Sensitization

After determining the end of convenient life, there is limited awareness on access to disposal. Also, only 2% of people think about the impact on the environment when disposing of their old electrical and electronic devices (Saphores *et al.*, 2012).

14.6.5 High-Cost of Setting Up Recycling Facility

Additionally, the study shows that progressive expertise in recycling developments is at a superior pecuniary drawback over basic procedures and are not economically viable in general (Gollakota *et al.*, 2020).

Formal reprocessing corporations in India are limited to the pre-processing of e-waste material, with the exception of a few, wherever e-waste, along with valuable metals, are sent to smelting refineries outside India (Deubzer, O. 2011; Schluep, M. 2014; Chaudhary, K., & Vrat, P. 2017). The stagnant sector in India has a long way to go in accepting state-of-the-art machinery for e-waste recycling because of problems in sourcing e-waste and partly because of high-end investments in superior and costly equipment. There is also a difficulty in making profitable technologies (Yamoah, S. B. 2014).

14.7 Future Scenarios of E-Waste

The E-Waste Problem Solving Initiative is a system of e-waste experts and a multi-participant platform to design policies that report all magnitudes of microelectronics in an increasingly digitized ecosphere. Independent processing pertains to an integrated and scientific methodology to generate a fundamental solution to address global e-waste challenges through the electronics lifecycle.

14.8 The Need for Scientific Acknowledgment and Research

Every year, e-waste destroys biodiversity around the world and serves as a major source of pollution. This is a very hot topic for scientists to observe and perceive the finer methods of recycling and disposal, as its effects in the developing world have been increasing from time to time to various substantial extents. The government should also encourage more research into hazardous waste management, sustainable environmental monitoring and development of standards, and regulation of hazardous waste disposal.

14.9 Conclusion

Altogether, electronics traded in the market will ultimately become outdated. The consequential e-waste, which is growing rapidly on a global scale, includes both materials that can be recovered as secondary raw materials as well as harmful resources and materials. Consequently, for a sustainable environment, e-waste needs to be fixed in the direction of appropriate end-of-life procedures, i.e., recycling and disposal. Currently, two-thirds of the e-waste generated in Europe are still being either landfilled or disseminated to developing countries, which repeatedly privatize suitable reprocessing and disposal services. Global solid waste management, which was previously a very challenging task, is becoming more and more problematic with the assault of e-waste. The problem is disturbing in several countries and almost all inventions are in the informal sector as there is currently no organized alternative available. Adequate legislation to treat e-waste and collection campaigns initiated by manufacturers is seen as the key to more efficient return and collection rates of end-of-life electronics and electrical equipment. Sequentially, the considerably increased volume of e-waste produces will make the recycling process more parsimoniously profitable. Improved revenue will encourage recycling companies to invest and improve novel and more effectual recycling technologies.

References

Ahirwar, R., & Tripathi, A. K. (2021). E-waste management: A review of recycling process, environmental and occupational health hazards, and potential solutions. Environmental Nanotechnology, Monitoring & Management, 15, 100409.

Ahmed, T., Liaqat, I., Murtaza, R., & Rasheed, A. (2019). Bioremediation Approaches for E-waste Management: A Step Toward Sustainable Environment. In Electronic Waste Pollution (pp. 267-290). Springer, Cham.A.Dubey, S.Narang, A.Srivastav, A.Kumar, V.Díaz , Woodhead Publishing, Science Direct,Artificial Intelligence for Renewable Energy Systems. Paperback ISBN: 9780323903967

A.Dubey, S.Narang, A.Srivastav, A.Kumar, V.Díaz , Woodhead Publishing, Science Direct, a Visualization Techniques for Climate Change with Machine Learning and Artificial Intelligence. ISBN: 9780323997140

Akram, R., Fahad, S., Hashmi, M. Z., Wahid, A., Adnan, M., Mubeen, M., ... & Nasim, W. (2019). Trends of electronic waste pollution and its impact on the global environment and ecosystem. Environmental Science and Pollution Research, 26(17), 16923-16938.

AMOLO, E. J. A. (2017). Assessing the sustainability of e-waste management in Kisumu city, Kenya (Doctoral dissertation, Maseno University).

Amuzu, D. (2015). Unravelling urban environmental (In) justice of e-waste processing activities in Agbogbloshie, Accra-Ghana (Master's thesis, NTNU).

Andeobu, L., Wibowo, S., & Grandhi, S. (2021). An assessment of E-waste generation and environmental management of selected countries in Africa, Europe and North America: A systematic review. Science of The Total Environment, 792, 148078.

Annamalai, J. (2015). Occupational health hazards related to informal recycling of E-waste in India: An overview. Indian journal of occupational and environmental medicine, 19(1), 61.

Ari, V. (2016). A review of technology of metal recovery from electronic waste. E-Waste in transition—From pollution to resource.

Ari, V. (2016). A review of technology of metal recovery from electronic waste. E-Waste in transition—From pollution to resource.

Bandyopadhyay, A. (2008). Indian initiatives on E-waste management—A critical review. Environmental Engineering Science, 25(10), 1507-1526.

Basiye, K. (2008). Extended Producer Responsibility for the Management of Waste from Mobile Phones.

Boubellouta, B., & Kusch-Brandt, S. (2021). Cross-country evidence on environmental Kuznets curve in waste electrical and electronic equipment for 174 countries. Sustainable Production and Consumption, 25, 136-151.

Chatterjee, S., & Kumar, K. (2009). Effective electronic waste management and recycling process involving formal and non-formal sectors. International Journal of Physical Sciences, 4(13), 893-905.

Chaudhary, K., & Vrat, P. (2017). Optimal location of precious metal extraction facility (PMEF) for E-waste recycling units in National Capital Region (NCR) of India. Opsearch, 54(3), 441-459.

Chaudhary, K., & Vrat, P. (2017). Overview and critical analysis of national law on electronic waste management. Envtl. Pol'y & L., 47, 181.

Chmielewski, A., & Szurgott, P. (2015). Modelling and simulation of repeated charging/discharging cycles for selected nickel-cadmium batteries. Journal of KONES, 22.

Davis, J. M., & Garb, Y. (2015). A model for partnering with the informal e-waste industry: Rationale, principles and a case study. Resources, Conservation and Recycling, 105, 73-83.

Deubzer, O. (2011). E-waste Management in Germany.

Dhas, D. B., Vetrivel, S. C., & Mohanasundari, M. (2021, November). E-waste management: An empirical study on retiring and usage of retiring gadgets. In AIP Conference Proceedings (Vol. 2387, No. 1, p. 130002). AIP Publishing LLC.

Dubey, A.K., Kumar, A., Ramirez, I.S., Marquez, F.P.G. (2022). A Review of Intelligent Systems for the Prediction of Wind Energy Using Machine Learning. In: Xu, J., Altiparmak, F., Hassan, M.H.A., García Márquez, F.P., Hajiyev, A. (eds) Proceedings of the Sixteenth International Conference on Management Science and Engineering Management – Volume 1. ICMSEM 2022. Lecture Notes on Data Engineering and Communications Technologies, vol 144. Springer, Cham. https://doi.org/10.1007/978-3-031-10388-9_35

Edge, J. L. (2011). Strategizing beyond the state: The global environmental movement and corporate actors (Doctoral dissertation).

Forti, V., Baldé, C. P., Kuehr, R., & Bel, G. (2020). The Global E-waste Monitor 2020. United Nations University (UNU), International Telecommunication Union (ITU) & International Solid Waste Association (ISWA), Bonn/Geneva/Rotterdam, 120.

Gollakota, A. R., Gautam, S., & Shu, C. M. (2020). Inconsistencies of e-waste management in developing nations–Facts and plausible solutions. Journal of Environmental Management, 261, 110234.

Grachev, V. A. (2019). Environmental effectiveness of energy technologies. GEOMATE Journal, 16(55), 228-237.

Gupta, S. (2011). E-waste management: teaching how to reduce, reuse and recycle for sustainable development-need of some educational strategies. Journal of Education and Practice, 2(3), 2222-1735.

Ilankoon, I. M. S. K., Ghorbani, Y., Chong, M. N., Herath, G., Moyo, T., & Petersen, J. (2018). E-waste in the international context–A review of trade flows, regulations, hazards, waste management strategies and technologies for value recovery. Waste Management, 82, 258-275.

Iles, A. (2004). Mapping environmental justice in technology flows: Computer waste impacts in Asia. Global Environmental Politics, 4(4), 76-107.

Isimekhai, K. A. (2017). Environmental risk assessment for an informal e-waste recycling site in Lagos State, Nigeria (Doctoral dissertation, Middlesex University).

Islam, M. T., & Huda, N. (2018). Reverse logistics and closed-loop supply chain of Waste Electrical and Electronic Equipment (WEEE)/E-waste: A comprehensive literature review. Resources, Conservation and Recycling, 137, 48-75.

Issac, M. N., & Kandasubramanian, B. (2021). Effect of microplastics in water and aquatic systems. Environmental Science and Pollution Research, 28(16), 19544-19562.

Jeyaraj, P. (2021). Management of E-waste in India–Challenges and recommendations. World Journal of Advanced Research and Reviews, 11(2), 193-218.

Kaur, P. J., Pant, K. K., Chauhan, G., & Nigam, K. D. P. (2020). Sustainable metal extraction from waste streams. John Wiley & Sons.

Kaya, M. (2016). Recovery of metals and nonmetals from electronic waste by physical and chemical recycling processes. Waste management, 57, 64-90.

Kaya, M. (2016). Recovery of metals and nonmetals from electronic waste by physical and chemical recycling processes. Waste management, 57, 64-90.

Khayanje, B. K. (2008). Extended Producer Responsibility for the Management of Waste from Mobile Phone. Unpublished Master of Science in Environmental Sciences, Policy & Management, IIIEE, Lund University.

Kubásek, M. (2013, October). Mapping of illegal dumps in the Czech Republic–Using a crowd-sourcing approach. In International Symposium on Environmental Software Systems (pp. 177-187). Springer, Berlin, Heidelberg.

Kumar, A., Dubey, A.K., Ramírez, I.S., Muñoz del Río, A., Márquez, F.P.G. (2022). A Review and Analysis of Forecasting of Photovoltaic Power Generation Using Machine Learning. In: Xu, J., Altiparmak, F., Hassan, M.H.A., García

Márquez, F.P., Hajiyev, A. (eds) Proceedings of the Sixteenth International Conference on Management Science and Engineering Management – Volume 1. ICMSEM 2022. Lecture Notes on Data Engineering and Communications Technologies, vol 144. Springer, Cham. https://doi.org/10.1007/978-3-031-10388-9_36

Kumar, A., & Gupta, V. (2021). E-waste: an emerging threat to "one health". In Environmental Management of Waste Electrical and Electronic Equipment (pp. 49-61). Elsevier.

Lazzarin, L., & Kusch, S. (2015). E-Waste management framework and the importance of producer responsibility and proactive hackerspaces. Proceedings EIIC, 188-192.

Magalini, F. (2016). Global challenges for e-waste management: the societal implications. Reviews on environmental health, 31(1), 137-140.

Mahmud, W., Ahmed, S., & Mahajan, S. (2008). Economic reforms, growth, and governance: The political economy aspects of Bangladesh's development surprise. Leadership and growth, 227.

Makhale, N. O. (2016). An investigation on attitudes and behaviours of residents towards recycling of municipal solid waste in Olievenhoutbosch, Centurion, in the Gauteng Province. University of Johannesburg (South Africa).

Méndez-Fajardo, S., Böni, H., Vanegas, P., & Sucozhañay, D. (2020). Improving sustainability of E-waste management through the systemic design of solutions: the cases of Colombia and Ecuador. In Handbook of Electronic Waste Management (pp. 443-478). Butterworth-Heinemann.

Murthy, V., & Ramakrishna, S. (2022). A Review on Global E-Waste Management: Urban Mining towards a Sustainable Future and Circular Economy. Sustainability, 14(2), 647.

Namias, J. (2013). The future of electronic waste recycling in the United States: obstacles and domestic solutions. Columbia University.

Needhidasan, S., Ramesh, B., & Prabu, S. J. R. (2020). Experimental study on use of E-waste plastics as coarse aggregate in concrete with manufactured sand. Materials Today: Proceedings, 22, 715-721.

Okereafor, U., Makhatha, M., Mekuto, L., Uche-Okereafor, N., Sebola, T., & Mavumengwana, V. (2020). Toxic metal implications on agricultural soils, plants, animals, aquatic life and human health. International journal of environmental research and public health, 17(7), 2204.

Osibanjo, O., & Nnorom, I. C. (2007). The challenge of electronic waste (e-waste) management in developing countries. Waste management & research, 25(6), 489-501.

Osibanjo, O., & Nnorom, I. C. (2007). The challenge of electronic waste (e-waste) management in developing countries. Waste management & research, 25(6), 489-501.

Osibanjo, O., & Nnorom, I. C. (2008). Material flows of mobile phones and accessories in Nigeria: Environmental implications and sound end-of-life management options. Environmental Impact Assessment Review, 28(2-3), 198-213.

Pahari, A. K., & Dubey, B. K. (2019). Waste from electrical and electronics equipment. In Plastics to Energy (pp. 443-468). William Andrew Publishing.

Pal, M., Sachdeva, M., Gupta, N., Mishra, P., Yadav, M., & Tiwari, A. (2015). Lead exposure in different organs of mammals and prevention by curcumin–nanocurcumin: a review. Biological trace element research, 168(2), 380-391.

Patil, R. A., & Ramakrishna, S. (2020). A comprehensive analysis of e-waste legislation worldwide. Environmental Science and Pollution Research, 27(13), 14412-14431.

Patil, S., Mathew, T., Sinha, N., Tewatia, C. R., Goswami, H., Thombre, M., & Vyas, M. (2021). The interface between law and technology.

Pinto, V. N. (2008). E-waste hazard: The impending challenge. Indian journal of occupational and environmental medicine, 12(2), 65.

Premalatha, M., Tabassum-Abbasi, Abbasi, T., & Abbasi, S. A. (2014). The generation, impact, and management of e-waste: State of the art. Critical Reviews in Environmental Science and Technology, 44(14), 1577-1678.

Premalatha, M., Tabassum-Abbasi, Abbasi, T., & Abbasi, S. A. (2014). The generation, impact, and management of e-waste: State of the art. Critical Reviews in Environmental Science and Technology, 44(14), 1577-1678.

Premalatha, M., Tabassum-Abbasi, Abbasi, T., & Abbasi, S. A. (2014). The generation, impact, and management of e-waste: State of the art. Critical Reviews in Environmental Science and Technology, 44(14), 1577-1678.

Rahman, M. S., & Alam, J. (2020). Solid Waste Management and Incineration Practice: A Study of Bangladesh. International Journal of Nonferrous Metallurgy, 9(01), 1.

Rajput, R., & Nigam, N. A. (2021). An overview of E-waste, its management practices and legislations in present Indian context. Journal of Applied and Natural Science, 13(1), 34-41.

Rathore, P.S., Chatterjee, J.M., Kumar, A. *et al.* Energy-efficient cluster head selection through relay approach for WSN. J Supercomput 77, 7649–7675 (2021). https://doi.org/10.1007/s11227-020-03593-4

Rao, C. S. (2007). Environmental pollution control engineering. New Age International.

Robinson, B. H. (2009). E-waste: an assessment of global production and environmental impacts. Science of the total environment, 408(2), 183-191.

Ryneal, M. (2016). Turning waste into gold: Accumulation by disposal and the political economy of e-waste urban mining (Doctoral dissertation, Northern Arizona University).

Saphores, J. D. M., Ogunseitan, O. A., & Shapiro, A. A. (2012). Willingness to engage in a pro-environmental behavior: An analysis of e-waste recycling based on a national survey of US households. Resources, conservation and recycling, 60, 49-63.

Sasaki, S. (2018, June). Issues with THAI WEEE (Waste Electrical and Electronic Equipment) recycling bill: comparison with other Asian countries. In The Proceeding of PIM 8th national and 1st international conference on challenges and opportunities of ASEAN: innovative, integrative and inclusive development (pp. L1-13).

Schluep, M. (2014). Case study e-waste management. E-Waste–Multi Country Case Study.

Schumacher, K. A., & Agbemabiese, L. (2019). Towards comprehensive e-waste legislation in the United States: design considerations based on quantitative and qualitative assessments. Resources, Conservation and Recycling, 149, 605-621.

Shrivastava, S., & Shrivastava, R. L. (2017). A systematic literature review on green manufacturing concepts in cement industries. International Journal of Quality & Reliability Management.

Singh, N., Li, J., & Zeng, X. (2016). Global responses for recycling waste CRTs in e-waste. Waste Management, 57, 187-197.

Stevens, A. The Global E-waste Monitor 2020.

Terazono, A., Murakami, S., Abe, N., Inanc, B., Moriguchi, Y., Sakai, S. I., ... & Williams, E. (2006). Current status and research on E-waste issues in Asia. Journal of Material Cycles and Waste Management, 8(1), 1-12.

Valone, T. F. (2021). Linear global temperature correlation to carbon dioxide level, sea level, and innovative solutions to a projected 6 C warming by 2100. Journal of Geoscience and Environment Protection, 9(03), 84.

Vats, M. C., & Singh, S. K. (2014). Status of e-waste in India-A review. transportation, 3(10).

Yamoah, S. B. (2014). E-waste in developing country context-Issues, challenges, practices, opportunities: Addressing the WEEE Challenge in Ghana. Master of Science), Aalborg University.

You, X., Snowdon, M. R., Misra, M., & Mohanty, A. K. (2018). Biobased poly (ethylene terephthalate)/poly (lactic acid) blends tailored with epoxide compatibilizers. ACS omega, 3(9), 11759-11769.

15
Impact of E-Waste on Reproduction

Adrija Roy, Sayantika Mukherjee*, Dipanwita Das and Amrita Saha

Department of Environmental Science, Amity Institute of Environmental Sciences, Amity University Kolkata, Kolkata, India

Abstract

E-waste refers to all waste produced when disposing of electronic devices, particularly consumer electronics. Personal computers and wireless devices that are quickly abandoned by consumers raise serious environmental issues. Various studies have shown associations between exposure to e-waste and physical health outcomes, including thyroid function, reproductive health, lung function, growth, and changes to cell functioning. Several researchers have investigated the outcomes of pregnancies in communities that have been exposed to e-waste. In most investigations, there have been consistent effects of exposure with increases in spontaneous abortions, stillbirths, premature deliveries, lower birthweights, and birth durations, despite diverse exposure settings and toxins that were examined. Increased exposure to polycyclic aromatic hydrocarbons 37 and persistent organic pollutants, such as polybrominated diphenyl ethers 36, polychlorinated biphenyls 37, and perfluoroalkyl, have been linked to adverse birth outcomes. The absence of a link between metal exposure and poor birth outcomes is the key exclusion to these effects. Numerous studies revealed that children's body-mass index, weight, and height were all much lower than adults'.

Keywords: E-waste, reproductive health, poor birth outcomes, still births, abortion

15.1 Introduction

The negative consequences for health and ecology of exposure to human consumption waste have long been recognized. Recently, a hazardous waste product originating from electrical and electronic equipment (EEE) was

---

*Corresponding author: sgmukherjee@kol.amity.edu

Abhishek Kumar, Pramod Singh Rathore, Ashutosh Kumar Dubey, Arun Lal Srivastav, T. Ananth Kumar and Vishal Dutt (eds.) Sustainable Management of Electronic Waste, (293–300) © 2024 Scrivener Publishing LLC

recognized. These products contain expensive components that have economic value when recycling, but EEE also includes potentially hazardous substances which may be directly released or generated during recycling. This creates what is called electronic waste. This formation and release of common hazardous by-products accumulates in the so-called "informal" electronic waste sector recycling where modern industrial processes cannot be used and occupational safety is often insufficient. Unprotected exposure to e-waste is discouraged by each person. Of the exposed groups, children are particularly vulnerable to many of the components in e-waste.

Electronic garbage, or e-waste, refers to outdated, end-of-life appliances that have been discarded by their original users, including computers, laptops, TVs, DVD players, refrigerators, freezers, mobile phones, and MP3 players. If not correctly treated, the numerous toxic components found in e-waste could have an adverse effect on both the environment and human health. To combat the ever-growing threat posed by e-waste to the environment and human health, numerous organisations, bodies, and governments of various nations have embraced and/or developed environmentally sound solutions and strategies for managing e-waste.

E-waste poses a serious risk to human health and the environment, particularly in Asian nations. Only 25% of electronic waste is recycled in official facilities with acceptable worker safety. Most e-waste pollutants released harmful chemicals into the air, water, and soil. It's crucial to manage e-waste effectively and use eco-remediation technology. Recycling electronic garbage has emerged as a serious environmental health concern. E-waste recycling practises that involve human disassembly, open burning to extract heavy metals, and open disposal of remaining fractions cause pernicious chemicals to escape into the environment. Lead (Pb), cadmium (Cd), chromium (Cr), manganese (Mn), nickel (Ni), mercury (Hg), arsenic (As), copper (Cu), zinc (Zn), aluminium (Al), and cobalt (Co) are heavy metals derived from electronic waste (e-waste) and they differ in their chemical composition, reaction properties, distribution, metabolism, excretion, and biological transmission [1].

Earlier research demonstrated that exposure to heavy metals has harmful effects on children's health, including lower birth weight, shorter anogenital distance, lower Apgar scores, lower current weight, decreased lung function, decreased levels of hepatitis B surface antibodies, increased prevalence of attention-deficit hyperactivity disorder, and increased DNA and chromosome damage. Acute and chronic consequences of heavy metal exposure on children's health might range from minor upper respiratory irritation to chronic respiratory, cardiovascular, neurological, urinary, and reproductive disease, as well as aggravated symptoms and disease of pre-existing conditions. Chronic respiratory, cardiovascular, neurological, urinary, and reproductive disease, as well as the aggravated symptoms and disease of pre-existing conditions [2]. For several reasons, the practice of rich countries exporting e-waste to developing nations has become widespread. The export of e-waste to less developed and regulated nations is encouraged by high labour costs and strict environmental restrictions for the disposal of hazardous waste in developed countries. E-waste importation for recycling can have some immediate financial advantages. However, many underdeveloped nations lack the tools, infrastructure, and resources required for ethical e-waste recycling and disposal. China, which is rapidly industrialising economically, as well as other Asian nations at a turning point in their industrial development, like India and Vietnam, which are home to sizable and expanding EEE manufacturing industries and assembling plants, present a different scenario: in addition to managing their own e-waste, these nations currently manage e-waste flow from the industrialised world. Both the "advanced" countries' e-waste stream and the business in developing and transitional nations are rising rapidly because of the pressing desire to enhance revenues as well as flaws in the law and its enforcement. Risky recycling and/or disposal techniques are the consequence, with negative effects on the health of humans, animals, and the environment.

15.2 Literature Review

Recycling of electronic waste is essential, but it must be done in a controlled and safe manner. When possible, electronic waste should be repaired and repurposed as a whole unit as opposed to being disassembled.

When refurbishment is not an option, electronic waste should be disassembled by skilled, well-protected, and paid workers at technologically cutting-edge recycling facilities in both developed and developing nations. The foundation of any e-waste legislation should be founded on a few key ideas. [3] First, there should be no distinction in acceptable risk criteria for hazardous, secondary e-waste compounds in developing and developed nations. However, given the physical distinctions and obvious vulnerabilities of children, the permissible criteria for children and adults should differ. Although effective, it is not practical to completely eradicate hazardous components from EEE. Even if there is a need for research, it is equally important to establish and conduct educational and awareness programs about the potential dangers of recycling e-waste. These initiatives are crucial in emerging nations. It is imperative to fight toward eliminating child labour and improving working conditions for all e-waste employees. Interventions ought to be tailored to the regional culture, geography, and constraints of the most vulnerable communities. All countries should adopt policies that would encourage the safe, regulated, and compensated recycling of e-waste [4]. The resulting toxic waste has had negative effects on the socioeconomic, ecological, and physical surroundings. There is severe pollution of the Earth's lithosphere, hydrosphere, biosphere, and atmosphere. People and other members of biodiversity are susceptible to lethal diseases like cancer, abnormalities of reproduction, neurological damage, endocrine disturbances, asthmatic bronchitis, and mental retardation. For humanity, current situations of unchecked e-waste generation have reached alarming levels. To protect nature and natural resources from further degradation, the public and private sectors, civil society, NGOs, industrialists, and the business community must all pay immediate attention to this. Studies done on birth outcomes related to informal e-waste recycling in Guiya, China proved that e-waste recycling should be a major thing in our life.

15.3 Discussion

It is essential to recycle e-waste at all costs because it seriously disintegrates childbirth (A. Chan *et al.*, 2009). Batteries, circuit boards, plastic housings, cathode ray tubes, activated glass, and lead capacitors are only a few examples of the parts of electrical and electronic equipment that are categorised as e-waste. Although e-waste is handled informally in many places, it has been stated that China, Ghana, Nigeria, India, Thailand, the Philippines, and Vietnam are among the countries where informal recycling occurs in considerable volumes. Due to its rapid growth and informal processing in developing nations, electronic garbage (or "e-waste") is a rising environmental and health concern on a global scale. E-waste consists of old, broken, or abandoned electrical or electronic equipment that may also include persistent organic pollutants (POPs) such as polybrominated diphenyl ethers (PBDEs), polychlorinated biphenyls (PCBs), polyvinyl chloride (PVC), and polycyclic aromatic hydrocarbons (PAHs), as well as heavy metals (lead [Pb], cadmium [Cd], mercury [Hg]). E-waste recycling activities have seriously contaminated local air, dust, soil, and water due to a lack of proper recycled technology and strict environmental regulations [5]. Guiyu, a small rural hamlet in South China, is now a location with significant dangerous material contamination because of e-waste commerce and recycling. Locals recycle e-waste using antiquated, unregulated techniques that cause significant environmental pollution. The negative impacts of improper recycling on the environment and human health are a subject of growing concern. Since the 1990s, nearly 6000 family workshops process roughly 1.6 million tonnes of e-waste annually using incredibly archaic techniques like incinerating e-waste in backyards or communities to recover valuable metals, creating significant pollution including heavy metals (such as Cd, Co, Cr, Cu, Ni, Pb, and Zn) and POPs (e.g., dioxins, furans, PBDEs, and PAHs). These pollutants were discovered in extraordinarily high concentrations, occasionally approaching the highest observed values in the literature. Guiyu, a small rural town in South China, has become a location with significant hazardous levels, sometimes exceeding the highest known levels detected in environmental, biota, and human samples because of e-waste trade and recycling. Previous research in Guiyu discovered a significant frequency of skin damage, headaches, vertigo, nausea, chronic gastritis, and gastroduodenal ulcers among those working in the recycling of e-waste. Blood lead, levels in children are more than 10 g/dL in over 70% of them. Pb, Cd, Cr, Ni, and PBDE levels were greater in the blood and placenta of newborns. Many harmful substances can enter the blood of the fetus through the placental barrier. Infants and developing fetuses are especially susceptible to toxicants. Asthma, diabetes, and cancer are a few of the adult disorders linked to periconceptional and prenatal environmental exposure. Many harmful substances can enter the blood of the fetus through the placental barrier. Infants and developing fetuses are especially susceptible to toxins. Asthma, diabetes, and cancer are a few of the adult disorders linked to preconceptionally and prenatal environmental exposure (Zhang et al., 2013). E-waste exposure's impacts on fetal human development are not well understood. To evaluate the effects of informal e-waste recycling on newborn development and health from maternal exposure, we will give epidemiologic data of birth outcomes and levels of CBPb in the typical e-waste recycling area of Guiyu, China. This is the first investigation on the effects of e-waste exposure during pregnancy [6]. The consequences of exposure to a particular component or element cannot be considered in isolation because pollutants are emitted as a mixture. The interactions between the chemical components of e-waste, however, require a more in-depth comprehension. Exposure to e-waste is a complicated process in which there are several pathways and sources of exposure, various durations of exposure, and potential inhibitory, synergistic, or additive effects of numerous chemical exposures. The exposures involved should be considered since e-waste exposure is a special variable in and of itself. Additional populations at risk for exposure include children, fetuses, pregnant women, the elderly, individuals with disabilities, workers in the unregulated e-waste recycling industry, and other vulnerable groups [7][8][9][10][11].

15.4 Conclusion

Due to additional exposure pathways (such as breastfeeding and placental exposures), high-risk behaviours (such as hand-to-mouth activities in early childhood and high risk-taking behaviours in adolescence), and their evolving physiology, children are a particularly susceptible group (e.g., high intakes of air, water, and food and low rates of toxin elimination). Additionally, if recycling is being done in their homes, the children of e-waste recycling employees could bring home contamination from their parents' skin and clothing as well as direct high-level exposure [12][13][14]. Additional populations at risk for exposure include children, fetuses, pregnant women, the elderly, individuals with disabilities, workers in the unregulated e-waste recycling industry, and other vulnerable groups. The management of e-waste in China has received a lot of attention over the past ten years. Import restrictions and pollution control measures have been changed in policy. Both laboratory and large-scale industrial operations have been used for research. An incentive mechanism has been set up to encourage ethical e-waste recycling. However, there is still a problem with how e-waste is handled in China and the problem hasn't been fixed. There hasn't been a lot of work done to investigate this matter further or to analyse ongoing issues from the standpoint of sustainable development. Hazardous e-wastes, including those from PCBs, CRT glass, and brominated flame retardant (BFR) plastics, have become problematic and probably flow to small or backyard recyclers without environmentally sound management. Traditional technologies are still being used to recover precious metals, such as the cyanide method of gold hydrometallurgy, from e-waste. While recovery rates of precious metals from e-waste are above 50%, they have encountered some challenges from environmental considerations. Worse, many critical metals contained in e-waste are lost because the recovery rates are less than 1%. On the other hand, this implies that there is an opportunity to develop the urban mine of critical metals from e-waste.

References

1. Huang, J., Nkrumah, P. N., Anim, D. O., & Mensah, E. (2014). E-waste disposal effects on the aquatic environment: Accra, Ghana. Reviews of environmental contamination and toxicology, 19-34.
2. Darshan, P. C., & Lakshmi, S. V. (2016). Global and Indian e-waste management–methods & effects. Retrieved from.
3. Hussain, M., & Mumtaz, S. (2014). E-waste: impacts, issues and management strategies. Reviews on environmental health, 29(1-2), 53-58.
4. Karim, S. M., Shariful Islam Sharif, Md Anik, and Anisur Rahman. "Negative Impact and Probable Management Policy of E-Waste in Bangladesh." arXiv preprint arXiv:1809.10021 (2018).
5. Li, W., & Achal, V. (2020). Environmental and health impacts due to e-waste disposal in China–A review. Science of the Total Environment, 737, 139745.
6. Igharo, Osaretin Godwin, John I. Anetor, Oladele Osibanjo, Humphrey Benedo Osadolor, Emmanuel C. Odazie, and Zedech Chukwuemelie Uche. "Endocrine disrupting metals lead to alteration in the gonadal hormone levels in Nigerian e-waste workers." Universa Medicina 37, no. 1 (2018): 65-74.
7. Khan, S. A. (2016). E-products, E-waste and the Basel Convention: regulatory challenges and impossibilities of international environmental law. Review of European, Comparative & International Environmental Law, 25(2), 248-260.
8. Kiddee, P., Pradhan, J. K., Mandal, S., Biswas, J. K., & Sarkar, B. (2020). An overview of treatment technologies of e-waste. Handbook of Electronic Waste Management, 1-18.
9. Dixit, S., & Singh, P. (2022). Investigating the disposal of E-Waste as in architectural engineering and construction industry. Materials Today: Proceedings, 56, 1891-1895.
10. Zeng, X., Xu, X., Boezen, H. M., & Huo, X. (2016). Children with health impairments by heavy metals in an e-waste recycling area. Chemosphere, 148, 408-415.
11. Bao, S., Pan, B., Wang, L., Cheng, Z., Liu, X., Zhou, Z., & Nie, X. (2020). Adverse effects in Daphnia magna exposed to e-waste leachate: Assessment based on life trait changes and responses of detoxification-related genes. Environmental Research, 188, 109821.
12. Orisakwe, O. E., Frazzoli, C., Ilo, C. E., & Oritsemuelebi, B. (2019). Public health burden of e-waste in Africa. Journal of Health and Pollution, 9(22).
13. Zhang, Y., Xu, X., Chen, A., Davuljigari, C. B., Zheng, X., Kim, S. S., ... & Huo, X. (2018). Maternal urinary cadmium levels during pregnancy associated with risk of sex-dependent birth outcomes from an e-waste pollution site in China. Reproductive toxicology, 75, 49-55.
14. Xu, Xijin, Hui Yang, Aimin Chen, Yulin Zhou, Kusheng Wu, Junxiao Liu, Yuling Zhang, and Xia Huo. "Birth outcomes related to informal e-waste recycling in Guiyu, China." Reproductive Toxicology 33, no. 1 (2012): 94-98.

16

Challenges in Scale-Up of Bio-Hydrometallurgical Treatment of Electronic Waste: From Laboratory-Based Research to Practical Industrial Applications

Ana Cecilia Chaine Escobar[1], Andrew S. Hursthouse[1] and Eric D. van Hullebusch[2*]

[1]*School of Computing, Engineering & Physical Sciences, University of the West of Scotland, Paisley, UK*
[2]*Université Paris Cité, Institut de Physique du Globe de Paris, CNRS, Paris, France*

Abstract

E-waste is rapidly growing in volume as a result of increasing consumption and reduced life expectancy of electronic products. Energy-intensive and cost-inefficient methods like pyrometallurgy and hydrometallurgy following chemical and/or physical separation steps are generally implemented for metal recovery from e-waste. As an environmentally sounder option, biohydrometallurgy was developed using microorganisms and their metabolic products to solubilize and recover metals from insoluble matrices like e-waste. Bioleaching applications to e-waste are still at the early stages of development and a techno-economic analysis indicates it is not yet viable. This review assesses peer-reviewed data gathered to establish the Technology Readiness Level of biohydrometallurgy for material recovery from e-waste at a pilot scale allowing the conclusion that bioleaching at the commercial scale currently faces diverse operational challenges that hamper its scale-up and industrial implementation.

Keywords: E-waste, recycling, metal extraction, biohydrometallurgy, bioleaching, technology readiness level

---

*Corresponding author: vanhullebusch@ipgp.fr

Abhishek Kumar, Pramod Singh Rathore, Ashutosh Kumar Dubey, Arun Lal Srivastav, T. Ananth Kumar and Vishal Dutt (eds.) Sustainable Management of Electronic Waste, (301–340) © 2024 Scrivener Publishing LLC

16.1 Introduction

E-waste includes any electrical and electronic equipment that is at the end of its life and has been disposed of with no intention of being reused in the future. This waste stream is growing yearly in volume at a rate of 16 to 28% [1] due to the increased consumption of electronic products and programmed obsolescence.

E-waste contains a wide variety of materials which are valuable and, in many cases, hazardous with the potential to cause severe environmental and health impacts. Pyrometallurgy is the main industrial technique used for e-waste recycling, particularly, Printed Circuit Boards (PBCs). However, it requires high energy inputs and produces harmful gaseous emissions (dioxins, furans) that require off-gas treatment, making pyrometallurgy a relatively costly process. Alternatively, hydrometallurgy is a simpler process using high volumes of concentrated acid/bases which produce effluents also resulting in a costly operation. As a more environmentally friendly option with low energy requirements, biohydrometallurgy has been developed using microorganisms and their metabolic products for the solubilization of metals from insoluble matrixes such as e-waste, presenting a promising technology.

This review focuses on establishing the Technology Readiness Level (TRL) of biohydrometallurgy for base and critical metals recovered from e-waste.

16.2 Methodology

To understand the current state of knowledge and to highlight the importance of further research, a qualitative action research method was adopted [2]. To clearly identify the processes included in the review, a conceptual system flow map was developed (Figure 16.1).

16.3 Results

16.3.1 Pre-Treatment

Dismantling is the first step in e-waste recycling where the separated parts can be reused as secondary resources or further processed [3]. Disassembling can be mechanical (automatic or semiautomatic) or manual.

Biohydrometallurgy for E-Waste: Scale-Up Challenges 303

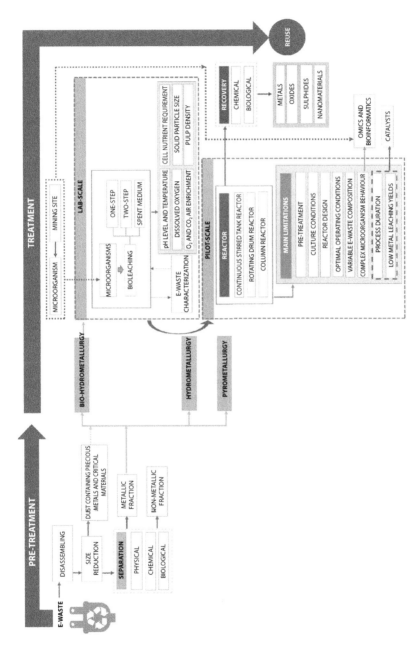

Figure 16.1 Conceptual system process flow map for biohydrometallurgical recovery of critical elements from E-waste.

Automatic dismantling has not yet emerged as an economically viable technique due to barriers like product variety, low volumes collected, and designs becoming increasingly intricate. Issues associated with the logistics of item return result in a variation of collected volumes, which also have a detrimental impact moving towards automatic disassembling. Consequently, manual dismantling is still more efficient [4], [5].

The disassembled parts are then physically processed for size reduction using different devices (mills, hammer crusher, cutters, and shredders). Particle-size is a very important factor in biohydrometallurgical e-waste recycling, as smaller particle sizes result in easier and more effective processing and extraction steps.

As part of the recycling process, it is important that different fractions are separated for efficient material recovery. Separation can be based on physical, chemical, or biological techniques.

16.3.1.1 Physical

Classification by physical properties can be divided into three techniques: particle shape/density, magnetic, and electrostatic separation [6] (Figure 16.2).

16.3.1.2 Chemical

Chemical techniques include pyrolysis (process carried out usually in fixed bed reactors at temperatures ranging between 200-700°C) for recycling the metal fractions, and systems using supercritical fluids and solvents for removing plastic layers.

Different studies report on the application of different physical or chemical techniques, including [4], [5], [7], [8], [9], and [10].

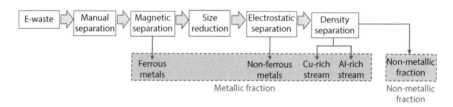

Figure 16.2 Physical separation techniques in separation process.

Table 16.1 Studies on bioflotation (Source: [12]).

Biological organism	Aim
Rhodococcus strains	Flotation of copper oxide minerals
	Collector for hematite
	Separation of apatite and quartz
Saccharomyces cerevisiae	Separation of quartz from hematite
Paenibacillus polymyxa	Selective separation of pyrite and galena from quartz and calcite
Bacillus subtilis	Separation of pyrite from galena

16.3.1.3 Biological: Bioflotation

Generally applied for mineral beneficiation, in bio-flotation biological organisms (bacteria, algae, fungi, yeast, and biomass) and their biomolecules interact with solid substrates by adhering to them and providing hydrophobicity. Table 16.1 presents the microorganisms studied for bio-flotation.

This technique could be a sounder and more cost-effective alternative to traditional chemical flotation [11].

Bioflotation is usually mediated by polymers either by biofilm formation or attachment of cells to the surface of the solid. Both processes are carried out by Extracellular Polymeric Substances (EPS) (Figure 16.3).

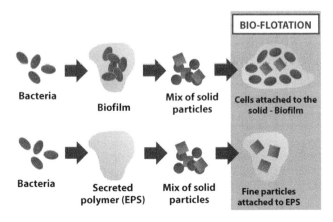

Figure 16.3 Bioflotation by biofilm and EPS formation (Adapted from [12]).

Other biomolecules have also been studied, particularly biosurfactants which in comparison to chemical surfactants, are less harmful, are effective at extreme conditions, and biodegradable [12].

Properties of bacteria related to the flotation process have been studied, however, the mechanisms behind bacterial adhesion and detachments are still to be determined [11]. Bioflotation is still in early development stages, thus, further research needs to be developed to scale the process up. The majority of studies have been carried out at lab scale and scaling up is difficult since physical characteristics such as shear force or solid fraction are different according to the working scale [13].

16.3.2 Treatment: Metal Extraction

Metallurgical methods for material extraction include pyrometallurgy, hydrometallurgy, and recently, biohydrometallurgy.

Table 16.2 Comparative characteristics of pyro- and hydrometallurgy as metal extraction methods.

	Pyrometallurgy	Hydrometallurgy
General Description	Involves a succession of incineration, smelting, and refining operations using high temperatures and high-temperature gas-phase reactions	Selective dissolution and precipitation of metals using strong acids and bases; often followed by separation and purification stages
Energy Requirements	High	Medium
Emissions	Greenhouse Gases (CO, SO_2 and NO_x), potentially dioxins and furans	Toxic chemicals
Cost Effectiveness	Low [14]	Higher than pyrometallurgy [9], [15]
Limitation	Only efficient for the recovery of Cu, leaving Al, Fe, and other elements (i.e., Gallium and Indium, [16]) concentrated in the slag	Mainly focused on precious metals and copper recovery [9]

Table 16.2 describes the main characteristics of the two traditional extraction methods.

16.3.2.1 Biohydrometallurgy: Bioleaching

Biohydrometallurgy consists of recovering metals from solid waste by biological means using weak organic acids or strong inorganic acids produced by microorganisms [17]. These processes are more environmentally friendly than hydrometallurgy and have lower energy requirements [18].

Microbial processes capable of dissolving or extracting metals from a solid phase are known as bioleaching. Microorganisms carry out this process by secreting organic or inorganic acids, oxidation/reduction of metal reactions, and generating complexing or chelating agents [19].

Table 16.3 presents the classification of microorganisms involved in bioleaching.

Table 16.3 Bioleaching mechanisms (Source [27]).

Bioleaching mechanism		Microorganism
Autotrophic Leaching	Fixing CO_2 and energy recovery by oxidation of Fe^{2+} or sulphur compounds; as a result of the oxidation, sulphides solubilize and the pH of the environment decreases, enhancing the solubilization of other metal compounds	Chemolithotrophic bacteria/Acidophilic bacteria Sulphur-oxidizing bacteria (*Acidithiobacillus thiooxidans*) Iron- and sulphur-oxidizing bacteria (*Acidithiobacillus ferrooxidans*) Iron oxidizing bacteria (*Leptospirillum ferrooxidans*)
Heterotrophic Bacteria	Heterotrophic microorganisms need a source of organic C for energy supply; the metabolic products (organic or inorganic acids, or cyanide) interact with the metals.	Heterotrophic microorganisms: *Chromobacterium violaceum* *Aspergillus* sp. *Penicillium* sp.
Heterotrophic Fungi		

16.3.3 E-Waste Bioleaching Process

16.3.3.1 Preparation

Generally, the desired components are manually separated and ground to obtain a homogeneous powder. To obtain a specific particle size, the mixture is sieved and, in some cases, dried before starting the bioleaching process [16].

Prior to bioleaching, samples are usually characterized using X-Ray Diffraction (XRD) [20], X-Ray Fluorescence (XRF) [21], [22], [23], Inductively Coupled Plasma Spectrometry (ICP) after sample acid digestion [24], and Laser Induced Breakdown Spectroscopy (LIBS) [25].

16.3.3.2 Microorganism Adaptation to E-Waste Phase

In e-waste, metals like Pb, Au, Pt, and Ag are partly responsible for the mixture's alkalinity increasing the pH of the medium. Furthermore, bioleaching can be inhibited by high metal concentrations, directly affecting the bacterial community activity and adding e-waste to the medium can end up having toxic effects (particularly bacteriostatic and bactericidal). Therefore, it is important that the microorganisms undergo an adaptation process before the bioleaching process to build resistance towards the innate toxicity of e-waste. This adaptation involves sub-culturing and gradual addition of waste, increasing its concentration. Gradually, bacteria start adapting and gaining resistance. It is also important that they are able to continue functioning under acidic conditions and in the presence of other potentially inhibitory components such as inorganic ions and organic plastics [17].

16.3.3.3 Bioleaching Methods

Bioleaching can be carried out through three basic strategies, as presented in Table 16.4.

The low and varied metal concentrations and toxicities in e-waste result in a decreased efficiency of the bioleaching process [26]. For some specific metals, strategies have been developed to solve this issue, nevertheless, applicability of comparable strategies for other critical materials should be further investigated.

16.3.3.4 Technical Considerations in the Bioleaching Process

The variables that affect bioleaching and its required technical considerations are present in Table 16.5.

16.3.4 E-Waste Bioleaching Up-Scaling

16.3.4.1 Bioleaching Process Scale-Up: Challenges

A variety of well-established methods can be used for metal extraction from e-waste, however, as the metal concentrations are low and toxic components are present, bioleaching application has been hindered [26]. For instance, when treating e-waste, the non-metallic fraction will not be dissolved and rare earth elements (REE) recovery is also challenging, as they are embedded in different materials [33].

Table 16.4 Bioleaching strategies (Source [8]).

Bioleaching method	Description
One-step	E-waste is added to the culture medium, therefore bacterial growth takes place in its presence. Microorganisms could suffer the inhibition of growth and metabolic activity.
Two-step	In the first step, the medium is inoculated and pre-cultured without the addition of e-waste. Leaching substrate is added in a second stage after microorganisms have reached the logarithmic growth phase. Better leaching times and higher recovery rates are obtained.
Spent-medium	After the colony has achieved the maximum cell density and metabolite production, cells are separated by filtration and centrifugation, leaving a cell-free metabolite solution. The metabolite is used for bioleaching, with the following advantages: - Microorganisms do not come in contact with the e-waste, enabling biomass recycling - Metabolite production is optimized - Higher waste concentrations and pulp densities may be used - No risk of microbial contamination of the waste

A number of studies have been carried out and their results are presented in Table 16.6.

16.3.4.2 Pre-Treatment

There are several physical and mechanical methods to treat e-waste to separate metallic from non-metallic parts. However, physical methods result in the loss of metals that are strongly adhered to plastics and generate fine materials that produce severe environmental impacts. Chemical substances covering the metal fractions (i.e., organic compounds and plastics)

Table 16.5 Physicochemical variables affecting the bioleaching process.

Variable	Impact	Reference
pH	Affects the bioleaching process differently depending on the microorganisms being used and the metal ion's chemistry	[27]
Temperature	Has a positive effect on bioleaching with higher temperatures, resulting in higher dissolution and metal recovery rates	[28]
Aeration and Dissolved Oxygen (DO) Concentration	The main final electron acceptor in the majority of bioleaching microorganisms is dissolved oxygen used for energy generation by oxidation of Fe^{2+} and/or S^0. Different microorganism may have a different affinity towards oxygen, which allows them to adapt to the oxygen availability. Once the stationary phase ends, the oxygen consumption ends. If anaerobic conditions are imposed during the logarithmic and stationary phases, it prevents cellular cyanide synthesis. An increase in aeration rate results in a reduction of the metal dissolution due to excessive turbulence and to cellular attrition which has a negative effect on bacterial growth.	[27] [29] [30]

(Continued)

Table 16.5 Physicochemical variables affecting the bioleaching process. (*Continued*)

Variable	Impact	Reference
O_2 and CO_2 Air Enrichment	Most leaching activity is carried out by aerobic autotrophs, therefore, O_2 and CO_2 are essential nutrients for their growth and survival. During the bioleaching process, O_2 may be used directly as an oxidiser of the energy substrate or as an electron acceptor. The O_2 molecule is reduced to H_2O or it may play a secondary role in the re-oxidation of Fe^{2+}/Fe^{3+}. It is well known that forced aeration is the main source of O_2 and CO_2 in bioleaching systems, both for sustaining microorganism activity, as well as for posterior metal recovery. Currently, it is a biotechnological and operational challenge to identify the most efficient and reliable method for the monitoring of O_2 and CO_2 availability in industrial bioleaching processes, as well as its effects on microorganism consortia. Different microorganisms have different physiologic requirements, therefore the level of O_2 and CO_2 depend on the microorganisms used, the matter that is being oxidized, and the leaching material.	[30] [31]

(*Continued*)

Table 16.5 Physicochemical variables affecting the bioleaching process. (*Continued*)

Variable	Impact	Reference
Cell Nutrient Requirements	<u>Sulphur</u>: In e-waste metal bioleaching processes using S^{2-}/S^{o} as substrates, critical acid-mediated reactions are carried out by sulphur-oxidizing acidophiles. However, there is little information about the proper use of sulphur sources, since the levels of sulphur consumption, its mobilization, and the role that soluble intermediates can have on mobilization and oxidation, are not yet well known. On the other hand, the efficiency of bioleaching processes using S^{o} as a substrate is known to be limited by its rate of oxidation. This is linked to several factors including the type of S^{o} used, its concentration, and the size of the particles. <u>Iron</u>: In metal bioleaching, bacteria maintain high redox potential by continuously oxidizing Fe^{2+} to Fe^{3+}. Since Fe^{3+} is a strong oxidizing agent, it is capable of attacking and oxidizing metals, converting them to their soluble form, and reducing itself to Fe^{2+}. As Fe^{2+} is a source of energy used by bacteria to grow in bioleaching systems, the increase in Fe^{2+} generation sustains a continuous metal bioleaching process. Therefore, the higher the initial concentrations of Fe^{2+} are, the faster the metal leaching (assuming that bioleaching is carried out by indirect mechanisms without contact).	[27] [30] [32]

(*Continued*)

Table 16.5 Physicochemical variables affecting the bioleaching process. (*Continued*)

Variable	Impact	Reference
Solid Particle Size and Pulp Density	Different particle sizes of e-waste may have different impacts on the bioleaching yields depending on the medium and the target metals. For instance, descent yields can be related to the particle-particle collision effect, which disrupts the cell, reduces air diffusion and forms a thick slurry, increasing the apparent viscosity of the medium. On the other hand, higher yields may be observed due to galvanic interaction between the metals present in the pulp. Additionally, they could be related to the preference of bacterial cells to attach to a specific site of a specific metal found in large particles in low concentrations.	[27]

do not allow for complete bioleaching since microorganisms are either not able to reach the metals or are affected by their toxicity. Therefore, they must be removed prior to bioleaching.

It has been reported that e-waste is naturally alkaline, resulting in an increase of pH in the medium they are treated in, however, the electronic components responsible for its alkalinity have yet to be identified. Various studies ([28], [34], [35], [36]) address the potential to increase the solid concentration during bioleaching.

16.3.4.2.1 Jarosite Precipitation

In the bioleaching process, Fe^{3+} is oxidized to Fe^{2+}. However, Fe^{3+} concentration in the medium increases when a considerable amount of bioleaching time has passed. Further, accumulated Fe^{3+} precipitates resulting in the formation of jarosite [32]. Sulphates are consumed in this process (Equation 16.1).

$$A^+ + 3Fe^{3+} + 2SO_4^{2-} + 6H_2O \rightarrow AFe_3(SO_4)_2(OH)_6 + 6H^+ \quad (16.1)$$

Jarosite is a basic hydrous sulphate of potassium and iron with a chemical formula of $AFe_3(SO_4)_2(OH)_6$, where A can be K,

Table 16.6 Work summary.

Source	Work summary						
[41]	Matrix	Mobile phone PCBs					
	Microorganism	*Acinetobacter* sp. Cr B2 (Inoculum size 9% v/v)					
	Process	**Particle Size:** 3.79 mm **Reactor:** 1.5 L; Pulsed plate column; Operated on sequential batch mode with repeated leaching in fresh bioleaching media every 24 h					
	Yield	Cu: 63.5%					
	Operation parameters						
	pH	T(°C)	Pulp density (g/L)	Aeration (mL/min)	Frequency (s^{-1})	Amplitude (cm)	Time (days)
	12	25	6.7 (10 g/stage)	1500	0.2	6.5	6

(*Continued*)

Biohydrometallurgy for E-Waste: Scale-Up Challenges 315

Table 16.6 Work summary. (*Continued*)

Source	Work summary					
[42]	Matrix	PCB scrap				
	Microorganism	Moderate thermophiles and mesophiles (Inoculum: Acid mine drainage)				
	Process	**Particle Size:** 74 μm and 1 mm collected and homogenized (agitator) **Reactor:** 3L (CSTR)				
	Yield	Zn: 85.23% Cu: 76.59% Al: 70.16%				
	Operation parameters					
	pH	T(°C)	Pulp density (g/L)	Aeration (mL/min)	Agitation (r/min)	Time (days)
	2	45±0.2	10 / 30 / 50 / 80	400	400	7
	Matrix	PCB scrap				

(*Continued*)

Table 16.6 Work summary. *(Continued)*

Source	Work summary		
[43]	Microorganism	*Sulfobacillus thermosulfidooxidans* with *Sulfobacillus acidophilus*		
	Process	**Particle Size:** >8+3mm **Reactor:** Short isothermal columns of 58 cm (height) and 13 cm (internal diameter). Operated on closed loop. Clean air blown into the base of each column at a rate of 1.7-1.8 nm³/m²/h.		
	Yield	Zn: 74% Al: 68% Cu: 85% Ni: 78%		
	Operation parameters			
	pH	T(°C)	Pulp density (g/L)	Time (days)
	2	45±2	10	165

(Continued)

Table 16.6 Work summary. (Continued)

Source	Work summary		
[30]	Matrix	PCB scrap	
	Microorganism	*Sulfobacillus thermosulfidooxidans* strain RDB	
	Process	**Particle Size:** 105 μm (Washed with saturated NaCl solution and rinsed with double-distilled H_2O; Dried to constant weight) 9K growth medium **Reactor:** 2L baffled glass reactor containing 4 titanium impellers pitched downward (45°)	
	Yield	Optimal yield* (Medium supplemented with S°, 25% O_2 + 0.03% CO_2): Al: 91% Cu: 95% Zn: 96% Ni: 94%	
	Operation parameters		
	pH	T(°C)	Pulp density (g/L)	Aeration (mL/min)	Agitation (r/min)	Time (days)
	2	45	10	*0.5	*280	* 15 days

(Continued)

Table 16.6 Work summary. *(Continued)*

Source	Work summary		
[39]	Matrix	PCB scrap	
	Microorganism	Acidophilic moderately thermophilic bacteria (*Sulfobacillus thermosulfidooxidans* and *Thermoplasma acidophilum*)	
	Process	**Particle Size:** 100-120 μm (Washed with saturated NaCl solution, rinsed and dried to constant weight) **Reactor:** Columns of 58 cm of height and 13 cm of internal diameter (approximately 8L capacity)	
	Yield	Zn: 80% Al: 64% Cu: 86% Ni: 74%	
	Operation parameters		
	pH	T(°C)	Pulp density (g/L)	Aeration (L/h)	Time (days)
	2	45±0.2	10	150-200	280

(Continued)

Table 16.6 Work summary. (Continued)

Source	Work summary					
[28]	Matrix	PCB scrap				
	Microorganism	*Sulfobacillus thermosulfidooxidans*				
	Process	**Particle Size:** 20 mm; Subjected to pre-weakening process (jaw crusher); Material over 20mm was chemically pre-treated using an aqueous solution of diethylene glycol and KOH **Reactor:** Rotating drum reactor; Perforated drum with 10 mm openings fitted with a gas sparger to provide aeration (air)				
	Yield	Cu: 85%				
	Operation parameters					
	pH	T(°C)	Pulp density (g/L)	Fe^{2+} conc. (g/L)	Rotation speed (min^{-1})	Time (days)
	1.75	50±1	25	5.0	80	7

(*Continued*)

Table 16.6 Work summary. (*Continued*)

Source	Work summary		
[38]	Matrix	Medium grade spent PCB	
	Microorganism	*Leptospirillum ferriphilum*	
	Process	Particle size: 750 μm Reactor: Double stage bioreactor: First stage: Bubble column (200 mL), Second stage: Continuous stirred tank reactor (2.25 L)	
	Yield	Cu: 96% Zn: 85% Co: 93% Pb: 12% Fe: 59% Ni: 73% Al: 54%	
	Operation parameters (second stage)		

pH	T(°C)	Pulp density (g/L)	Air flow rate (L/h)	Stirring speed (rpm)	Time (days)
1.1	36	1-1.8	60 (1% CO_2 enrich)	600	2

(*Continued*)

Table 16.6 Work summary. (*Continued*)

Source	Work summary		
[34]	Matrix	Waste PCB	
	Microorganism	*Leptospirillum ferriphilum*	
	Process	**Size of PCB:** 16cm x 19cm **Reactor:** Indigenously designed reactor of Length: 24 cm, Width: 17.5 cm, Height: 24.5 cm (10,29 L)	
	Yield	Cu: 99%	
	Operation parameters		
	pH	T(°C)	Pulp density (g/L)	Rotation speed (min⁻¹)	Time (days)
	1.4-2.43	30±2	125	180	8

(*Continued*)

Table 16.6 Work summary. (Continued)

Source	Work summary					
[44]	Matrix	Electronic scrap shredding dust				
	Microorganism	*Thiobacillus thiooxidans* and *Thiobacillus ferrooxidans*				
	Process	**Reactor:** Aerated and stirred 1-5 L flasks				
	Yield	Al, Cu, Ni, Zn: >90%				
	Operation parameters					
	pH	T(°C)	Pulp density (g/L)	Rotation speed (min^{-1})	Time (days)	
	2.5-2.7	30	5 & 10	150	17	
[44]	Matrix	Electronic scrap shredding dust				
	Microorganism	*Aspergillus niger* and *Penicillium simplicissimum*				
	Process	**Reactor:** Aerated and stirred 1-5 L flasks				
	Yield	Cu, Sn: > 65% Al, Ni, Pb, Zn: > 95%				

(*Continued*)

Table 16.6 Work summary. (Continued)

Source	Work summary				
[46]	**Operation parameters**				
	pH	T(°C)	Pulp density (g/L)	Rotation speed (min^{-1})	Time (days)
	Refer to report	30	10	150	21
	Matrix	E-waste (IT equipment, consumer appliances and small electronic equipment) shredding dust			
	Microorganism	*Acidithiobacillus thiooxidans* (First step) & *Pseudomonas putida* (Second step)			
	Process	Reactor: Brunswich Innova 2000: USA			
	Yield	First Step: Ce, Eu, Ne: > 99%; La, Y: 80%		Second Step: Au: 48%	
	Operation parameters				
	pH	T(°C)	Pulp density (g/L)	Rotation speed (min^{-1})	Time
	Refer to report	30	First step: 0.5; 1; 2 Second step: 1	150	First step: 8 days Second step: 3 hours

NH$_4$, Na, Ag, or Pb. This sulphate mineral is formed in ore deposits by the oxidation of iron sulphides ([27], [37]).

Jarosite precipitation is enhanced by high concentration of Fe^{2+} ions, pH, and temperature, therefore it can be prevented by maintaining low pH of the medium and ambient temperatures. If temperatures are high, jarosite still precipitates even under acidic conditions (pH<2) [27].

This precipitates directly and negatively affects the effectiveness of metal removal as jarosite precipitates on the surface of the solid waste and consumes ferric ions from the solution, which reduces the oxidation-reduction potential, both promoting metal passivation and hampering their solubilization [37].

In experimental copper bioleaching from waste PCBs studies by [38], it was observed that the leaching rate of copper decreased with the increase of incubation time. This might be related to jarosite precipitation, producing a decrease of Fe^{3+} concentration in solution and solids passivation. For bioleaching, scale-up is important to enhance Fe^{3+} utilization by bacteria which would yield higher metal removal efficiencies and lower operating costs. Thus, jarosite precipitation has to be reduced to a minimum by the addition of Fe^{2+}, maintaining low levels of pH and operating at ambient temperatures. pH values below 1.8 are suggested to effectively control precipitation rates.

16.3.4.2.2 Oxide Precipitation

Several studies report low Pb and Sn dissolution yields which could be explained by the oxide precipitation phenomena of PbSO$_4$ and SnO [38]. Further actions need to be designed in order to prevent the precipitation or recovery of the metals from these oxides.

16.3.4.3 Culture Conditions

The focus of research has been on identifying and providing details of the different mechanisms that are part of the bioleaching process.

16.3.4.3.1 Microorganisms

Different microorganisms are capable of leaching different metals/materials under different conditions. Therefore, it is necessary to find the most appropriate microorganisms for the waste that is being treated in order to design a suitable bioleaching process.

Considering the diverse composition of e-waste, no studies were found focusing on assessing the best microorganisms under the same operating conditions. For instance, [39] proved that a mixed culture of *Acidophilic chemolithotrophic* and heterotrophic bacteria were more efficient for metal bioleaching than when used individually at laboratory scale.

Two different possible approaches for the design of consortia for ore bioleaching in a stirred tank reactor were assessed in [40]. The "top-down" approach (performed in a CSTR) considers inoculation with a broad range of microorganisms of which only the more adaptable survive. On the other hand, the "bottom-up" approach uses a specific microorganism' consortia logically designed with a small number of microorganisms carefully selected based on their physiological features related to their ability for treating the target mineral. Both approaches would use microorganisms isolated from biomining plants or from environmental samples.

Considering the consortia may need a long time to adapt to the CSTR conditions, it is likely that while some microorganisms will more easily adapt, others will either fail or need a longer period to adapt.

16.3.4.3.2 Reactor Design

Few studies have been reported on e-waste bioleaching using different reactor designs. Therefore, it is still not possible to fully understand bioleaching at a large scale [34].

Selecting the most suitable reactor for biohydrometallurgy is based on the physicochemical and biological features of the system. Characteristics such as rheological and chemical behaviour of the matrices in an acid media, oxygen and CO_2 transfer, and the microorganisms used are to be considered for the process design.

<u>Continuous Stirred Tank Reactors (CSTR)</u>

In general, CSTR include continuous aeration circuits and agitation for sustaining the finely ground materials in suspension, favouring the O_2 and CO_2 transfer necessary for microorganisms' activity and growth. The operation mode is a continuous flow and nutrients are added.

CSTR provide higher mass transfer and mixing capacities than column reactors and allow for better control of process variables with shorter and more efficient bioleaching processes. CSTR are usually used for bioleaching of metal sulphides (cobalt, zinc, copper, nickel) and precious metals (gold and silver). Its continuous operation mode allows for an uninterrupted selection of the microorganisms capable of more efficiently growing in the tanks. The more efficient the microorganisms, the less they will be washed out, resulting in the dominant population in the bioreactor so is not necessary to sterilize the system as continuous microorganism selection catalyses the metal dissolution.

It is important to consider that although CSTR operations do not require a sterilization process, these reactors have higher construction and operations costs.

The main disadvantage of CSTRs is low pulp densities with a maximum of 20% w/v. However, it has been reported that by using thermophilic bacteria, 30% w/v pulp densities could be used [47].

Rotating-Drum Reactors

Rotating-drum reactors have higher pulp densities and less global energy requirements than CSTRs and produce a lower degree of particle collision with a reduced effect on the microbial cells. Additionally, they allow the treatment of high solid loads with no negative effects on the ferrous ions' bio-oxidation [27].

A rotating-drum reactor could be a suitable alternative to stirred tank reactors as they allow treatment of higher pulp densities, lower impact on cells, and the reduction of energy consumption [28].

Column Reactor

The use of column design for bioleaching has lower operating costs than commonly used CSTR due to a reduction in the energy requirements as there is no agitation involved. In addition, microorganisms are less subjected to mechanical stress due to the absence of propellers.

i. E-Waste Composition

The complex composition of e-waste makes this waste physiochemically singular, consequently, single techniques are not sufficient for precious and rare metals recovery being necessary to perform a combination of processes. In this regard, integration of mechanical processes as part of a pre-treatment stage can improve the leaching efficiency of metals from e-waste [44]. As for the microorganisms used, the consortia should be adapted to the varying compositions of different e-wastes which would not be practical or suitable at an industrial scale.

ii. Process Duration

Biohydrometallurgical process are usually limited by their duration. However, there is potential for optimization. The process duration has a direct impact on its profitability, i.e., higher profit for shorter times. The two main limitations for using bioleaching in an industrial scale are slow dissolution kinetics and low metal leaching yields. According to reports, obtaining 60% metal recovery takes long periods of time ranging from

300 to 900 days on laboratory and industrial scales, respectively [47]. This problem could be solved by introducing catalysts which would lower the activation energy and increase the reaction's rate.

Different catalysts were tested and proved to be effective, including surfactants, metal ions (Cu^{2+}, Ag^+, Hg^{2+}, Bi^{3+}, Co^{2+}), and non-metallic carbonaceous materials. Metal ions have been proven to be excellent catalysts in contrast with non-metallic compounds, which, according to research, only produce low bioleaching yields (as low as 12.5% and require amounts of catalyst as large as 2.5 kg/kg of ore [48], [49].

16.3.5 Omics and Bioinformatics

The term omics refers to a group of novel molecular technologies including genomics, metabolomics, proteomics, and genetics. The objective behind applying omics and bioinformatics to bioleaching is to determine the complex ways in which microorganisms contribute to the process and predict metabolic models. The omics tools are used to study the mechanisms behind the adaptation of microorganisms to their environment and it allows the identification of the steps needed in order to genetically/chemically manipulate capacity and efficiency of the metal removal in the system [17].

Current techniques capable of genome sequencing are of great aid to studying the whole consortium microbiome, however, genetic modification applied to biomining microorganisms, even though available, still needs further development. This will eventually allow for the modification of consortia in order to increase their metal removal efficiencies and better-controlled bioleaching [50].

Additionally, by partially or fully sequencing genomes, biodiversity within leaching environments has been successfully determined, thus, new microorganisms are studied that could potentially be applied to e-waste bioleaching. To date, genomics has had a significant impact on determining leaching biodiversity as well as developing metabolic models. In the near future, metagenomics will provide new important information by studying the microbial dynamics throughout the bioleaching stages with the use of breakthrough sequencing technology and metabolic operating models, which will in turn allow for an optimization of the design and control of bioleaching.

Research in this area is key to achieving economical bioleaching processes at an industrial scale that are applicable for all metals.

16.3.6 Metal/Metalloids Recovery Techniques

Selective and economically viable metal recovery from low-metal concentration leachates is one of the main challenges behind commercial scaling of e-waste recycling. Different chemical technologies can be applied including chemical precipitation, cementation, solvent extraction, electrowinning, ion exchange, and adsorption. However, recent developments are focused on greener processes such as biological technologies including biosorption, bioprecipitation, bioaccumulation, bioelectrochemical systems, and biomineralization. These recovery techniques are discussed below.

16.3.6.1 Biological Recovery Techniques: Immobilization Mechanisms

Biological techniques for the recovery of critical metals from e-waste are currently at a low TRL. Many studies have been presented on the development of novel green strategies for metal recovery from e-waste using biomaterials [50]. Research includes both mobilization and immobilization processes (see Figure 16.4).

i. Biosorption
Being a physicochemical and metabolism-independent process, biosorption involves the interaction between ions in solution and the charge surface of microorganisms' cell walls (bacteria, actinomycetes, fungi, algae,

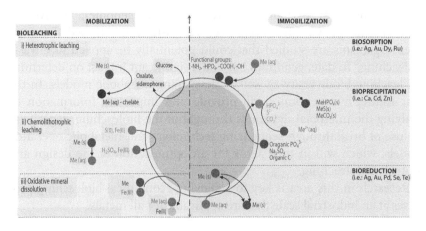

Figure 16.4 Microbial mechanisms for metal recovery (Adapted from [52]).

yeast, and biowaste). This is a competitive process in comparison to chemical ones, due to its high efficiency in removing metals from liquid matrices, lower operating costs, and less generation of waste sludge to be managed.

For this extracellular process cells are not required to be alive, therefore the process can be performed under extreme temperature and pH conditions (high and low, respectively), resulting in an advantage from an industrial point of view as the conditions and control of the process should be less difficult to manage. Another advantage is that recovered materials can be easily retrieved by centrifugation and filtration. The main disadvantages are the low volume of metals/metalloids recovered and the long process duration.

Applying biosorption to metal recovery from e-waste is a viable option. Each biosorbent has a set of properties depending on its functional groups which determine its biosorption capacity.

As the biosorption process takes place in an aqueous phase, its application to e-waste is hindered since metals have to be first leached out of the solid. In addition, to maximize the capacity of the biosorbent it is necessary to chemically modify its surface. These disadvantages result in a process that can only be applied at laboratory scale with further investigation needed for its scale-up and commercialization [14].

ii. Bioprecipitation
Precipitation occurs by disturbance of the ionic equilibrium of metals in an aqueous solution as a result of the addition of certain chemicals that alter the solution's pH. Therefore, the main factors determining the efficiency of precipitation for metal removal are the pH and metal concentration in solution. In bioprecipitation, the chemicals interacting with the metals present in the solution are special proteins excreted by organisms. This interaction yields metal compounds that are insoluble and, therefore, precipitate. Either cells or excreted polymers can act as nucleation surfaces. Examples of bacteria-excreted proteins used for bioprecipitation are metallothioneins and phytochelatins [53].

Bioprecipitation allows for selective metal recovery and, by adjusting the pH of the system, the precipitate can also be selected to be sulphides or carbonates.

iii. Bioelectrochemical Systems
Bioelectrochemical systems (BES) include microbial electrolysis cells (MECs) and microbial fuel cells (MFCs) used for extracting and recovering metals (Figure 16.5).

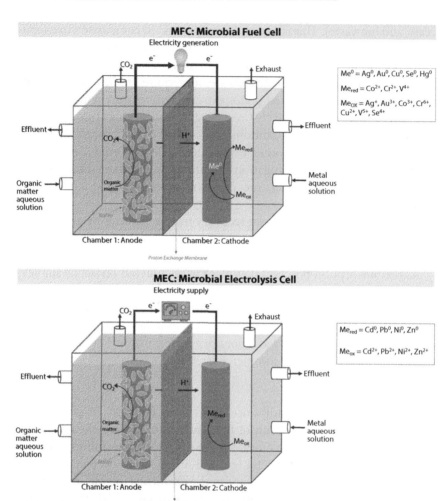

Figure 16.5 Microbial fuel cell & microbial electrolysis cell models (Adapted from [52]).

In BES, microbial colonies need a solid electron acceptor (bioanode) or an electron donor (biocathode). By delivering electrons to the cathode through the biological reactions in the anode, metals/metalloids precipitation is promoted as they are used as terminal electron acceptors in the cathode. The driving force for the electrochemical reduction process and the metal/metalloid recovery in the cathode is provided by the oxidation of organic matter on the biofilms located on the anode. Depending on the speciation of the metal/metalloid and the solution's chemistry, the reduced metals can either precipitate, deposit on the cathode, or remain solubilized.

In MFCs, reducing equivalents are released from organic matter by microbial and electrochemical processes and chemical energy is converted to electrical energy via a succession of microbial metabolic redox reactions. When the redox potential of a cathodic reaction is similar or higher than that of the anode, metal ions are spontaneously reduced in the cathode. When the redox potential of the metal/metalloid is lower than that of the anode, the electron flow from the anode to the cathode is not thermodynamically favourable and additional energy supply is required to achieve the reduction of metal ions in the cathode [51].

BES has potential for metal recovery from e-waste leachate, however most reported studies were performed on synthetic solutions. Therefore, further research focused on e-waste leachate is necessary to fully understand the mechanisms behind these systems. Determining the optimum operating capacities is key in order to improve biorecovery rates and upscale for commercial use.

iv. Bioreduction (Biomineralization)
Living organisms produce mineral crystals as part of a process for overcoming metal toxicity. Even though the process is widely applied to waste treatment and nanomaterial synthesis, the basic mechanisms are yet not completely understood [26].

Despite the need for further investigation of biomineralization mechanisms, a number of innovative critical metal biomineralization processes have been studied and reported ([54], [55], [56]) as its application has potential for the production of nanomaterials.

In particular, in e-waste recycling, biomineralization is starting to rise as a viable platinum group metals (PGMs) recovery technology [56].

16.3.7 Recovered Materials

16.3.7.1 Rare Earth Elements (REEs)

E-waste could pose as a viable option for sourcing, however, as they are present only in small quantities in marketed products (per unit), they are not usually targeted for recovery. This adds to its relatively low prices in comparison to those of precious metals which result in uneconomic recycling processes.

Biohydrometallurgy could be a potential alternative for REEs recovery through mobilization, mineralization, and metal reduction [57]. In general, the overall bioleaching mechanisms can be assumed to be the same as for any other metal, nevertheless, mechanisms of REEs-microorganism interactions are still not well known.

Studies by [58], [59], [60], [61], and [45] report on bioleaching for REEs recovery. A study by [62] reports on a process designed for the extraction and purification of REE from complex metal mixtures such as e-waste using lanmodulin, a natural occurring protein which is an REE-specific macromolecule tolerant to industrial conditions including acidity, high temperature, and concentrations of competing ions. This is a promising extraction technology.

16.3.7.2 Nanomaterials

Biosynthesis of metallic nanomaterials has recently drawn attention due to the better superior biocompatibility of biologically produced nanoparticles. Metallic nanoparticles of several metals have been biologically synthesised including Au, Ag, Pl, Al, Zn, Ti, Pd, Fe, and Cu. Other nanoparticles, such as oxides and sulphides, have also been studied due to their wide range of important applications from medicine to the energy industry. Highly pure nanometals can be recovered from e-waste. Nanoparticle recovery dates back to 2015 when abiotic synthesis of magnetite nanoparticles was carried out from iron ore tailings [63]. Obtaining nanoparticles from mining waste has received attention in recent years ([62], [64], [65]), however, it remains a challenge due to the complexity of the matrix.

16.4 Economic Feasibility

E-waste contains a variety of valuable materials, including about 69 elements from the periodic table that are potentially recoverable, although it is not economically feasible in every case. According to estimates, on a global scale, e-waste is worth at least 57 billion dollars annually [66]. The majority of the total weight of raw materials present in e-waste is due to Fe, Al, and Cu [67].

Globally, only 17% of this waste is being managed appropriately. The destination of the remaining 83% is still adequately confirmed and its disposal and environmental and health impacts vary depending on the geographical location. For the recycled component, techniques currently used include pyrometallurgy and hydrometallurgy, which are not environmentally sound. With the current recycling rates, the value of raw materials recovered amounts to approximately 9.4 billion dollars annually [66].

16.5 Conclusions

This review indicates that research on the upscaling of biohydrometallurgical processes for the recovery of metals from e-waste is still limited.

Pyrometallurgy is predominant in the e-waste recycling industry with highly negative environmental impacts, including the risk of dioxin emission and formation of toxic materials. Hydrometallurgy is a cleaner technology; however, large volumes of chemicals (high operating costs) are used as lixiviants, producing large volumes of sludge with high treatment costs.

Environmentally improved techniques are being developed by the e-waste recycling industry, including biohydrometallurgy in the recovery of metals from solid waste by biological means using metal extracting agents produced by microorganisms. These processes use weak organic acids and have low energy and temperature requirements.

Although bioleaching has been widely researched for the recovery of metals from complex resources and low-grade ores and is currently applied to industrial scale ore treatment, only a few studies report on bioleaching of different e-waste using different bioreactor configurations. Thus, it is difficult to reach firm conclusions on viability since studies were carried out under different operating conditions. Even though most of them were performed using PCB scraps, it is important to reflect that the composition of PCBs varies between sources.

Research on the upscaling of bioleaching is not deeply developed and it has mainly focused on specific metals and e-wastes (i.e., PCBs). Nevertheless, a number of studies have been carried out, mainly at bench scale, as summarised in Table 16.6.

Bioleaching efficiency for metal recovery from e-waste is associated with operating conditions that must be considered when designing the process. A vast number of limitations related to the working conditions of the microorganisms have been elucidated. This is why, when considering the wide variety of microorganisms recently discovered, it is likely that better technologies will be delivered for efficient up-scaling of the e-waste bioleaching process. Some mechanistic aspects are still not well understood, and it has still to be determined how to up-scale the progress made in the laboratory to translate into full-scale operation. It is expected that further developments in biotechnological processes, such as genetic engineering, will allow for the improvement of bioleaching efficiency from a long-term perspective.

As e-waste contains a wide variety of components, different materials can be recovered by biohydrometallurgical processes including oxides, sulphides, and metals. Biohydrometallurgy could be an alternative for REEs recovery through mobilization, mineralization, and metal reduction. A reduced number of studies on bioleaching of REEs from secondary sources were identified. Highly pure nanometals can be recovered from e-waste as well, however, it is a challenging process due to the complexity of the matrix with little detailed information being reported.

Technical and economic considerations were made in order to analyse the method's feasibility at an industrial scale, concluding that bioleaching has low operation costs in comparison with the physicochemical techniques, increasing its economic feasibility. Pre-treatment stages involving dismantling and size reduction are well resolved in the e-waste recycling industry, however when looking at biological separation methods, although it has been studied, they are still in early development stages. As for bioleaching as a treatment process, there are different factors hindering its development to pilot and industrial scales, such as bacterial culture conditions, reactor design and operation parameters, and jarosite or oxide precipitation. In some cases, strategies have been developed for the related challenges. However, further investigation is needed. Last but not least, recovery biological processes of critical metals are currently at low TRL and mechanisms behind recovery of materials such as REEs and nanomaterials are still not well known. In conclusion, current bioleaching at the commercial scale faces diverse operational challenges that hamper its scale-up and industrial implementation.

References

1. Pahari, A. K. and Dubey, B. K. (2019) Waste From Electrical and Electronics Equipment. In: Plastics to Energy. Elsevier Inc., pp.443–468.
2. Dawood, I. and Underwood, J. (2010) Research Methodology Explained. PM-05 - Advancing Project Management for the 21st Century "Concepts, Tools & Techniques for Managing Successful Projects". Vol.29–31.
3. Awasthi, A. K. and Li, J. (2017) An overview of the potential of eco-friendly hybrid strategy for metal recycling from WEEE. Resources, Conservation and Recycling. Vol.126, pp.228–239.
4. Gu, F., Summers, P. A. and Hall, P. (2019) Recovering materials from waste mobile phones: Recent technological developments. Journal of Cleaner Production. Vol.237, p.117657.

5. Işıldar, A., Rene, E. R., van Hullebusch, E. D. and Lens, P. N. L. (2018) Electronic waste as a secondary source of critical metals: Management and recovery technologies. Resources, Conservation and Recycling. Vol.135, pp.296–312.
6. Vidyadhar, A. (2016) A Review of Technology of Metal Recovery from Electronic Waste. In: Mihai, F.-C. (ed.). E-Waste in Transition - From Pollution to Resource. InTech, pp.121–158.
7. Kaya, M. (2016) Recovery of metals and nonmetals from electronic waste by physical and chemical recycling processes. Waste Management. Vol.57, pp.64–90.
8. Li, S., Zhong, H., Hu, Y., Zhao, J., He, Z. and Gu, G. (2014) Bioleaching of a low-grade nickel–copper sulfide by mixture of four thermophiles. Bioresource Technology. Vol.153, pp.300–306.
9. Hsu, E., Barmak, K., West, A. C. and Park, A.-H. A. (2019) Advancements in the treatment and processing of electronic waste with sustainability: a review of metal extraction and recovery technologies. Green Chemistry. Vol.21 (5), pp.919–936.
10. Islam, A., Ahmed, T., Awual, Md. R., Rahman, A., Sultana, M., Aziz, A. A., Monir, M. U., Teo, S. H. and Hasan, M. (2020) Advances in sustainable approaches to recover metals from e-waste-A review. Journal of Cleaner Production. Vol.244, p.118815.
11. Kim, G., Park, K., Choi, J., Gomez-Flores, A., Han, Y., Choi, S. Q. and Kim, H. (2015) Bioflotation of malachite using different growth phases of *Rhodococcus opacus*: Effect of bacterial shape on detachment by shear flow. International Journal of Mineral Processing. Vol.143, pp.98–104.
12. Pollmann, K., Kutschke, S., Matys, S., Raff, J., Hlawacek, G. and Lederer, F. L. (2018) Bio-recycling of metals: Recycling of technical products using biological applications. Biotechnology Advances. Vol.36 (4), pp.1048–1062.
13. Kim, G., Choi, J., Silva, R. A., Song, Y. and Kim, H. (2017) Feasibility of bench-scale selective bioflotation of copper oxide minerals using *Rhodococcus opacus*. Hydrometallurgy. Vol.168, pp.94–102.
14. Ramanayaka, S., Keerthanan, S. and Vithanage, M. (2020) 2 - Urban mining of E-waste: treasure hunting for precious nanometals. In: Prasad, M. N. V., Vithanage, M., and Borthakur, A. (eds.). Handbook of Electronic Waste Management. Butterworth-Heinemann, pp.19–54.
15. Khan, A., Inamuddin and Asiri, A. M., eds. (2020) E-waste Recycling and Management: Present Scenarios and Environmental Issues. Cham: Springer International Publishing.
16. Avarmaa, K.; Yliaho, S.; Taskinen, P. (2018) Recoveries of rare elements Ga, Ge, In and Sn from waste electric and electronic equipment through secondary copper smelting. Waste Management. Vol.71, pp.400-410
17. Baniasadi, M., Vakilchap, F., Bahaloo-Horeh, N., Mousavi, S. M. and Farnaud, S. (2019) Advances in bioleaching as a sustainable method for metal recovery

from e-waste: A review. Journal of Industrial and Engineering Chemistry. Vol.76, pp.75–90.
18. Kaya, M. (2018) 3 - Current WEEE recycling solutions. In: Waste Electrical and Electronic Equipment Recycling. Woodhead Publishing, pp.33–93.
19. Yoshida, T. (2017) Applied Bioengineering: Innovations and Future Directions. WILEY-VCH.
20. Rajeshkumar, S., Bharath, L. V. and Geetha, R. (2019) Chapter 17 - Broad spectrum antibacterial silver nanoparticle green synthesis: Characterization, and mechanism of action. In: Shukla, A. K. and Iravani, S. (eds.). Green Synthesis, Characterization and Applications of Nanoparticles. Elsevier, pp.429–444.
21. Barhoum, A., García-Betancourt, M. L., Rahier, H. and Van Assche, G. (2018) Chapter 9 - Physicochemical characterization of nanomaterials: polymorph, composition, wettability, and thermal stability. In: Emerging Applications of Nanoparticles and Architecture Nanostructures. Elsevier, pp.255–278.
22. Kamberović, Ž., Korać, M., Ivšić, D., Nikolić, D. and Ranitović, M. (2009) Hydrometallurgical process for extraction of metals from electronic waste - Part I: Material characterization and process option selection. Metalurgija - Journal of Metallurgy. Vol.15 (4), pp.231–243.
23. Aldrian, A., Ledersteger, A. and Pomberger, R. (2015) Monitoring of WEEE plastics in regards to brominated flame retardants using handheld XRF. Waste Management. Vol.36, pp.297–304.
24. Kumar, A., Holuszko, M. E. and Janke, T. (2018) Characterization of the non-metal fraction of the processed waste printed circuit boards. Waste Management. Vol.75, pp.94–102.
25. Aguirre, M. A., Hidalgo, M., Canals, A., Nóbrega, J. A. and Pereira-Filho, E. R. (2013) Analysis of waste electrical and electronic equipment (WEEE) using laser induced breakdown spectroscopy (LIBS) and multivariate analysis. Talanta. Vol.117, pp.419–424.
26. Yu, Z., Han, H., Feng, P., Zhao, S., Zhou, T., Kakade, A., Kulshrestha, S., Majeed, S. and Li, X. (2020) Recent advances in the recovery of metals from waste through biological processes. Bioresource Technology. Vol.297, p.122416.
27. Habibi, A., Kourdestani, S. S. and Hadadi, M. (2020) Biohydrometallurgy as an environmentally friendly approach in metals recovery from electrical waste: A review. Waste Management & Research. Vol.38 (3), pp.232–244.
28. Rodrigues, M. L. M., Leão, V. A., Gomes, O., Lambert, F., Bastin, D. and Gaydardzhiev, S. (2015) Copper extraction from coarsely ground printed circuit boards using moderate thermophilic bacteria in a rotating-drum reactor. Waste Management. Vol.41, pp.148–158.
29. Xingyu, L., Rongbo, S., Bowei, C., Biao, W. and Jiankang, W. (2009) Bacterial community structure change during pyrite bioleaching process: Effect of pH and aeration. Hydrometallurgy. Vol.95 (3), pp.267–272.

30. Ilyas, S. and Lee, J. (2014a) Bioleaching of metals from electronic scrap in a stirred tank reactor. Hydrometallurgy. Vol.149, pp.50–62.
31. Cortés, M., Marín, S., Galleguillos, P., Cautivo, D. and Demergasso, C. (2019) Validation of Genetic Markers Associated to Oxygen Availability in Low-Grade Copper Bioleaching Systems: An Industrial Application. Frontiers in Microbiology. Vol.10.
32. Priya, A. and Hait, S. (2018) Extraction of metals from high grade waste printed circuit board by conventional and hybrid bioleaching using *Acidithiobacillus ferrooxidans*. Hydrometallurgy. Vol.177, pp.132–139.
33. Işıldar, A., Van Hullebusch, E. D., Lenz, M., Du Laing, G., Marra, A., Cesaro, A., Panda, S., Akcil, A., Kucuker, M. A. and Kuchta, K. (2019) Biotechnological strategies for the recovery of valuable and critical raw materials from waste electrical and electronic equipment (WEEE) – A review. Journal of Hazardous Materials. Vol.362, pp.467–481.
34. Sodha, A. B., Tipre, D. R. and Dave, S. R. (2020) Optimisation of biohydrometallurgical batch reactor process for copper extraction and recovery from non-pulverized waste printed circuit boards. Hydrometallurgy. Vol.191, p.105170.
35. Zhu, P., Chen, Y., Wang, L., Qian, G., Zhang, W. J., Zhou, M. and Zhou, J. (2013) Dissolution of brominated epoxy resins by dimethyl sulfoxide to separate waste printed circuit boards. Environmental Science & Technology. Vol.47 (6), pp.2654–2660.
36. Tatariants, M., Yousef, S., Sidaraviciute, R., Denafas, G. and Bendikiene, R. (2017) Characterization of waste printed circuit boards recycled using a dissolution approach and ultrasonic treatment at low temperatures. RSC Advances. Vol.7 (60), pp.37729–37738.
37. Hubau, A., Minier, M., Chagnes, A., Joulian, C., Silvente, C. and Guezennec, A.-G. (2020) Recovery of metals in a double-stage continuous bioreactor for acidic bioleaching of printed circuit boards (PCBs). Separation and Purification Technology. Vol.238, p.116481.
38. Xiang, Y., Wu, P., Zhu, N., Zhang, T., Liu, W., Wu, J. and Li, P. (2010) Bioleaching of copper from waste printed circuit boards by bacterial consortium enriched from acid mine drainage. Journal of Hazardous Materials. Vol. 184 (1-3), pp. 812-818
39. Ilyas, S., Ruan, C., Bhatti, H. N., Ghauri, M. A. and Anwar, M. A. (2010) Column bioleaching of metals from electronic scrap. Hydrometallurgy. Vol.101 (3), pp.135–140.
40. Bryan, C. G., Joulian, C., Spolaore, P., El Achbouni, H., Challan-Belval, S., Morin, D. and d'Hugues, P. (2011) The efficiency of indigenous and designed consortia in bioleaching stirred tank reactors. Minerals Engineering. Vol.24 (11), pp.1149–1156.
41. Jagannath, A., Shetty K., V. and Saidutta, M. B. (2017) Bioleaching of copper from electronic waste using Acinetobacter sp. Cr B2 in a pulsed plate column

operated in batch and sequential batch mode. Journal of Environmental Chemical Engineering. Vol.5 (2), pp.1599–1607.
42. Xia, M.-C., Wang, Y.-P., Peng, T.-J., Shen, L., Yu, R.-L., Liu, Y.-D., Chen, M., Li, J.-K., Wu, X.-L. and Zeng, W.-M. (2017) Recycling of metals from pretreated waste printed circuit boards effectively in stirred tank reactor by a moderately thermophilic culture. Journal of Bioscience and Bioengineering. Vol.123 (6), pp.714–721.
43. Ilyas, S., Lee, J. and Chi, R. (2013) Bioleaching of metals from electronic scrap and its potential for commercial exploitation. Hydrometallurgy. Vol.131–132, pp.138–143.
44. Brandl, H., Bosshard, R. and Wegmann, M. (2001) Computer-munching microbes: metal leaching from electronic scrap by bacteria and fungi. Hydrometallurgy. Vol.59 (2), pp.319–326.
45. Marra, A., Cesaro, A., Rene, E. R., Belgiorno, V. and Lens, P. N. L. (2018) Bioleaching of metals from WEEE shredding dust. Journal of Environmental Management. Vol.210, pp.180–190.
46. Mahmoud, A., Cézac, P., Hoadley, A. F. A., Contamine, F. and D'Hugues, P. (2017) A review of sulfide minerals microbially assisted leaching in stirred tank reactors. International Biodeterioration & Biodegradation. Vol.C (119), pp.118–146.
47. Pathak, A., Morrison, L. and Healy, M. G. (2017) Catalytic potential of selected metal ions for bioleaching, and potential techno-economic and environmental issues: A critical review. Bioresource Technology. Vol.229, pp.211–221.
48. Zhang, R.; Wei, D.; Shen, Y.; Liu, W.; Lu, T.; Han, C. (2016) Catalytic effect of polyethylene glycol on sulfur oxidation in chalcopyrite bioleaching by *Acidithiobacillus ferrooxidans*. Mineral Engineering. Vol. 95, pp. 74-78.
49. Dong, H.; Zhao, J.; Chen, J; Wu, Y.; Li, B. (2015) Recovery of platinum group metals from spent catalysts: A review. International Journal of Mineral Processing. Vol. 145, pp. 108-113.
50. Gopikrishnan, V., Vignesh, A., Radhakrishnan, M., Joseph, J., Shanmugasundaram, T., Doble, M. and Balagurunathan, R. (2020) Microbial leaching of heavy metals from e-waste. In: Biovalorisation of Wastes to Renewable Chemicals and Biofuels. Elsevier, pp.189–216.
51. Kucuker, M. A. and Kuchta, K. (2018) Biomining – Biotechnological Systems for the Extraction and Recovery of Metals from Secondary Sources. Global NEST Journal. Vol.20 (4), pp.737–742.
52. Nancharaiah, Y. V., Venkata Mohan, S. and Lens, P. N. L. (2015) Metals removal and recovery in bioelectrochemical systems: A review. Bioresource Technology. Vol.195, pp.102–114.
53. Hashmi, M. Z. and Varma, A., eds. (2019) Electronic Waste Pollution: Environmental Occurrence and Treatment Technologies. Cham: Springer International Publishing.

54. Ali, J.; Ali, N.; Wang, L.; Waseem, H.; Pan, G. Revisiting the mechanistic pathways for bacterial mediated synthesis of noble metal nanoparticles. J Microbiol Methods 2019, 159, 18-25, doi:10.1016/j.mimet.2019.02.010.
55. Foulkes, J.M.; Deplanche, K.; Sargent, F.; Macaskie, L.E.; Lloyd, J.R. A Novel Aerobic Mechanism for Reductive Palladium Biomineralization and Recovery by *Escherichia coli*. Geomicrobiology Journal 2016, 33, 230-236, doi:10.1080/01490451.2015.1069911.
56. Rene, E. R., Sahinkaya, E., Lewis, A. and Lens, P. N. L., eds. (2017) Sustainable Heavy Metal Remediation. Cham: Springer International Publishing.
57. Ilyas, S. and Lee, J. (2014b) Biometallurgical Recovery of Metals from Waste Electrical and Electronic Equipment: a Review. ChemBioEng Reviews. Vol.1 (4), pp.148–169.
58. Auerbach, R., Bokelmann, K., Stauber, R., Gutfleisch, O., Schnell, S. and Ratering, S. (2019) Critical raw materials – Advanced recycling technologies and processes: Recycling of rare earth metals out of end-of-life magnets by bioleaching with various bacteria as an example of an intelligent recycling strategy. Minerals Engineering. Vol.134, pp.104–117.
59. Maneesuwannarat, S., Vangnai, A. S., Yamashita, M. and Thiravetyan, P. (2016) Bioleaching of gallium from gallium arsenide by *Cellulosimicrobium funkei* and its application to semiconductor/electronic wastes. Process Safety and Environmental Protection. Vol.99, pp.80–87.
60. Hopfe, S., Flemming, K., Lehmann, F., Möckel, R., Kutschke, S. and Pollmann, K. (2017) Leaching of rare earth elements from fluorescent powder using the tea fungus Kombucha. Waste Management. Vol.62, pp.211–221.
61. Jowkar, M. J., Bahaloo-Horeh, N., Mousavi, S. M. and Pourhossein, F. (2018) Bioleaching of indium from discarded liquid crystal displays. Journal of Cleaner Production. Vol.180, pp.417–429.
62. Deblonde, G. J.-P., Mattocks, J. A., Park, D. M., Reed, D. W., Cotruvo, J. A. and Jiao, Y. (2020) Selective and Efficient Biomacromolecular Extraction of Rare-Earth Elements using Lanmodulin. Inorganic Chemistry. [Online]. Available: https://pubs.acs.org/doi/10.1021/acs.inorgchem.0c01303 [Accessed 27 Jul 2020].
63. Wong-Pinto, L., Menzies, A. and Ordóñez, J. I. (2020) Bionanomining: biotechnological synthesis of metal nanoparticles from mining waste— opportunity for sustainable management of mining environmental liabilities. Applied Microbiology and Biotechnology. Vol.104 (5), pp.1859–1869.
64. Pat-Espadas, A. M., Cervantes, F. J. (2018) Microbial recovery of metallic nanoparticles from industrial wastes and their environmental applications. Journal of Chemical Technology & Biotechnoloby. Vol. 93 (11), pp. 3091-3112.
65. Bindschedler, S., Vu Bouquet, T.Q.T., Job, D., Joseph, E., Junier, P. (2017) Chapter Two – Fungal Biorecovery of Gold from E-waste. Advances in Applied Microbiology. Vol. 99, pp. 53-81.

66. Balde, C.P., D'Angelo, E., Luda, V., Deubzer, O., Kuehr, R. (2022). Global Transboundary E-waste Flows. United Nations Institute for Training and Research.
67. Forti, V., Baldé, C. P., Kuehr, R. and Bel, G. (2020) The Global E-waste Monitor 2020 p.120.

17

Current Advances in Recycling of Electronic Wastes

Kumar Sagar Maiti, Irin Khatun, Serma Rimil Hansda and Dipankar Ghosh*

Microbial Engineering & Algal Biotechnology Laboratory, Department of Biosciences, JIS University Agarpara, Kolkata, India

Abstract

In the previous few decades, the global demand for electrical and electronics instrumentalities has been growing exponentially in developed and developing nations. It tends to lower the life span of these electrical and electronics materials. Moreover, ameliorating the demands of these electrical and electronics materials make the current situation worse. As a result, enormous quantities of electrical and electronic wastes (e-waste) have regularly been generated in developed and developing nations. It ultimately causes extensive troubles associated with e-waste disposal, recycling, and management to harness environmental ecological dynamics and health hazards. Therefore, a paradigm shift has been required which can spark the interest in tine-waste recycling processes and their upgradation. There are diverse ranges of e-waste recycling technologies, including conventional (physical and chemical treatments) and modern (biological or microbial treatments) approaches to combat environmental pollution and community health hazards. However, conventional and modern techniques suffer from multiple lacunae related to the efficacies of e-waste recycling techniques. To this end, the current literature review deals with a general outline of e-waste generation and categorization with special emphasis on e-waste recycling processes in great detail.

Keywords: Electronic wastes, e-waste classification, e-waste disposal, e-waste recycling, environmental management, health hazards

---

*Corresponding author: dghosh.jisuniversity2@gmail.com; d.ghosh@jisuniversity.ac.in

Abhishek Kumar, Pramod Singh Rathore, Ashutosh Kumar Dubey, Arun Lal Srivastav, T. Ananth Kumar and Vishal Dutt (eds.) Sustainable Management of Electronic Waste, (341–374) © 2024 Scrivener Publishing LLC

17.1 Introduction

The current growth in living standards has increased the formation of domestic or home waste, which includes biomass leftovers derived from food waste, grass, garden waste, and paper (Rodríguez-Padrón et al., 2019), as well as glass, plastics, metals, and electronic waste (e-waste) [Xu et al., 2019]. Electronic waste (e-waste) is the most rapidly increasing type of solid waste on the planet. This waste stream is closely linked to the usage of electrical and electronic equipment (EEE), which has increased by 2.5 million metric tonnes (Mt) each year as a result of worldwide economic and technical progress. In the last five years, the global quantity of e-waste has climbed by more than 9.0 Mt, reaching a total of 53.6 Mt in 2019, with Africa (2.9 Mt), the Americas (13.1 Mt), Asia (24.9 Mt), Europe (12 Mt), and Oceania (12 Mt) accounting for the majority of the growth (0.7 Mt) (Adrian et al., n.d.). E-waste is typically defined as any device (or component) that has hit the end of its useful life and has a plug, cell, or wire. E-waste refers to a wide range of items that are categorised according to their functional resemblance, material composition, overall weight, and obsolescence characteristics. There are approximately 54 product-centric categories that can be divided into six classes (Forti V, 2018). Temperature exchange equipment (air conditioners, heating pumps, fridge, ceiling fans, etc.), small equipment (toaster ovens, recording devices, video cameras, electric cameras radio sets, coffee machines, blenders, etc.), large equipment (washing machines, dishwashing machines, and photovoltaic panels), screens (TVs, monitors, laptop computers, and tablet computers), lamps (fluorescent lamps, low-pressure sodium lamps, high-intensity discharge lamps, and LED lamps.), and small IT and telecommunication equipment (mobile phones and tablets, Global Positioning System (GPS) devices, pocket calculators, routers, personal computers, printers, and telephones) are examples of such compounds (Figure 17.1) [Cui and Roven, 2011].

E-waste is made up of a variety of metals and non-metals, giving it a distinct and diversified makeup. 30 percent of organics (flame retardants, polymers, and glass fibre), 30 per cent of ceramics (mica, silica, and alumina), and 40 percent of inorganics are common E-waste composition patterns (e.g., non-ferrous and ferrous metals). Base metals (e.g., copper, aluminium, tin, and iron), noble metals (e.g., palladium, silver, and gold), heavy metals (e.g., beryllium, zinc, chromium, lead, mercury, cadmium, and nickel), and scarce earth metals (e.g., tantalum, gallium, and platinum groups) are all found in e-waste [Kaya, 2016a]. Heavy metals and persistent organics are released when WEEE is improperly handled, posing

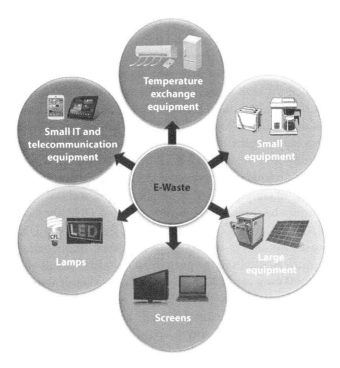

Figure 17.1 General overview on classification of e-waste.

a threat to human health and the environment. E-waste is responsible for 70% of all harmful and dangerous substances found in the environment today. Heavy metals including cadmium, mercury, lead, and beryllium, as well as harmful PVC plastics like brominated flame retardants, are among them. POPs (persistent organic pollutants) are a major component of e-waste and then bioaccumulate and biomagnify as they move through the food web [Abalansa et al., 2021]. Because of technological advancements, consumer preferences for many electronic devices such as smartphones, tablets, laptops, and gaming devices, as well as their quick expiration and global consumption increased by around 2.5 Mt per year. This massive volume growth poses a significant hazard to the surrounding people since it contains poisonous metallic and non-metallic substances that can harm human health. Deoxyribonucleic acid damage as a result of mutation, cardiovascular problems, skin disorders, cancer, hearing problems, neurological disorders, and severe learning and respiratory impacts are among the side effects. As a result, the growth of e-waste is a major worry and appropriate management is required to protect human health and preserve the ecosystem's long-term viability [Rautela et al., 2021]. Heavy metals

(Co, Cr, Cu, Fe, Se, and Zn) were discovered at higher amounts in the goods, PAH metabolites in urine, and polychlorinated biphenyls in mother's milk in e-waste workers, according to two separate but similar studies [Feldt et al., 2014]. Contact of pregnant women or children with e-waste has been associated with some neuro-developmental abnormalities and/or fetal disturbances [Asante et al., 2011]. Pregnancy-related concerns like unexpected abortions, stillbirths, and premature babies, as well as lower birth weights and lengths, have all been connected to e-waste exposure at e-waste processing plants [Wu et al., 2012]. Blood chromium concentrations have been linked to decreased vital capacity in schoolchildren in the Chinese city of Guiyu, as well as weight, height, and BMI [Zheng et al., 2013]. According to several studies, the reported DNA damage is due to oxidative stress. According to several studies, the observed DNA damage among e-waste workers is the result of increased frequencies of micronucleated binucleated cells in the peripheral blood as a result of their exposure to e-waste (Wang et al., 2011). Lead, a major chemical in e-waste, has been connected to delayed puberty in girls, while polychlorinated biphenyls, tetra-chloro-dibenzodioxin (TCDD), and perfluoroalkyl in e-waste have been linked to reduced sperm count and quality [McAuliffe et al., 2012]. In addition to human health, inappropriate e-waste landfilling has a significant impact on soil fertility. Heavy metal ions included in dumped e-waste may seep out and permeate through the soil, eventually reaching water bodies or groundwater as a result of such landfilling. Furthermore, hazardous metals pollute water and have an impact on aquatic environments [Ali and Khan, 2018]. Heavy metal ions are also stubborn compounds that can infiltrate the soil-crop-food chain, posing a greater hazard to public health [Abdu, Abdullahi, and Abdulkadir, 2017]. Furthermore, the release of dioxins into the environment during the open combustion process degrades the air quality and inhaling dioxin-contaminated air causes a variety of health problems, including birth defects, brain, bone, renal, and liver damage. As a result, appropriate solutions for recovering or recycling e-waste are urgently needed to protect the ecosystem and its inhabitants.

In today's world of rapidly changing technology, a considerable amount of e-waste has gathered as a result of increased production of EEE (Electrical and Electronic Equipment) that becomes e-waste when its useful life is through. For environmentally sound e-waste management, effective techniques are required at both the production and post-use stages. Recycling is an important technique for reducing the growing piles of e-waste, yet, poor recycling practices are contributing to higher levels of harmful chemicals in the ecosystem. The present global e-waste management trend (informal processing of 82.6 percent of global e-waste in 2019) is unsustainable and a complete solution requires a diverse strategy. In this direction, a better understanding of the e-waste recovery and recycling process is critical (Ahirwar and Tripathi, 2021a). For both local and international e-waste recycling, technological advancements and recycling processes have been devised. Recycling e-waste is required to reduce environmental impact, effectively manage existing natural resources, and, most importantly, save money. E-waste and other urban ores must be recycled and reused if resource efficiency is to be increased [Rene *et al.*, 2021]. Recycling of E-waste is carried out by informal or formal processes.

17.1.1 Informal Recycling

Informal recycling, often known as backyard recycling, refers to the unscientific and primitive methods used to recover resources from E-waste. In this instance, E-waste management tasks (such as shipping, processing, sorting, repairing, refurbishing, and dismantling) are performed inefficiently by untrained individuals [Adrian *et al.*, 2020]. In nations like India, Ghana, and China, around 82.6 percent of e-waste created globally gets recycled informally [Adrian *et al.*, 2020]. Rag pickers in the informal recycling industry play a significant part in minimising e-waste's environmental impact [Chi *et al.*, 2011]. Temperature-control equipment, heavy equipment, screens, and information-technology equipment are the main E-waste categories managed by the informal sectors [Baldé, Wang, and Kuehr, 2015]. More than 80% of e-waste is exported from Japan, Australia, New Zealand, South Korea, North America, and Europe to nations like China, Pakistan, India, Nigeria, Ghana, and others in Southeast Asia. The Basel Action Network (BAN) has improved surveillance and reporting of unlawful E-waste trade [Visvanthan, Yin, and Karthikeyan, 2010]. Local rag pickers collect the e-waste trash, which is then traded in slums and backyards to individuals looking to recover rich precious metals. Volumes are reduced by large-scale manual dismantling and primitive recycling procedures (e.g., melting, chipping, heating, stripping, and physical dismantling). Landfilling and depumping of post-recovery residue (unwanted materials) and effluent in neighbouring drains, rivers, lakes, or oceans are common. Backyard e-waste recycling has negative environmental consequences, polluting soil (through toxic element leaching and effluent discharge), water (acidic effluent discharge, washing of circuit boards, and disposing of waste residue into nearby drainage systems), and air (open circuit board burning, emission of toxic fumes, and suspended particulate matter) [Chi *et al.*, 2011]. Air pollution occurs as a result of the production of hydrocarbons during the informal combustion of economically valuable metals such as Cu, Al, Au, Pb, Hg, Cd, Pd, and Pt. Cu is one of these metals that RoHS (Restriction of Certain Detrimental Substances) considers as very poisonous and hazardous to the ecological system. Open combustion, which releases dangerous substances into the environment, is favoured for recovering Cu through informal ways [Song et al., 2015]. Similarly, to recycle metals such as Ag, Au, and Pd, various very toxic chemicals such as HCl, HNO3, H2SO4, and cyanide are used, which harm the environment severely. Furthermore, extraction efficiency is less than 80%, but formal e-waste processing can achieve an extraction efficiency of greater than 95% [Matsukami et al., 2015].

17.1.2 Formal Recycling

Reusable elements such as plastics and metals are recycled by informal e-waste recycling facilities. Collecting, pre-processing, post-processing, and disposing are the main processes in formal e-waste recycling. E-waste can be collected and bought from rag pickers through driving initiatives and take back schemes. By grouping components of a similar type and fractioning refined metals, pre-processing renders e-waste more homogeneous [Ashiq, Kulkarni, and Vithanage, 2019]. Heat, chemical, and metallurgical methodologies (e.g., hydrometallurgy, pyrometallurgy, electrometallurgy, and biometallurgy) or a combination of two or more methods are used to extract valuable elements (e.g., precious metals) from homogeneous groups after pre-processing [Arya and Kumar, 2020]. To separate the metallic and non-metallic fractions of waste printed circuit boards, a variety of mechanical processes (such as shredding/crushing, magnetic separation, eddy current separation, ball milling, and gravity separation) have been used [Kaya, 2016a]. The creation of large volumes of post-processing by-products (e.g., trash residues and acidic effluents) that must eventually be handled is one of the key issues connected with current recycling procedures (e.g., landfill or incineration) [A.Dubey et al. 2022].

Electronic and electrical equipment have become vital in this age of information and communication technology. Many of these devices become obsolete within a few years of their manufacturing due to the fast-changing nature of technology. These characteristics, combined with an increase in demand owing to people's increased spending power around the world, result in massive amounts of e-waste being generated. The growing amount of electronic waste indicates a serious environmental problem that must be addressed immediately. The accumulation of e-waste could have serious effects on human health and the environment's long-term viability. Valorization of e-waste could also provide economic benefits by allowing valuable and scarce materials to be recovered. Several ways for recovering metals from e-waste have been described, but there is still a long way to go in terms of developing ecologically acceptable and efficient protocols in this area [Hsu *et al.*, 2019a].

17.2 E-Waste: A General Description and Classification and Issues on the Environment and Health

An electrical-powered appliance, which can be of any size or function and is no longer desired by the consumer, is referred to as e-waste. E-waste is made up of abandoned electronic appliances, with computers and mobile phones accounting for a disproportionate amount due to their short lifespan. In recent years, there has been a growing recognition of our impact on the environment as a result of our lifestyle, with the need to adopt a more sustainable approach to our consumption patterns emerging as particularly important. This trend pertains to industrial areas that influence consumer behaviour, particularly the electronic industry, where short life cycles and fast-evolving technologies have resulted in growing e-waste amounts [Gaidajis, Angelakoglou, and Aktsoglou, 2010]. The concern over e-waste stems from the fact that its amount is increasing every year as a result of rising consumption and shorter lifespan. E-waste has been the fastest-growing garbage in the last decade [Robinson, 2009]. The Asia-Pacific region is considered one of the fastest developing regions in the world, and the electrical and electronics business, which has seen enormous transformations as a result of growing technological and commercial advances, is one of the industries that have been fitted from these circumstances. Electrical and electronic equipment has become vital in today's world, enhancing living standards while also containing hazardous compounds that harm human health and the environment and contributing to the climate problem [Halim and Suharyanti, 2020]. In India, the majority of discarded electronic equipment arises in households because individuals do not know how to dispose of them properly. This ever-increasing garbage is quite complicated, as well as a rich source of metals like gold, silver, and platinum. Silver and copper are two metals that can be reclaimed and brought back to life [Monika and Kishore, 2010]. It is a multidimensional waste stream comprised of both scarce and economically valuable constituents. Effective waste management can reduce criminal and civil responsibility concerns, operational costs, and the requirement for transport and disposal [Mor *et al.*, 2021]. WEEE (waste from electronic and electrical pieces of equipment) is divided into three categories, which account for about 90% of total trash generated, with 2% accounting for big household appliances, 34% for ICT equipment, and 14% for consumer electronics. The continent of Asia produced the most electronic garbage, followed by Europe and the Americas. Interestingly, despite being the world's second-most-populous continent, Africa produced one of the lowest levels of e-waste. Although the African continent produces the least amount of e-waste in comparison to other continents due to slow technological growth and limited access to energy, it suffers from other types of pollution caused by traffic emissions, oil spills, heavy metals, refuse dumps, dust, and open burnings and incineration, all of which contribute significantly to environmental contamination in Africa [Andeobu, Wibowo, and Grandhi, 2021].

17.2.1 E-Waste Classification

Electronic trash is defined by the Organisation for Economic Co-operation and Development (OECD) as any device that is powered by electricity and has reached the end of its useful life. So, this isn't just about cell phones. Let's have a look at some of the different categories of WEEE (Waste Electrical and Electronic Equipment) that exist today, as defined by an EU directive:

- Type 1: **Major appliances** (refrigerators, washing machines, dryers, etc.)
- Type 2: **Small appliances** (vacuum cleaners, irons, blenders, fryers, etc.)
- Type 3: **Computer and telecommunication appliances** (laptops, PCs, telephones, mobile phones, etc.)
- Type 4: **Consumer electronics** (video and audio equipment, musical instruments)
- Type 5: **Lighting devices** (incandescent light bulbs, fluorescent tubes, gas-discharge lamps, etc.)
- Type 6: **Electrical and electronic tools** (drills, saws, gardening devices, etc.)
- Type 7: **Toys, leisure** (electronic toys, models, sports equipment)
- Type 8: **Medical devices** (all medical equipment except implants)
- Type 9: **Monitoring devices** (detectors, thermostats, laboratory equipment, etc.)
- Type 10: **Vending machines**

Gold and silver are employed as contacts, bonding wires, and switches in the electronics industry, whereas palladium is used in computer hard disc drives. In the sector of electronics, global demand for gold, silver, and palladium in 2015 was 254 tonnes, 12.816 tonnes, and 40.18 tonnes, respectively [Srivastava, 2020].

17.2.2 Issues on Environment and Health

According to current projections, the global pace of e-waste growth is over three times that of municipal solid garbage. E-waste has become a severe concern in China and other Asian emerging countries, as one of the largest sources of heavy metals and organic contaminants in municipal garbage and the fastest increasing waste stream [Isimekhai et al., 2017]. These countries not only generate massive volumes of domestic e-waste as a result of their high consumption rates of electrical and electronic (EE) equipment, but they also import massive amounts of used information technology (IT) gadgets from other countries. India is a growing country and the demand for electrical devices has risen in recent decades as the population has grown and lifestyles have changed. E-waste generation in India is increasing at a rate of 15% per year and is predicted to reach 8,000,000 tonnes per year in 2021. According to a report by the Central Pollution Control Board (CPCB), 65 cities in India generate more than 60-70 percent of total e-waste, which comes from 10 states. Maharashtra, Tamilnadu, Andhra Pradesh, Uttar Pradesh, West Bengal, Delhi, Karnataka, Gujarat, Madhya Pradesh, and Punjab are among the e-waste generating states in India [Sankhla et al., 2016]. Globally, 536 million metric tonnes (Mt) of e-waste were produced in 2019, according to the Global E-waste Monitor. By 2030, this quantity is expected to rise to 747 Mt. Asia (249) produced the most e-waste in 2019, followed by the Americas (131 Mt), Europe (120 Mt), Africa (29 Mt), and Oceania (07 Mt). An estimated 80% of e-waste from industrialised countries is illegally shipped to LMICs such as China, India, Nigeria, Brazil, Ghana, and Pakistan, where labour costs and disposal are low and restrictions are lax or non-existent [Parvez et al., 2021]. A successful take-back program that encourages manufacturers to design less wasteful goods, include less harmful components that are easier to disassemble reuse, and recycling could help reduce waste. This is due to the discharge of recyclable wastes in rivers, such as acids and sludges. To meet the population's needs, water is now being carried from faraway towns. The incineration of e-waste can release harmful fumes and gases into the atmosphere, contaminating it. Landfills that are not adequately regulated can pose a threat to the environment. When certain electronic devices, such as circuit breakers, are broken, the mercury will leak. PCBs (polychlorinated biphenyls) from condensers are the same way [Sankhla et al., 2016]. Metals including lead, cadmium, mercury, and nickel, as well as organic substances like flame retardants, chlorofluorocarbons, polycyclic aromatic hydrocarbons (PAHs), polybrominated diphenyl ethers (PBDEs), and polychlorinated dibenzo-p-dioxins and furans (PCDD/Fs), are all found in e-waste. E-waste recycling also recovers precious materials such as iron, aluminium, copper, silver, and rare earth metals, while prolonged exposure can be harmful. These environmental toxins pose a serious hazard to both human and environmental health [Parvez et al., 2021; Rathore et al., 2021]. Humans can be affected by e-waste in a variety of ways, including changes in thyroid function, negative neonatal outcomes, changes in cellular function and expression, psychological alterations in behaviour and temperament, and a decline in lung function. People who lived in an e-waste recycling town had significantly more DNA damage than those who lived in control communities. Studies on plant and bacterial cells have been undertaken to determine the effect of leachate on genetic material. Additionally, it has been established that leachate is genotoxic to mammalian cells. Leachates can be absorbed by cells, altering the pH both inside and outside, potentially affecting enzymes and changing the structure of DNA. Another theory is that the presence of free radicals in the cell reacts with lipids, causing peroxidation of the cell membrane in tissues and DNA breakage. Furthermore, free radicals cause base substitution by charging nucleic acid constituents such as purine and pyridine, which causes DNA to break and promote mutation [Ankit et al., 2021]. Because pollutants are emitted as a mixture, the consequences of exposure to a single component or element cannot be studied separately. Many pathways and sources of exposure, varied periods of exposure time, and possible inhibitory, synergistic, or additive effects of many chemical exposures are all key variables in e-waste exposure. Exposure to e-waste is a distinct variable in and of itself and the various exposures involved should be taken into account as a whole. There are three types of e-waste exposure sources: informal recycling, formal recycling, and exposure to hazardous e-waste chemicals still in the environment [Grant et al., 2013]. Burning e-waste on an open landfill to extract gold and other precious metals produces fine particulate matter, which causes cardiovascular and pulmonary problems in children in the region. Toxic particles are carried by the wind pattern of a certain place and reach the soil-crop-food pathway, damaging both people and animals when they enter the food chain. Mercury levels in motherboards are abnormally high and improper disposal could result in skin and respiratory illnesses [Kumar et al., 2022]. Drinking water contaminated with lead hurts the central and nervous systems, resulting in stunted brain growth, dwarfism, hearing loss, and blood cell formation and function problems. When the exhaust created by the combustion of the mother circuit board is inhaled, it causes lung cancer. Dermatological problems were also caused by it [Dharini *et al.*, 2017].

17.3 Conventional Approaches to E-Waste Recycling, Advantages and Disadvantages

E-waste contains rich secondary resources such as metals, polymers, glass, and rare earth elements. Traditional e-waste recycling initiatives, on the other hand, have mostly focused on collecting useful and easily extractable components like metals, which are present in relatively high concentration and extractable forms [Baccini and Brunner, 2012]. Despite the difficulties of good e-waste management, recycling e-waste has benefited the environment as well as human health. The conventional methods of e-waste collection and processes for recovering valuable metals from e-waste are briefly discussed in this section with potential merits and pitfalls (Table 17.1).

17.3.1 Physical Methods

Pre-treating e-waste to separate the metallic and non-metallic fractions is required before separation or extraction techniques. Physical separation techniques have minimal capital and operational expenses, but they lose a lot of valuable metal (10–35%) due to insufficient metal liberation [Kaya, 2016b]. Disassembly or dismantling is the first step in the pre-treatment process [Sethurajan *et al.*, 2019; Lu *et al.*, 2015]. The goal of disassembly is to reduce the harmful components of e-waste in the recycling process while also focusing on the reusable components [Gollakota, Gautam, and Shu, 2020]. There are two types of disassembly: selective and simultaneous disassembly, which takes place in stages for the first and only one step for the second. A significant aspect of selective disassembly is to look at and select specific components. During simultaneous disassembly, the e-waste is heated and de-soldered, followed by form and size identification [Hsu *et al.*, 2019b; Yaashikaa *et al.*, 2022]. Both of these processes have important advantages and disadvantages, for example, during simultaneous disassembly there is a high danger of destroying components when heating, additional sorting facilities are required, processing times are lengthened, and the cost is considerable [Chauhan *et al.*, 2018]. Physical methods include the following:

Table 17.1 E-waste recycling approaches with possible merits and demerits.

Methods	Advantages	Disadvantages
Simultaneous Disassembly	Minimal capital and operational costs	Higher danger of destroying components
Incineration	Common method and ease of operation	Toxic hazardous dioxins are discharged along with fly ash
Sieving	Allows shredded e-waste to be classified into different size fractions	Have to employ secondary extraction processes
Magnetic Separation	Prevalent in the faster separation of components	Sometimes aggregates of non-ferrous particles are attracted along with the ferrous substances
Pyrometallurgy	Can handle any type of scraps	Toxic byproducts are released like polybrominated dibenzodioxins (PBDDs), phenol, dibenzo-p-dioxin; difficulty in recovering pure metals
Hydrometallurgy	Release of fewer harmful gases, lower energy consumption	Toxic and flammable reagents are used in hydrometallurgical processes, which also produce large amounts of solid waste and effluents.
Biometallurgy	Lower costs and very specific extraction of metals by microbes	It is a time consuming process

17.3.1.1 Incineration

E-waste is heated in this approach to transform the calorific value into energy by releasing gas and eliminating the non-metallic component. This approach is still commonly used in Asia, Europe, and America due to its ease of use. Toxic hazardous components such as heavy metals, polybrominated/polychlorinated dibenzo-dioxins (PBDD/PCDD), and fly ash are discharged into the environment during the combustion process before being purified. Cadmium, lead, zinc, copper, and nickel, among other metallic compounds, melt at their melting points and are then discharged into the environment. Copper metal in PCBs acts as a catalyst for dioxin production during the process [Amato, Becci, and Beolchini, 2020]. In addition to gaseous emissions, the incineration process produces solid leftovers known as bottom ash, which will be treated or landfilled using different treatment methods. Because toxic elements are released during the process, incineration is not considered an environmentally beneficial method. The cost of building the incineration plant is similarly considerable [Yaashikaa et al., 2022].

17.3.1.2 Sieving

Plastics, polymers, and flame retardant compounds make up the majority of e-waste. Sieving is a size separation process that allows shredded e-waste to be classified into size fractions. To ensure consistent distribution of input to various physical or chemical processing procedures, the process is extremely desirable. Rotating screens and trammels are the most common screening technologies. To filter the dangerous metal fines indicated in the previous section, a secondary crushing through a dry air cyclone classifier with a dust collection system is used before proceeding to the shred and screening procedures [Kaya, 2016b].

17.3.1.3 Gravity Separation

Components are separated using this strategy based on their different specific gravity. Component separation is caused by the movement of materials in air and water to gravity and other factors. This separation is determined by the form, size, and density of the components. Because of its low density, aluminium is easily separated, whereas copper is easily separated due to its size [Amit Kumar, Holuszko, and Janke, 2018].

17.3.1.4 Magnetic Separation

Magnetic particles are separated from non-magnetic particles using this technology. Low-intensity magnetic drum separators can separate ferromagnetic metals like aluminium from non-magnetic waste items. Although a magnet can be utilised for this technique, the main disadvantage is particle agglomeration, which can cause nonferrous particles to attract ferrous particles, resulting in decreased productivity and efficiency [Wang et al., 2020].

17.3.1.5 Electrochemical Separation

Finally, electrochemical separation is a highly regarded technique based on the principle of electrical conductivity, in which the eddy current and magnetic field aid in the separation of Fe and non-Fe fractions, the separation of plastics from metal mixtures, and the identification of fine grind particles [Hsu et al., 2019b]. Electrochemical separation is a promising method of recovering basic elements from WEEE that is low-cost, environmentally friendly, and uses few chemicals [Lister, Wang, and Anderko, 2014]. Electrochemical reactions are heterogeneous in general and they occur when the charge is transferred across the electrode-electrolyte contact [Gollakota, Gautam, and Shu, 2020]. This approach is mostly used to recover noble metals like gold and silver, which are electropositive and electrodeposit more readily than other metals [Yaashikaa et al., 2022].

17.3.1.6 *Metallurgical Methods*

The next step in recycling e-waste is to enrich valuable metals by removing non-essential and non-recyclable components. It entails disassembling the EEE to collect individual components. E-waste contains different amounts of ferrous components (e.g. iron, steel), non-ferrous metals (e.g. silver, gold, copper, aluminium, platinum), plastics, glass, wood, rubber, and other materials, depending on the kind, manufacturer, and age of the equipment. These components are divided into batches of metals, plastics, ceramics, paper, and wood, as well as categories like capacitors, batteries, LCDs, and PCBs, before being shredded into smaller bits using mechanical processes. For these applications, industrial-scale granulation machines are utilised. By separating the processed e-waste into metallic and non-metallic fractions using magnetic, current-based, or density separation techniques, the e-waste is enhanced for recovery. Plastic from e-waste is usually recycled as a source of energy and to make plastic bottles. Hydrometallurgical, pyrometallurgical, and biometallurgical processes are used to recover metals from processed e-waste [Ahirwar and Tripathi, 2021; Thakur and Kumar, 2020].

17.3.1.7 *Pyrometallurgy*

Pyrometallurgy is a traditional thermo physical separation technique that involves a complex combination of internal processes such as smelting, drossing, sintering, melting, and gas-phase reactions [R. Li et al., 2019]. It is a high-temperature technique for extracting or purifying non-ferrous components from metallurgical materials. Pyrometallurgical processes in Cu and Pb smelters are used for the majority of e-waste recycling on an industrial scale, especially for Pb, Cu, and precious metals. The flexibility to handle any type of scrap is the main benefit of pyrometallurgical treatment. As a result, electronic scrap can be used as a source of raw materials in smelters to recover Cu, Au, and Ag [Kaya, 2016b]. Because of flame retardants and varied plastic blends, e-waste plastic parts are difficult to recycle. However, because of the energy released from the combustion of plastics and combustible elements in e-waste, smelting operations utilise less energy [van Schaik and Reuter, 2014; Abdelbasir et al., 2018; Dubey et al., 2022]. Pyrometallurgy begins with liberation, then separation, and lastly purification. E-waste is diverted to smelters, for example, valuable metals are isolated in a solvent metal phase, such as copper or lead, which is a vital stage in pyrometallurgy. In general, two types of smelting techniques are used: flash smelting, which uses oxygenated gas to promote autogenous conditions, and bath smelting, which involves a molten pool containing both slag and melt phases and relies solely on the roasting and smelting steps before being sent to the conversion process. Finally, these e-waste slags and smelt conversion processes are carried out in traditional lead smelting and copper smelting routes [Khaliq et al., 2014]. Pyrometallurgy is followed by hydrometallurgical and electrometallurgical processes to recover pure base and precious metals. Platinum group metals [e.g., Ruthenium (Ru), Rhodium (Rh), Platinum (Pt), and Palladium (Pd)] are recovered from e-waste using a combination of hydro-, pyro-, and electrometallurgical methods [Mohdee et al., 2018]. Some of the thermal plants available for formal processing by pyrometallurgical techniques are Aurubis smelter, Noranda smelter, Ronnskar smelter, and Umicore. After this procedure, some precious metals are left over, which can be recovered using other methods such as electro-refining and electro-winning [Thakur and Kumar, 2020].

17.3.1.8 Disadvantages of Pyrometallurgy

Pyrometallurgy has its own set of issues and constraints, particularly in terms of the environment, for example, the release of toxins like polybrominated dibenzodioxins (PBDDs), phenol, dibenzo-p-dioxin, biphenyl, anthracene, dibromobenzene, naphthalene, and polybrominated dibenzofurans (PBDFs) [Tue et al., 2013], in traditional and cost-effective smelters [Kaya, 2016b]. When flame retardants and polyvinyl chloride (PVC) included in e-waste are melted, it leads to the production of dangerous dioxins, which requires special emission controls to protect the environment [Abdelbasir et al., 2018]. Pyrometallurgy has not been proven to be acceptable for low-grade WEEE due to the high-grade feed required [Xia et al., 2018]. Smelting is also not recommended since it produces complicated slags of Al and Fe, which cannot be retrieved and are additional industrial waste to be dealt with. The method has considerable problems, including ineffective or partial metal recovery, a longer residence time for complete separation of valuables from e-waste, and an additional purification stage by electro-refining, among others, which adds to the process' complexity. In terms of mass balance, the pyrometallurgical process results in increased metal losses, which are typically lost in the form of slag, refractories, or dust [Gollakota, Gautam, and Shu, 2020; Tesfaye et al., 2017]. Toxic by-products and the difficulty of recovering pure metals are two other issues that limit the utilisation of pyrometallurgical processes [Kuyucak and Akcil, 2013].

17.3.1.9 Hydrometallurgy

Hydrometallurgical procedures are used due to the limits of pyrometallurgical processes. Some of the advantages that attract the use of hydrometallurgical processes include the release of fewer harmful gases, lower energy consumption, simplicity of execution in the laboratory, and lower operating expenses combined with a greater efficiency rate [Ni et al., 2013]. The selective process of metal extraction by a series of acid/alkaline leaches is known as hydrometallurgical treatment or leaching. To dissolve the valuables from the leachate, a variety of chemicals are utilised as leaching agents. The solutions are then subjected to decontamination processes like precipitation, solvent extraction, adsorption, and ion exchange to separate the concentrate from the metals [Birloaga et al., 2014; Coman, Robotin, and Ilea, 2013]. For the pure phases, the solutions are subjected to electro-refining, chemical reduction, and crystallisation. Under the impact of dissolving, desorption, and complexation, organic pollutants or radionuclides are released from the solid phase into the waste phase [Ghosh et al., 2015]. Chemical leaching is the technique of recovering precious metals such as Au and Cu by employing acids or ligand support complications. Solo chemical leaching, chemical leaching with ligand, chemical leaching with acids, and hydrometallurgical etching are all examples of leaching processes. Chemical leaching is divided into four categories: cyanide leaching, halide leaching, thiourea leaching, and thiosulphate leaching [Pant et al., 2012a]. For the recovery of pure base and precious metals from e-waste, several mineral acids and oxidants have been utilised, including aqua-regia ($3HCl:HNO_3$), HCl, H_2SO_4, HNO_3/H_2O_2, $HClO_4$, $NaClO_4$, and lixiviants such as cyanide, thiosulfate, thiourea, and halide [Ding et al., 2019]. Thiosulfate is most commonly used to dissolve gold metal. Copper ion and ammonia, in combination with thiosulfate solution, promote gold metal recovery while also acting as a catalyst [Thakur and Kumar, 2020].

Disadvantages of Hydrometallurgy

Toxic, very acidic/alkaline, or flammable reagents are used in hydrometallurgical processes, which also produce large amounts of solid waste and effluents. The halogenated flame retardants used in printed circuit boards produce hazardous dioxin, furans, and volatile metals, necessitating off-gas treatment in hydrometallurgy [Ahirwar and Tripathi, 2021; Shah, Tipre, and Dave, 2014]. While hydrometallurgy produces fewer hazardous gases than pyrometallurgy, it is plagued by acid solutions and toxic vapours, which are severe drawbacks [Priya and Hait, 2017].

17.3.1.10 Biometallurgy

Biometallurgical processes are common mineral separation procedures that are also thought to be an alternate technology for recovering metals from low-grade ores and concentrate [Borja *et al.*, 2016]. Biometallurgy occurs by bioleaching and biosorption. Bioleaching relies on microorganisms' ability to transform solid substances into soluble and extractable components that can be recovered. It is used in recovering base metals such as Cu, Ni, Zn, and Cr, as well as REEs, Au, and Ag. Algae, bacteria, yeasts, and fungus are used in biosorption to collect heavy and precious metals, prevalently gold and copper. These bacteria are utilised as adsorbents for precious metal biosorption, which is a complex process involving metal adsorption onto cell walls or cell-associated compounds [Namias *et al.*, 2013; R. Chauhan and Upadhyay, 2015]. Microbiological leaching is effective in the mobilisation of metals from e-waste in research conducted over the last 20 years. When compared to traditional methods, the biosorption process has several advantages, including lower operating costs, a smaller volume of chemical sludge to manage, and high effluent depollution efficiency [Abdelbasir *et al.*, 2018]. Numerous studies have shown that moderate thermophiles have a higher bioleaching capability than mesophilic and extreme thermophiles, hence many *Thiobacillus* bacteria and thermophilic fungi (e.g., *A. niger* and *P. simplicissimum*) have been employed to extract metals from e-waste and low-grade metal reservoirs [Jadhav, Su, and Hocheng, 2016]. Acidophilic bacteria, among the different bacterial strains available, also have a specialisation and importance in the e-waste leaching process. *Acidithiobacillus ferrooxidans*, *Acidithiobacillus thiooxidans*, *Leptospirillum ferrooxidans*, and *Sulfolobus sp.* are some of the common acidophilic strains and have a strong affinity to solubilize base metals like Cu, Ni, and Al [Liang, Mo, and Zhou, 2010; Hong and Valix, 2014]. The most widely utilised chemolithoautotrophic bacteria, such as *Acidithiobacillus ferrooxidans* and *Acidithiobacillus thiooxidans*, have a strong affinity to solubilize base metals (Cu, Ni, Al) [Ilyas, Lee, and Chi, 2013]. Bioleaching, as is well known, uses an acid media in conjunction with active oxidising agents to solubilize metals into ion form. As a result, bacterial activity promotes metal solubilization through two mechanisms: contact and non-contact [Borja et al., 2016]. The non-contact mechanism is the oxidation of minerals by oxidising agents, mainly Fe, and the contracting mechanism is the oxidation of minerals by the attachment of the microbe to the surface of the metal [Johnson, 2013].

As bacterial growth is inhibited by a high concentration of metals, scientists proposed a new idea of using consortia and a two-step bioleaching process. In this method, in the absence of e-waste, a bacterial culture is permitted to create maximum lixiviant before the leaching procedure. [Xiang et al., 2010] used a bacterial consortium from the genera *Acidithiobacillus sp.*, *Gallionella sp.*, and *Leptospirillum sp.* to optimise the conditions for optimal bacterial growth and maximum Cu metal solubilization. With an initial pH of 1.5, a Fe^{2+} content of 9 g/l, and a pulp density of 20 g/l, they were able to achieve 95 percent Cu leaching. They showed that employing a consortium reduces leaching time to 5 days as compared to using a single culture (i.e., 12 days). Biohydrometallurgical processes are (beneficial) in terms of reducing environmental footprint, cleaner alternative lixiviants, improved procedures for maximal lixiviant products, and reduced hazardous material generation.

Disadvantages of Biometallurgy

Biometallurgy generates a lesser amount of wastewater than other metallurgical processes. Because active bioagents are used in biometallurgy, it has a higher rate of metal extraction from depleted, lower-grade, or complex sources than pyrometallurgy and hydrometallurgy with the least energy requirement. However, when compared to the other two metallurgical processes, the bio-metallurgical approach is the most time-consuming, expertise-intensive, and slow process. It is essentially a set of hydrometallurgical techniques that use a variety of microorganisms (bacteria and fungus) to improve metal solubility and mobilisation from ores, deposits, and wastes [Priya and Hait, 2017].

17.4 Advances in Approaches for Improving E-Waste Recycling for Value-Added Materials and Biomaterials Generation

The existing methods for recycling e-waste are often primitive and sometimes result in waste streams that are more harmful than the original e-waste. As a result, developing alternative methods that allow for the efficient recovery of key metals from trash is extremely desirable (Ilankoon et al., 2018). Biological technologies offer obvious advantages because of their low cost and eco-friendliness in the extraction of Au, Ag, As, Co, Cu, Mn, Mo, and Ni. Furthermore, bio-based technologies may be more selective toward metals, giving them an advantage [Watling, 2014]. Improved biometallurgical methods have been a progressive step towards developing new extraction methods (Table 17.2). Metal extraction from e-waste has been accomplished using a variety of traditional technologies. However, such technologies are prohibitively expensive or might result in secondary contamination requiring additional treatment, whereas the biological approach to metal recycling from e-waste is environmentally beneficial. A hybrid technique combining biological and chemical leaching was used for the extraction of Zn, Cd, Pb, and Cu with EDTA with A. ferroxidans bacterial strain DSM 9103 [Pant et al., 2012b]. For example, utilising a mix of hydrometallurgical and biometallurgical processes, [Bhat, Rao, and Patil, 2012] suggested an integrated model for the recovery of Au and Ag from WEEE. They found that *Eichhornia* root biomass and waste tea powder were effective biosorbents for recovering leached silver-cyanide from electronic scrap and that the concentrated silver-cyanide recovered in the biosorption method may be used as an electroplating industry input material. Studies have shown that organic acids produced from A. niger (oxalic, malic, gluconic, and citric acid) acted as chelating agents for recovering metals from spent lithium-ion batteries [Bahaloo-Horeh, Mousavi, and Baniasadi, 2018]. According to [Sahni, Kumar, and Kumar, 2016], chemo-biohydrometallurgy can be used to recover metals from old SIM cards using microorganisms like *Chromobacterium violaceum* (e-waste). By employing acidic pre-treatment of SIM e-waste in two-step bioleaching, they were able to recover 13.79 percent of copper, 2.55 percent of silver, and 0.44 percent of the gold from untreated SIM trash, while roughly 72 percent of copper was mobilised from SIM e-waste. For gold recovery from waste PCBs, Sheel and Pant (2018) developed a modified technique involving a blend of *Lactobacillus acidophilus* (LA) and ammonium 350 thiosulfate (AT). This approach provides an excellent leaching solution for gold recovery, with the indicated mixture resulting in 85 percent gold extraction (using AT as leachate) [Sheel and Pant, 2018]. Even though cyanogenic bacteria have been frequently employed for precious metal bioleaching, new isolates may yield better results. Under ideal conditions, *P. balearica* SAE1 isolated from an e-waste recycling plant showed a leaching rate of 68.5 percent for gold and 33.8 percent for silver from a 10 g/L pulp density [Anil Kumar, Saini, and Kumar, 2018]. Microbial consortium (*Acidithiobacillus*

Table 17.2 Microbial regime involvement for E-waste recycling and recovery.

Organism	Function	Reference
Lysinibacillus sphaericus	At an initial concentration of 60 ppm, *Lysinibacillus sphaericus* encapsulated in alginate matrix recovers 100 percent of Au (III) after three hours.	[Páez-Vélez, Rivas, and Dussán, 2019]
Tepidimonas fonticaldi	Demonstrates a remarkable adsorption capability for gold (Au) (III).	[Han et al., 2017]
Saccharomyces cerevisiae	The silver (Ag) (I) binding capability of Saccharomyces cerevisiae cells is considerably increased when the CueR protein is overexpressed on the cell surface.	[Tao et al., 2016]
Pseudomonas balearica	Under ideal conditions, *P. balearica* SAE1 extracted from an e-waste recycling facility shows a leaching rate of 68.5 percent for Au and 33.8 percent for Ag from a 10 g/L pulp density.	[Anil Kumar, Saini, and Kumar, 2018]
Cellulosimicrobium funkei	For gallium extraction from thin-film GaAs solar cell waste, *Cellulosimicrobium funkei* has a 70% effective leaching rate.	[Maneesuwannarat et al., 2016]
Aspergillus genus	The production of citric and oxalic acids by the Aspergillus genus resulted in a high recovery of 90-95 percent from Zn-Mn and Ni-Cd batteries.	[Kim et al., 2016]
Penicillium expansum	After three weeks of cultivation, *Penicillium expansum* can bio-concentrate up to 390 ppm of Lanthanum and 1520 ppm of Terbium from WEEE.	[di Piazza et al., 2017]
Acidithiobacillus ferrooxidans	Nickel and gallium leaching rates are increased by *Acidithiobacillus ferrooxidans*. At ideal conditions, it also removes up to 80% of Cu and other base metals.	[Jowkar et al., 2018]

(Continued)

Table 17.2 Microbial regime involvement for E-waste recycling and recovery. (*Continued*)

Organism	Function	Reference
Chromobacterium violaceum	Recovery of gold from waste PCBs.	[J. Li, Liang, and Ma, 2015]
Acidithiobacillus ferrivorans and *Acidithiobacillus thiooxidans*, *Pseudomonas. fluorescens*, *Pseudomonas putida*	Improves gold recovery from waste PCBs by 1.6-fold.	[Işıldar *et al.*, 2016]
Phomopsis sp.	Without any pre-treatment, the *Phomopsis sp.* XP-8 fungus can selectively recover roughly 80% of Au from electronic wastewater.	[X. Xu *et al.*, 2019]
Shewanella algae	Recovers platinum (IV), palladium (II), and rhodium (III) simultaneously from the aqua regia leachate of used automotive catalysts.	[Saitoh, Nomura, and Konishi, 2017]
Acidocellaaromatica sp. and *Acidiphiliumcryptum* sp.	Recovers palladium (Pd) from acidic Pd (II) solutions.	[Okibe, Nakayama, and Matsumoto, 2017]
Pseudomonas aeruginosa	Extracellular bioreduction of silver along with nitrate reductase.	[J. Ali *et al.*, 2017]
Raoultellaornithinolytica sp. and *Raoultellaplanticola* sp.	When glucose is used as an electron source, *Raoultellaornithinolytica* sp. and *Raoultellaplanticola* sp. strain MoI show excellent molybdate reduction capacity.	[Sae ed, el Shatoury, and Hadid, 2019]
Bacillus sp.	In ideal conditions, a keratin-degrading *Bacillus* sp. strain can convert molybdate to molybdenum.	[Khayat *et al.*, 2016; S. Wu *et al.*, 2019]
Shinella sp.	In the reduction of Te(IV), *Shinella* sp. WSJ-2 is discovered.	[S. Wu *et al.*, 2019]
Lactobacillus acidophilus	*Lactobacillus acidophilus* along with ammonium thiosulfate achieves 85% gold recovery from waste PCBs.	[Sheel and Pant, 2018]

thiooxidans and *Leptospirillum ferriphilum*) has also been reported to be able to bioleach Cu, Mn, and Zn from spent batteries [Niu *et al.*, 2015].

Biotechnologies have a long history of being used to process low-grade ores. When compared to main ores, WEEE differs in terms of chemical composition, metal abundance, and complexity. In most WEEE mixtures, there are large amounts of ordinary metals and a lesser concentration of essential metals. Current WEEE recycling methods are insufficient to target key metals, which are usually found in low amounts [Ilyas *et al.*, 2017]. It also necessitates providing more energy to the bacteria. This particular difficulty necessitates the development of unique key metal recovery technologies from WEEE. Some of the key leaching mechanisms are not fully understood, so additional fundamental study on WEEE bioprocessing is needed. To progress further into full-scale applications, more research is required, including optimization of operational conditions and examination of environmental implications. In addition, in biotechnological solutions for metal recovery, incorporating scale-up studies with techno-economic assessment and environmental sustainability analysis considerations are significant considerations (Işıldar *et al.*, 2019).

17.5 Conclusion

Electronic and electrical equipment have become nearly vital in this age of information and communication technology. Many of these devices become obsolete within a few years of their manufacturing due to the fast-changing nature of technology. These characteristics, combined with an increase in demand owing to people's increased spending power around the world, result in massive amounts of e-waste being generated. Because 2 Mt of e-waste is generated every year around the world, it is a major concern for electrical and electronic equipment makers, regulatory organisations, and recyclers to work together to ensure that it is properly managed. The accumulation of e-waste could have serious effects on human health and the environment's long-term viability. Many countries are attempting to control municipal solid waste, but the E-waste management strategy is frequently overlooked. E-waste recycling and treatment facilities demand large capital investments, yet few countries set aside funds for this purpose. E-waste is processed and managed informally and illegally in most underdeveloped countries. Due to a lack of implementation of current legislation, large populations in developing nations are still unaware of E-waste, applicable policies, rules and regulations, and its handling. As a result, it is critical to evaluate existing practises, regulations, and recycling infrastructures, raise awareness and tackle the problem by utilising a cost-effective, safe, and environmentally-friendly (bio) technical method. Furthermore, end-users must be informed of the issues connected with inappropriate waste disposal and are required to participate in formal recycling programmes. These recycling processes assist not only the environment but also the economy of the country because of the retrieved metals. E-waste is a valuable economic resource. As a result, proper resource recovery should begin to ensure that resource usage does not harm the ecosystem or generate adverse health impacts. Hydrometallurgy is a potential method for recovering metals from WEEE at all concentration levels. Biohydrometallurgy has been demonstrated to be a versatile method for selectively recovering the metals of interest from e-waste utilising predefined/acclimated biocatalysts, making it an environmentally friendly and green technical choice. Hydrometallurgy, pyrometallurgy, and pyrolysis are utilised at the industrial scale where fast processing times are needed, but they would need post-treatment to deal with the hazardous chemical solution or gases, as well as to retrieve the final metal products from e-waste. Many countries have enacted legislation to encourage reuse, recycling, as well as other types of trash recovery. The analysis of e-waste revealed that it is highly varied and complicated in terms of component and material types. Copper and precious metals, on the other hand, account for greater than 80% of the value of most e-waste samples. This suggests that precious metals and copper recovery will likely continue to be a significant economic driver for a long period. The ladder of e-waste management advocates the use of the entire equipment first, followed by remanufacturing, material recovery via recycling procedures, and finally, incineration and landfilling as the last choice. The recycling of e-waste can be divided into three main stages: (a) preferential dismantling, targeting, and singling out of dangerous or precious components for special treatment; (b) mechanical and/or metallurgical applications to enhance preferable materials content; and (c) refining retrieved materials that have been retreated or purified using chemical (metallurgical) processing to be appropriate for use in their authentic implementation. However, the method is hampered by several constraints, including sluggish and low yielding outputs, a lack of space and equipment, and present insufficiency for handling complex metal mixtures. The technology is linked to two main areas of sustainability: environmental, resource, and health protection, as well as profitability measured by the value and use of recovered items. However, given the current situation in underdeveloped nations, a decentralised strategy is required as a top priority for improving e-waste handling from start to finish. Advancements in technology, as well as overcoming limitations in existing treatment alternatives, are required for a more sustainable, resilient ecosystem, as well as improved health, safety, and resource conservation.

Acknowledgement

The author would like to thank JIS University and the JIS Group of educational initiatives.

References

Abalansa, Samuel, BadrelMahrad, John Icely, and Alice Newton. "Electronic Waste, an Environmental Problem Exported to Developing Countries: The Good, the Bad and the Ugly." *Sustainability (Switzerland)* 13 (9), 2021. https://doi.org/10.3390/su13095302.

Abdelbasir, Sabah M., Saad S.M. Hassan, Ayman H. Kamel, and Rania Seif El-Nasr. "Status of Electronic Waste Recycling Techniques: A Review." *Environmental Science and Pollution Research*. Springer Verlag, 2018. https://doi.org/10.1007/s11356-018-2136-6.

Abdu, Nafiu, Aliyu A. Abdullahi, and Aisha Abdulkadir. "Heavy Metals and Soil Microbes." *Environmental Chemistry Letters*. Springer Verlag, 2017. https://doi.org/10.1007/s10311-016-0587-x.

Adrian, S, M BruneDrisse, Y Cheng, L Devia, O Deubzer, F Goldizen, J Gorman, et al. "Quantities, Flows, and the Circular Economy Potential The Global E-Waste Monitor 2020," 2020.

A.Dubey, S.Narang, A.Srivastav, A.Kumar, V.Díaz, Woodhead Publishing, Science Direct, Artificial Intelligence for Renewable Energy Systems. Paperback ISBN: 9780323903967

A.Dubey, S.Narang, A.Srivastav, A.Kumar, V.Díaz, Woodhead Publishing, Science Direct, a Visualization Techniques for Climate Change with Machine Learning and Artificial Intelligence. ISBN: 9780323997140

Ahirwar, Rajesh, and Amit K. Tripathi. "E-Waste Management: A Review of Recycling Process, Environmental and Occupational Health Hazards, and Potential Solutions." *Environmental Nanotechnology, Monitoring and Management*. Elsevier B.V., 2021. https://doi.org/10.1016/j.enmm.2020.100409.

Ali, Hazrat, and Ezzat Khan. "Bioaccumulation of Non-Essential Hazardous Heavy Metals and Metalloids in Freshwater Fish. Risk to Human Health." *Environmental Chemistry Letters*. Springer Verlag, 2018. https://doi.org/10.1007/s10311-018-0734-7.

Ali, Jafar, Naeem Ali, Syed Umair Ullah Jamil, Hassan Waseem, Kifayatullah Khan, and Gang Pan. "Insight into Eco-Friendly Fabrication of Silver

Nanoparticles by Pseudomonas Aeruginosa and Its Potential Impacts." *Journal of Environmental Chemical Engineering* 5 (4), 2017: 3266–72.

Amato, Alessia, Alessandro Becci, and Francesca Beolchini. "Sustainable Recovery of Cu, Fe and Zn from End-of-Life Printed Circuit Boards." *Resources, Conservation and Recycling* 158 (July), 2020: 104792. https://doi.org/10.1016/J.RESCONREC.2020.104792.

Andeobu, Lynda, Santoso Wibowo, and SrimannarayanaGrandhi. "A Systematic Review of E-Waste Generation and Environmental Management of Asia Pacific Countries." *International Journal of Environmental Research and Public Health*. MDPI, 2021. https://doi.org/10.3390/ijerph18179051.

Ankit, Lala Saha, Virendra Kumar, Jaya Tiwari, Sweta, Shalu Rawat, Jiwan Singh, and Kuldeep Bauddh. "Electronic Waste and Their Leachates Impact on Human Health and Environment: Global Ecological Threat and Management." *Environmental Technology and Innovation* 24 (November), 2021. https://doi.org/10.1016/j.eti.2021.102049.

Arya, Shashi, and Sunil Kumar. "Bioleaching: Urban Mining Option to Curb the Menace of E-Waste Challenge." *Bioengineered*. Taylor and Francis Inc., 2020. https://doi.org/10.1080/21655979.2020.1775988.

Asante, Kwadwo Ansong, Sam Adu-Kumi, Kenta Nakahiro, Shin Takahashi, TomohikoIsobe, AgusSudaryanto, GnanasekaranDevanathan, et al. "Human Exposure to PCBs, PBDEs and HBCDs in Ghana: Temporal Variation, Sources of Exposure and Estimation of Daily Intakes by Infants." *Environment International* 37 (5), 2011: 921–28. https://doi.org/10.1016/j.envint.2011.03.011.

Ashiq, Ahamed, Janhavi Kulkarni, and Meththika Vithanage. "Hydrometallurgical Recovery of Metals from E-Waste." In *Electronic Waste Management and Treatment Technology*, 225–46. Elsevier, 2019. https://doi.org/10.1016/B978-0-12-816190-6.00010-8.

Baccini, Peter, and Paul H Brunner. *Metabolism of the Anthroposphere: Analysis, Evaluation, Design*. MIT Press, 2012.

Bahaloo-Horeh, Nazanin, Seyyed Mohammad Mousavi, and MahsaBaniasadi. "Use of Adapted Metal Tolerant Aspergillus Niger to Enhance Bioleaching Efficiency of Valuable Metals from Spent Lithium-Ion Mobile Phone Batteries." *Journal of Cleaner Production* 197, 2018: 1546–57.

Baldé, C P, F Wang, and R Kuehr. "Transboundary Movements of Used and Waste Electronic and Electrical Equipment Estimates from the European Union Using Trade Statistics," 2015.

Bhat, Viraja, Prakash Rao, and Yogesh Patil. "Development of an Integrated Model to Recover Precious Metals from Electronic Scrap-A Novel Strategy for e-Waste Management." *Procedia-Social and Behavioral Sciences* 37, 2012: 397–406.

Birloaga, Ionela, VasileComan, Bernd Kopacek, and Francesco Vegliò. "An Advanced Study on the Hydrometallurgical Processing of Waste Computer Printed Circuit Boards to Extract Their Valuable Content of Metals."

Waste Management 34 (12), 2014: 2581–86. https://doi.org/10.1016/j.wasman.2014.08.028.

Borja, Danilo, Kim Anh Nguyen, Rene A. Silva, Jay Hyun Park, Vishal Gupta, Yosep Han, Youngsoo Lee, and Hyunjung Kim. "Experiences and Future Challenges of Bioleaching Research in South Korea." *Minerals*. MDPI AG, 2016. https://doi.org/10.3390/min6040128.

Chauhan, Garima, Prashant Ram Jadhao, K K Pant, and K D P Nigam. "Novel Technologies and Conventional Processes for Recovery of Metals from Waste Electrical and Electronic Equipment: Challenges & Opportunities–a Review." *Journal of Environmental Chemical Engineering* 6 (1), 2018: 1288–1304.

Chauhan, Ruchi, and Kanjan Upadhyay. "Removal of Heavy Metal from E-Waste: A Review." *IJCS* 3 (3), 2015: 15–21.

Chi, Xinwen, Martin Streicher-Porte, Mark Y.L. Wang, and Markus A. Reuter. "Informal Electronic Waste Recycling: A Sector Review with Special Focus on China." *Waste Management* 31 (4), 2011: 731–42. https://doi.org/10.1016/j.wasman.2010.11.006.

Coman, V, B Robotin, and P Ilea. "Nickel Recovery/Removal from Industrial Wastes: A Review." *Resources, Conservation and Recycling* 73, 2013: 229–38.

Cui, Jirang, and Hans Jørgen Roven. "Electronic Waste." In *Waste*, 281–96. Elsevier Inc., 2011. https://doi.org/10.1016/B978-0-12-381475-3.10020-8.

Dharini, K., J. Bernadette Cynthia, B. Kamalambikai, J. P. Arul Sudar Celestina, and D. Muthu. "Hazardous E-Waste and Its Impact on Soil Structure." In *IOP Conference Series: Earth and Environmental Science*. Vol. 80. Institute of Physics Publishing, 2017. https://doi.org/10.1088/1755-1315/80/1/012057.

Ding, Yunji, Shengen Zhang, Bo Liu, Huandong Zheng, Chein chi Chang, and Christian Ekberg. "Recovery of Precious Metals from Electronic Waste and Spent Catalysts: A Review." *Resources, Conservation and Recycling*. Elsevier B.V., 2019. https://doi.org/10.1016/j.resconrec.2018.10.041.

Dubey, A.K., Kumar, A., Ramirez, I.S., Marquez, F.P.G. (2022). A Review of Intelligent Systems for the Prediction of Wind Energy Using Machine Learning. In: Xu, J., Altiparmak, F., Hassan, M.H.A., García Márquez, F.P., Hajiyev, A. (eds) Proceedings of the Sixteenth International Conference on Management Science and Engineering Management – Volume 1. ICMSEM 2022. Lecture Notes on Data Engineering and Communications Technologies, vol 144. Springer, Cham. https://doi.org/10.1007/978-3-031-10388-9_35

Feldt, Torsten, Julius N. Fobil, Jürgen Wittsiepe, Michael Wilhelm, Holger Till, Alexander Zoufaly, Gerd Burchard, and Thomas Göen. "High Levels of PAH-Metabolites in Urine of e-Waste Recycling Workers from Agbogbloshie, Ghana." *Science of the Total Environment* 466–467, 2014: 369–76. https://doi.org/10.1016/j.scitotenv.2013.06.097.

Gaidajis, G, K Angelakoglou, and D Aktsoglou. "E-Waste: Environmental Problems and Current Management Engineering Science and Technology Review." *Journal of Engineering Science and Technology Review*. Vol. 3, 2010. www.jestr.org.

Ghosh, B., M. K. Ghosh, P. Parhi, P. S. Mukherjee, and B. K. Mishra. "Waste Printed Circuit Boards Recycling: An Extensive Assessment of Current Status." *Journal of Cleaner Production*. Elsevier Ltd, 2015. https://doi.org/10.1016/j.jclepro.2015.02.024.

Gollakota, Anjani R.K., Sneha Gautam, and Chi Min Shu. "Inconsistencies of E-Waste Management in Developing Nations – Facts and Plausible Solutions." *Journal of Environmental Management*. Academic Press, 2020. https://doi.org/10.1016/j.jenvman.2020.110234.

Grant, Kristen, Fiona C. Goldizen, Peter D. Sly, Marie Noel Brune, Maria Neira, Martin van den Berg, and Rosana E. Norman. "Health Consequences of Exposure to E-Waste: A Systematic Review." *The Lancet Global Health* 1 (6), 2013. https://doi.org/10.1016/S2214-109X(13)70101-3.

Halim, L., and Y. Suharyanti. "E-Waste: Current Research and Future Perspective on Developing Countries." *International Journal of Industrial Engineering and Engineering Management* 1 (2), 2020: 25. https://doi.org/10.24002/ijieem.v1i2.3214.

Han, Yin-Lung, Jen-Hao Wu, Chieh-Lun Cheng, Dillirani Nagarajan, Ching-Ray Lee, Yi-Heng Li, Yung-Chung Lo, and Jo-Shu Chang. "Recovery of Gold from Industrial Wastewater by Extracellular Proteins Obtained from a Thermophilic Bacterium TepidimonasFonticaldi AT-A2." *Bioresource Technology* 239, 2017: 160–70.

Hong, Y, and M Valix. "Bioleaching of Electronic Waste Using Acidophilic Sulfur Oxidising Bacteria." *Journal of Cleaner Production* 65, 2014: 465–72.

Hsu, Emily, KatayunBarmak, Alan C. West, and Ah Hyung A. Park. "Advancements in the Treatment and Processing of Electronic Waste with Sustainability: A Review of Metal Extraction and Recovery Technologies." *Green Chemistry*. Royal Society of Chemistry, 2019a. https://doi.org/10.1039/c8gc03688h.

———. "Advancements in the Treatment and Processing of Electronic Waste with Sustainability: A Review of Metal Extraction and Recovery Technologies." *Green Chemistry*. Royal Society of Chemistry, 2019b. https://doi.org/10.1039/c8gc03688h.

Ilyas, Sadia, Min-Seuk Kim, Jae-Chun Lee, Asma Jabeen, and Haq Nawaz Bhatti. "Bio-Reclamation of Strategic and Energy Critical Metals from Secondary Resources." *Metals* 7 (6), 2017: 207.

Ilyas, Sadia, Jae-chun Lee, and Ru-an Chi. "Bioleaching of Metals from Electronic Scrap and Its Potential for Commercial Exploitation." *Hydrometallurgy* 131, 2013: 138–43.

Isimekhai, Khadijah A, HemdaGarelick, John Watt, and Diane Purchase. "Heavy Metals Distribution and Risk Assessment in Soil from an Informal E-Waste Recycling Site in Lagos State, Nigeria." *Environmental Science and Pollution Research* 24 (20), 2017: 17206–19.

Işıldar, Arda, Jack van de Vossenberg, Eldon R Rene, Eric D van Hullebusch, and Piet N L Lens. "Two-Step Bioleaching of Copper and Gold from Discarded Printed Circuit Boards (PCB)." *Waste Management* 57, 2016: 149–57.

Jadhav, U, C Su, and H Hocheng. "Leaching of Metals from Printed Circuit Board Powder by an Aspergillus Niger Culture Supernatant and Hydrogen Peroxide." *RSC Advances* 6 (49), 2016: 43442–52.

Johnson, D. Barrie. "Development and Application of Biotechnologies in the Metal Mining Industry." *Environmental Science and Pollution Research* 20 (11), 2013: 7768–76. https://doi.org/10.1007/s11356-013-1482-7.

Jowkar, Mohammad Javad, Nazanin Bahaloo-Horeh, Seyyed Mohammad Mousavi, and Fatemeh Pourhossein. "Bioleaching of Indium from Discarded Liquid Crystal Displays." *Journal of Cleaner Production* 180, 2018: 417–29.

Kaya, Muammer. "Recovery of Metals from Electronic Waste by Physical and Chemical Recycling Processes Project Number and Title: Tubitak/216M405-Recovery of Nickel and Cobalt from Laterite Leach Solution by Synergistic Solvent Extraction (SSX) Method View Project Integrated Eco-Technology for a Selective Recovery of Base and Precious Metals in Cu and Pb Mining by-Products (MINTECO) View Project," 2016a. https://www.researchgate.net/publication/295605709.

———. "Recovery of Metals and Nonmetals from Electronic Waste by Physical and Chemical Recycling Processes." *Waste Management*. Elsevier Ltd, 2016b. https://doi.org/10.1016/j.wasman.2016.08.004.

Khaliq, Abdul, Muhammad Akbar Rhamdhani, Geoffrey Brooks, and Syed Masood. "Metal Extraction Processes for Electronic Waste and Existing Industrial Routes: A Review and Australian Perspective." *Resources*. MDPI AG, 2014. https://doi.org/10.3390/resources3010152.

Khayat, MohdEzuan, Mohd Fadhil Abd Rahman, Mohd Shukri Shukor, Siti Aqlima Ahmad, Nor AripinShamaan, and MohdYunusShukor. "Characterization of a Molybdenum-Reducing Bacillus Sp. Strain Khayat with the Ability to Grow on SDS and Diesel." *RendicontiLincei* 27 (3), 2016: 547–56.

Kim, Min-Ji, Ja-Yeon Seo, Yong-Seok Choi, and Gyu-Hyeok Kim. "Bioleaching of Spent Zn–Mn or Ni–Cd Batteries by Aspergillus Species." *Waste Management* 51, 2016: 168–73.

Kumar, Amit, Maria E. Holuszko, and Travis Janke. "Characterization of the Non-Metal Fraction of the Processed Waste Printed Circuit Boards." *Waste Management* 75 (May), 2018: 94–102. https://doi.org/10.1016/J.WASMAN.2018.02.010.

Kumar, Anil, Harvinder Singh Saini, and Sudhir Kumar. "Bioleaching of Gold and Silver from Waste Printed Circuit Boards by Pseudomonas Balearica SAE1 Isolated from an E-Waste Recycling Facility." *Current Microbiology* 75 (2), 2018: 194–201.

Kumar, A., Dubey, A.K., Ramírez, I.S., Muñoz del Río, A., Márquez, F.P.G. (2022). A Review and Analysis of Forecasting of Photovoltaic Power Generation Using Machine Learning. In: Xu, J., Altiparmak, F., Hassan, M.H.A., García Márquez, F.P., Hajiyev, A. (eds) Proceedings of the Sixteenth International Conference on Management Science and Engineering Management – Volume 1. ICMSEM 2022. Lecture Notes on Data Engineering and

Communications Technologies, vol 144. Springer, Cham. https://doi.org/10.1007/978-3-031-10388-9_36

Kuyucak, Nural, and Ata Akcil. "Cyanide and Removal Options from Effluents in Gold Mining and Metallurgical Processes." *Minerals Engineering* 50, 2013: 13–29.

Li, Jingying, Changjin Liang, and Chuanjing Ma. "Bioleaching of Gold from Waste Printed Circuit Boards by ChromobacteriumViolaceum." *Journal of Material Cycles and Waste Management* 17 (3), 2015: 529–39.

Li, Ronghua, Hongxia Deng, Xiaofeng Zhang, Jim J. Wang, Mukesh Kumar Awasthi, Quan Wang, Ran Xiao, Baoyue Zhou, Juan Du, and Zengqiang Zhang. "High-Efficiency Removal of Pb(II) and Humate by a CeO_2–MoS_2 Hybrid Magnetic Biochar." *Bioresource Technology* 273 (February), 2019: 335–40. https://doi.org/10.1016/j.biortech.2018.10.053.

Liang, Guobin, Yiwei Mo, and Quanfa Zhou. "Novel Strategies of Bioleaching Metals from Printed Circuit Boards (PCBs) in Mixed Cultivation of Two Acidophiles." *Enzyme and Microbial Technology* 47 (7), 2010: 322–26. https://doi.org/10.1016/j.enzmictec.2010.08.002.

Lister, Tedd E., Peiming Wang, and Andre Anderko. "Recovery of Critical and Value Metals from Mobile Electronics Enabled by Electrochemical Processing." *Hydrometallurgy* 149 (October), 2014: 228–37. https://doi.org/10.1016/J.HYDROMET.2014.08.011.

Lu, Chenyu, Lin Zhang, Yongguang Zhong, Wanxia Ren, Mario Tobias, Zhilin Mu, Zhixiao Ma, Yong Geng, and Bing Xue. "An Overview of E-Waste Management in China." *Journal of Material Cycles and Waste Management*. Springer Japan, 2015. https://doi.org/10.1007/s10163-014-0256-8.

Maneesuwannarat, Sirikan, Alisa S Vangnai, Mitsuo Yamashita, and PaitipThiravetyan. "Bioleaching of Gallium from Gallium Arsenide by CellulosimicrobiumFunkei and Its Application to Semiconductor/Electronic Wastes." *Process Safety and Environmental Protection* 99, 2016: 80–87.

Matsukami, Hidenori, Nguyen Minh Tue, Go Suzuki, Masayuki Someya, Le Huu Tuyen, Pham Hung Viet, Shin Takahashi, Shinsuke Tanabe, and Hidetaka Takigami. "Flame Retardant Emission from E-Waste Recycling Operation in Northern Vietnam: Environmental Occurrence of Emerging Organophosphorus Esters Used as Alternatives for PBDEs." *Science of the Total Environment* 514 (May), 2015: 492–99. https://doi.org/10.1016/j.scitotenv.2015.02.008.

McAuliffe, Megan E., Paige L. Williams, Susan A. Korrick, Larisa M. Altshul, and Melissa J. Perry. "Environmental Exposure to Polychlorinated Biphenyls and p,p´-DDE and Sperm Sex-Chromosome Disomy." *Environmental Health Perspectives* 120 (4), 2012: 535–40. https://doi.org/10.1289/ehp.1104017.

Mohdee, Vanee, KreangkraiManeeintr, ThanapornWannachod, SuphotPhatanasri, and UraPancharoen. "Optimization of Process Parameters Using Response Surface Methodology for Pd(II) Extraction with Quaternary Ammonium

Salt from Chloride Medium: Kinetic and Thermodynamics Study." *Chemical Papers* 72 (12), 2018: 3129–39. https://doi.org/10.1007/s11696-018-0542-3.

Monika, and Jugal Kishore. "E-Waste Management: As a Challenge to Public Health in India." *Indian Journal of Community Medicine* 35 (3), 2010: 382–85. https://doi.org/10.4103/0970-0218.69251.

Mor, Rahul S., Kuldip Singh Sangwan, Sarbjit Singh, Atul Singh, and Manjeet Kharub. "E-Waste Management for Environmental Sustainability: An Exploratory Study." In *Procedia CIRP*, 98:193–98. Elsevier B.V., 2021. https://doi.org/10.1016/j.procir.2021.01.029.

Namias, Jennifer, Nickolas J Themelis, Phillip J Mackey, and P J Mackey. "THE FUTURE OF ELECTRONIC WASTE RECYCLING IN THE UNITED STATES: Obstacles and Domestic Solutions," 2013. http://www.epa.gov/wastes/conserve/materials/ecycling/docs/fullbaselinereport2011.pdf.

Ni, Kun, Yonglong Lu, Tieyu Wang, Kurunthachalam Kannan, Jorrit Gosens, Li Xu, Qiushuang Li, Lin Wang, and Shijie Liu. "A Review of Human Exposure to Polybrominated Diphenyl Ethers (PBDEs) in China." *International Journal of Hygiene and Environmental Health* 216 (6), 2013: 607–23.

Niu, Zhirui, Qifei Huang, Jia Wang, Yiran Yang, Baoping Xin, and Shi Chen. "Metallic Ions Catalysis for Improving Bioleaching Yield of Zn and Mn from Spent Zn-Mn Batteries at High Pulp Density of 10%." *Journal of Hazardous Materials* 298, 2015: 170–77.

Okibe, Naoko, Daisuke Nakayama, and Takahiro Matsumoto. "Palladium Bionanoparticles Production from Acidic Pd (II) Solutions and Spent Catalyst Leachate Using Acidophilic Fe (III)-Reducing Bacteria." *Extremophiles* 21 (6), 2017: 1091–1100.

Páez-Vélez, Carolina, Ricardo E Rivas, and Jenny Dussán. "Enhanced Gold Biosorption of LysinibacillusSphaericus CBAM5 by Encapsulation of Bacteria in an Alginate Matrix." *Metals* 9 (8), 2019: 818.

Pant, Deepak, Deepika Joshi, Manoj K. Upreti, and Ravindra K. Kotnala. "Chemical and Biological Extraction of Metals Present in E Waste: A Hybrid Technology." *Waste Management* 32 (5), 2012a: 979–90. https://doi.org/10.1016/j.wasman.2011.12.002.

———. "Chemical and Biological Extraction of Metals Present in E Waste: A Hybrid Technology." *Waste Management* 32 (5), 2012b: 979–90. https://doi.org/10.1016/j.wasman.2011.12.002.

Parvez, Sarker M., Farjana Jahan, Marie Noel Brune, Julia F. Gorman, Musarrat J. Rahman, David Carpenter, Zahir Islam, *et al.* "Health Consequences of Exposure to E-Waste: An Updated Systematic Review." *The Lancet Planetary Health*. Elsevier B.V., 2021. https://doi.org/10.1016/S2542-5196(21)00263-1.

Piazza, Simone di, Grazia Cecchi, Anna Maria Cardinale, Cristina Carbone, Mauro Giorgio Mariotti, Marco Giovine, and MircaZotti. "Penicillium Expansum Link Strain for a Biometallurgical Method to Recover REEs from WEEE." *Waste Management* 60, 2017: 596–600.

Priya, Anshu, and Subrata Hait. "Comparative Assessment of Metallurgical Recovery of Metals from Electronic Waste with Special Emphasis on Bioleaching." *Environmental Science and Pollution Research* 24 (8), 2017: 6989–7008. https://doi.org/10.1007/s11356-016-8313-6.

Rautela, Rahul, Shashi Arya, Shilpa Vishwakarma, Jechan Lee, Ki Hyun Kim, and Sunil Kumar. "E-Waste Management and Its Effects on the Environment and Human Health." *Science of the Total Environment*. Elsevier B.V., 2021. https://doi.org/10.1016/j.scitotenv.2021.145623.

Rene, Eldon R., ManivannanSethurajan, Vinoth Kumar Ponnusamy, Gopalakrishnan Kumar, Thi Ngoc Bao Dung, KathirvelBrindhadevi, and ArivalaganPugazhendhi. "Electronic Waste Generation, Recycling and Resource Recovery: Technological Perspectives and Trends." *Journal of Hazardous Materials* 416 (August), 2021. https://doi.org/10.1016/j.jhazmat.2021.125664.

Rathore, P.S., Chatterjee, J.M., Kumar, A. et al. Energy-efficient cluster head selection through relay approach for WSN. J Supercomput 77, 7649–7675 (2021). https://doi.org/10.1007/s11227-020-03593-4

Robinson, Brett H. "E-Waste: An Assessment of Global Production and Environmental Impacts." *Science of the Total Environment*, 2009. https://doi.org/10.1016/j.scitotenv.2009.09.044.

Saeed, Ali Mohamed, EinaselShatoury, and Reem Hadid. "Production of Molybdenum Blue by Two Novel Molybdate-reducing Bacteria Belonging to the Genus Raoultella Isolated from Egypt and Iraq." *Journal of Applied Microbiology* 126 (6), 2019: 1722–28.

Sahni, Aditya, Anil Kumar, and Sudhir Kumar. "Chemo-Biohydrometallurgy—a Hybrid Technology to Recover Metals from Obsolete Mobile SIM Cards." *Environmental Nanotechnology, Monitoring & Management* 6, 2016: 130–33.

Saitoh, Norizoh, Toshiyuki Nomura, and Yasuhiro Konishi. "Biotechnological Recovery of Platinum Group Metals from Leachates of Spent Automotive Catalysts." In *Rare Metal Technology 2017*, 129–35. Springer, 2017.

Sankhla, Mahipal Singh, Mayuri kumari, Manisha Nandan, ShriyashMohril, Gaurav Pratap Singh, Bhaskar Chaturvedi, and Dr. Rajeev Kumar. "Effect of Electronic Waste on Environmental & Human Health- A Review." *IOSR Journal of Environmental Science, Toxicology and Food Technology* 10 (09), 2016: 98–104. https://doi.org/10.9790/2402-10090198104.

Schaik, Antoinette van, and Markus A. Reuter. "Material-Centric (Aluminum and Copper) and Product-Centric (Cars, WEEE, TV, Lamps, Batteries, Catalysts) Recycling and DfR Rules." In *Handbook of Recycling: State-of-the-Art for Practitioners, Analysts, and Scientists*, 307–78. Elsevier Inc., 2014. https://doi.org/10.1016/B978-0-12-396459-5.00022-2.

Sethurajan, Manivannan, Eric D van Hullebusch, Danilo Fontana, Ata Akcil, HaciDeveci, Bojan Batinic, João P Leal, *et al.* "Recent Advances on Hydrometallurgical Recovery of Critical and Precious Elements from End of Life Electronic Wastes - a Review." *Critical Reviews in Environmental Science* and *Technology* 49 (3), 2019: 212–75. https://doi.org/10.1080/10643389.2018.1540760.

Shah, Monal B, Devayani R Tipre, and Shailesh R Dave. "Chemical and Biological Processes for Multi-Metal Extraction from Waste Printed Circuit Boards of Computers and Mobile Phones." *Waste Management & Research* 32 (11), 2014: 1134–41.

Sheel, Anvita, and Deepak Pant. "Recovery of Gold from Electronic Waste Using Chemical Assisted Microbial Biosorption (Hybrid) Technique." *Bioresource Technology* 247, 2018: 1189–92.

Song, Mengke, Chunling Luo, Fangbai Li, Longfei Jiang, Yan Wang, Dayi Zhang, and Gan Zhang. "Anaerobic Degradation of Polychlorinated Biphenyls (PCBs) and Polychlorinated Biphenyls Ethers (PBDEs), and Microbial Community Dynamics of Electronic Waste-Contaminated Soil." *Science of the Total Environment* 502 (January), 2015: 426–33. https://doi.org/10.1016/j.scitotenv.2014.09.045.

Srivastava, Rajeev. "A Literature Review and Classification of E-Waste Management Research." *The Journal of Solid Waste Technology and Management* 46 (2), 2020: 258–73.

Tao, Hu-Chun, JieSu, Peng-Song Li, He-Ran Zhang, and Guo-Yu Qiu. "Specific Adsorption of Silver (I) from the Aqueous Phase by Saccharomyces Cerevisiae Cells with Enhanced Displaying CueR Protein." *Journal of Environmental Engineering* 142 (9), 2016: C4015018.

Tesfaye, Fiseha, Daniel Lindberg, Joseph Hamuyuni, PekkaTaskinen, and Leena Hupa. "Improving Urban Mining Practices for Optimal Recovery of Resources from E-Waste." *Minerals Engineering* 111 (September), 2017: 209–21. https://doi.org/10.1016/j.mineng.2017.06.018.

Thakur, Pooja, and Sudhir Kumar. "Metallurgical Processes Unveil the Unexplored 'Sleeping Mines' e- Waste: A Review." *Environmental Science and Pollution Research*. Springer, 2020. https://doi.org/10.1007/s11356-020-09405-9.

Tue, Nguyen Minh, Shin Takahashi, Annamalai Subramanian, Shinichi Sakai, and Shinsuke Tanabe. "Environmental Contamination and Human Exposure to Dioxin-Related Compounds in e-Waste Recycling Sites of Developing Countries." *Environmental Science: Processes & Impacts* 15 (7), 2013: 1326–31.

Visvanthan, C., Nang Htay Yin, and Obuli P. Karthikeyan. "Co-Disposal of Electronic Waste with Municipal Solid Waste in Bioreactor Landfills." *Waste Management* 30 (12), 2010: 2608–14. https://doi.org/10.1016/j.wasman.2010.08.006.

Wang, Qin, Baogui Zhang, Shaoqi Yu, Jingjing Xiong, Zhitong Yao, Baoan Hu, and Jianhua Yan. "Waste-Printed Circuit Board Recycling: Focusing on Preparing Polymer Composites and Geopolymers." *ACS Omega* 5 (29), 2020: 17850–56. https://doi.org/10.1021/ACSOMEGA.0C01884/ASSET/IMAGES/ACSOMEGA.0C01884.SOCIAL.JPEG_V03.

Watling, Helen R. "Review of Biohydrometallurgical Metals Extraction from Polymetallic Mineral Resources." *Minerals* 5 (1), 2014: 1–60.

Wu, Kusheng, Xijin Xu, Lin Peng, Junxiao Liu, Yongyong Guo, and Xia Huo. "Association between Maternal Exposure to Perfluorooctanoic Acid (PFOA) from Electronic Waste Recycling and Neonatal Health Outcomes." *Environment International* 48 (November), 2012: 1–8. https://doi.org/10.1016/j.envint.2012.06.018.

Wu, Shijuan, Tengfei Li, Xian Xia, Zijie Zhou, Shixue Zheng, and Gejiao Wang. "Reduction of Tellurite in Shinella Sp. WSJ-2 and Adsorption Removal of Multiple Dyes and Metals by Biogenic Tellurium Nanorods." *International Biodeterioration & Biodegradation* 144, 2019: 104751.

Xia, Mingchen, Peng Bao, Ajuan Liu, Mingwei Wang, Li Shen, Runlan Yu, Yuandong Liu, Miao Chen, Jiaokun Li, and Xueling Wu. "Bioleaching of Low-Grade Waste Printed Circuit Boards by Mixed Fungal Culture and Its Community Structure Analysis." *Resources, Conservation and Recycling* 136, 2018: 267–75.

Xiang, Yun, Pingxiao Wu, Nengwu Zhu, Ting Zhang, Wen Liu, Jinhua Wu, and Ping Li. "Bioleaching of Copper from Waste Printed Circuit Boards by Bacterial Consortium Enriched from Acid Mine Drainage." *Journal of Hazardous Materials* 184 (1–3), 2010: 812–18.

Xu, Chunping, Mahmoud Nasrollahzadeh, Maurizio Selva, Zahra Issaabadi, and Rafael Luque. "Waste-to-Wealth: Biowaste Valorization into Valuable Bio(Nano)Materials." *Chemical Society Reviews*. Royal Society of Chemistry, 2019. https://doi.org/10.1039/c8cs00543e.

Xu, Xiaoguang, Ying Yang, Xixi Zhao, Haobin Zhao, Yao Lu, Chunmei Jiang, Dongyan Shao, and Junling Shi. "Recovery of Gold from Electronic Wastewater by Phomopsis Sp. XP-8 and Its Potential Application in the Degradation of Toxic Dyes." *Bioresource Technology* 288, 2019: 121610.

Yaashikaa, P. R., B. Priyanka, P. Senthil Kumar, S. Karishma, S. Jeevanantham, and SravyaIndraganti. "A Review on Recent Advancements in Recovery of Valuable and Toxic Metals from E-Waste Using Bioleaching Approach." *Chemosphere* 287 (January), 2022. https://doi.org/10.1016/j.chemosphere.2021.132230.

Zheng, Guina, Xijin Xu, Bin Li, Kusheng Wu, TaofeekAkangbeYekeen, and Xia Huo. "Association between Lung Function in School Children and Exposure to Three Transition Metals from an E-Waste Recycling Area." *Journal of Exposure Science and Environmental Epidemiology* 23 (1), 2013: 67–72. https://doi.org/10.1038/jes.2012.84.

18

E-Waste: The Problem and the Solutions

Krati Taksali[1*] and Pramod Singh Rathore[2]

[1]Parul University, Vadodara, India
[2]Department of Computer and Communication Engineering, Manipal University Jaipur, Jaipur, India

Abstract

Throughout the ages, the surplus of our daily routine has been disposed in the easiest and least expensive way, which leads to "no value" for the stuff we waste.

E-Waste is a now a part of this surplus. It's an innovative addition to our ever-growing hazardous solid waste. Electronics and electrical equipment are part of e-waste. Many countries, especially developing countries like India, are facing infinite challenges for the management of e-waste which is imported illegally or generated internally. India is also one of the countries fighting towards e-waste management. In India, existing e-waste management practices, rules, and regulations are not so good and can affect the environment as well as human health. On top of that, policies are not appropriately followed or implemented. At the time of study, it was noticed that there is a crucial need to mark the issues and challenges faced in e-waste management to avoid the consequences of these in future.

Keywords: E-waste, electronic garbage, CRT, recycle, electronic trash, RDED, dispose, cutting-edge battery

18.1 Introduction

There has been a rise in the use of any kind of electronic device or apparatus in this age of modernization and information technology. Since everyone is constantly upgrading to newer technology, e-waste is increasing in the solid waste stream at an alarming rate [1].

Corresponding author: kratitaksali05@gmail.com

Abhishek Kumar, Pramod Singh Rathore, Ashutosh Kumar Dubey, Arun Lal Srivastav, T. Ananth Kumar and Vishal Dutt (eds.) Sustainable Management of Electronic Waste, (375–396) © 2024 Scrivener Publishing LLC

E-waste, short for electronic garbage, is produced when electronic devices become obsolete or are no longer needed. For example, when they are no longer needed, electronic devices such as computers, mobile phones, CDs, monitors, TVs, WAs, and ACs are considered e-waste. As technology advances and newer models become available, it will soon be time to upgrade this equipment.

E-waste contains a wide variety of compounds, some of which are useful but many of which are hazardous and can have a bad effect on the environment or human health. Materials including metal, polymer, CRT, circuit board, wire, and so on make up the bulk of electronic waste. It is possible to recover Ag, CU, Au, and Pt from garbage by scientific processing methods [2][3]. When discarded improperly, e-waste can pose health risks due to the presence of harmful compounds such LC, Hg, Ni, Se, Ba, Cr, Co, and Pb.

E-waste management is a top issue in wealthy nations, but it is a growing problem in developing nations due to a number of factors including a lack of investment, technically trained people, and resources, which developed nations have successfully overcome. There are several harmful and hazardous compounds that may be found in the tons of electronic garbage that are sent throughout the world every year. Most of which end up in landfills, incinerated, or sent to foreign recyclers. The majority of recyclers send their e-waste to developing nations like China, Africa, and India [4].

E-waste disassembly entails more than just taking apart electronics; it also involves ripping them to shreds or even setting them on fire. A wide variety of serious inflammatory, respiratory, and skin illnesses may be brought on by the smoke's inclusion of several dangerous chemical particles. Similarly, precious metals like gold, copper, and platinum are extracted from circuits by burning them, but the PVC used to coat the wires emits smoke and may cause cancer of the lungs or skin when inhaled or absorbed via the skin.

The Basel Action Network (BAN) is an international treaty with the goal of reducing and controlling the regulation of hazardous and toxic substances across countries. In fact, it is acknowledged within the convention itself that some e-waste is being shipped without proper authorization. There are around 50 million tons of electronic garbage produced annually throughout the world. Personal electronic gadgets such as desktops, notebooks, cellphones, tablets, and displays account for 50% of this total. Despite the fact that 66% of the world's population lives in a region with e-waste rules and regulations, only 20% of it gets recycled. This leaves over 4Cr tons of e-waste vulnerable to being trafficked illegally, burnt for resource extraction, or handled poorly [5][6][7][8].

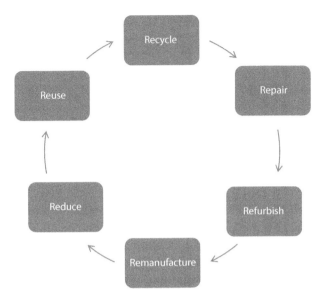

Figure 18.1 Worldwide problem of electronic garbage.

Worldwide problem of electronic garbage is shown in Figure 18.1. More than a hundred million computers and other electronic devices are thrown away in the United States every year, with only around twenty percent being recycled in an effective manner. Reports from the Basel Action Network (BAN) show that anything from 50-80% of American e-waste ends up in (In)India, (CN)China, (Pak)Pakistan, (Tw)Taiwan, and a few nations in (SA)Africa. Simply said, the cost of labor associated with the recycling process is lower in these nations. In addition, the United States allows the export of electronic trash. For environmental and safety reasons, (CN)China has outlawed the import of e-waste. Given the lack of accountability on the part of the national government and the e-industry, users, and recyclers, the United States (US) has allowed the export of e-waste [9].

About 160,000,000 electronic devices are discarded each year in (CN) China. (CN)China has been the leading e-waste dumping country in previous years. The number of individuals learning how to properly dispose of electronic garbage is in the millions.

Electronic trash generation is rising at a rate of 5–10% annually worldwide, according to studies. Each year, 146,000 tons of electronic garbage is produced in India. Still, this number only accounts for domestic e-waste production and does not include e-waste that has been imported (either legally or illegally) from elsewhere.

The reason for this is that many nations export electrical and electronic trash (WEEE) to India. In contrast to the (EU's) European Union's aim of

4 kg per capita, CH(Switzerland) has recycled 11 kg per capita of e-waste, making it the first country in the world to do so (EU).

Members of the European Union are required to meet strict assembly, regaining, and reuse quotas as stated in the EU WEEE rules. Therefore, it requires all member nations to collect at least 4 kg/capita annually. Increasing the availability of recyclable resources and decreasing the need for virgin material in new products are 2 goals of these collection and weight-based recycling targets.

It has been stated that 1/3 of e-waste in the EU is properly gathered and disposed of. Most electronic garbage in South Korea was collected by manufacturers when the EPR plan was implemented in 2003. Reuse and recycling accounted for 12% and 69% of e-waste produced within the same time frame. The remaining 19% was sent to a dumping area or incinerator facilities [10].

The management of e-waste in developing countries is poor due to the neglectable or nonexistent enforcement of the existing law framework, a low level of perception and sensitization, and sparse occupational safety for those involved in these actions. This is in contrast to the European Union and Japan, both of which have well-developed initiatives at all levels that are aimed at changing the etiquettes of consumers. Therefore, in order to lessen the risk of harm to human health and the environment, developing countries need to devise effective strategies to encourage the reuse, renovation, or reprocessing of e-waste in facilities that are specifically designed for that purpose [11].

The top countries that get electronic garbage from developed nations are (Cn)China, (PE)Peru, Ghana, (NI)Nigeria, (In)India, and (Pak)Pakistan (Mmereki, 2016). The mission of the BAN is to promote eco-friendly practices for the disposal of electronic waste to protect the Earth against the commercialization of hazardous waste. Organizations working for environmental protection in the US include the BAN, the (Silicon Valley Toxic Coalition) SVTC, and the (Electronics Take-Back Coalition) ETBC, all of which work together. All three groups are working toward the same goal of advancing comprehensive national strategies for handling hazardous waste. E-Stewards is a relatively new program that audits and certifies recyclers and takeback programs so that discerning customers may choose which ones actually live up to their claims.

18.2 India's Electronic Waste Crisis

Two of the fastest-growing sectors in the economy are those related to information technology and telecommunications. Computers in India have become increasingly common, with 95 per 1,000 people using one in 2011. Instead of throwing away their old phones, Indians usually give it to a friend or family member who will use it until it breaks, at which point it is sold at a flea market or given to the Kabadiwallas. In 2014, India discarded 1.7 Mt of electronic and e-equipment, making it the sixth largest producer of electron-waste worldwide. In the casual economy of (In)India, unskilled workers gather, transport, sort, dismantle, recycle, and dispose of electronic garbage by hand. Because of a lack of education and awareness, e-waste is often discarded in regular trash and then sorted by rag pickers. The valuable and reusable components found in old electronics are often discarded along with the rest. This electronic garbage is collected by rag pickers, who then make a living by selling it to scrap merchants. To recycle and handle e-waste, recyclers employ antiquated and sometimes dangerous technology and tools. About 12,500,000 metric tons (MT) of e-garbage is generated annually in India. On the Environmental Performance Index, India comes in at position 155, much below the average of a ranking of 67. Additionally, it has a low ranking in a number of other categories, including health hazards (127), air quality (174), and water and sanitation (124). India's standing in these fields would also be enhanced by the implementation of ESM practices for electronic waste [12][13].

Numerous industrialized countries ship their electronic garbage to India as a dumping place. A breakdown by country of (In)India's e-waste imports is shown in the chart below. When broken down by origin, the United States(US) accounts for 42% of (In)India's e-waste imports, followed by China (30%), Europe (18%), and other nations (10%) such as Taiwan, South Korea, Japan, etc.

As per the report that was given at the World Economic Forum (WEF) in 2018, India ranked 177/180 on the Environmental Performance Index, making it one of the five nations that performed the lowest overall. This was connected to the high death toll caused by air pollution and the ineffectiveness of legislation intended to protect environmental health. Additionally, In(India) recycles < 2% of the entire amount of electronic trash that it produces annually in an official capacity, which places it sixth in the world after the USA, (Cn)China, (Jp)Japan, and (De)Germany respectively. As of the year 2018, India receives almost 2,000,000 metric tons of waste from e-goods each year, most of which is generated locally. Dumping waste in open dumpsites is a common practice despite the fact that it can lead to a variety of problems, including groundwater contamination, bad health, and more. According to the findings of the research project titled Electronic Waste Management in India, which was carried out by the ASSOCHAM and KPMG, computers account for around 70 percent of all e-waste.

This included cell phones (12%), e-appliances (8%), and medical gadgets (7%), with the remainder of the e-waste originating from residences [14] [15].

The informal economy plays a significant role in all phases of the e-waste life cycle: collection, transportation, processing, and recycling. When it comes to regulation, this industry is both well-connected and unrestricted. In many cases, resources that could be salvaged are not. Environmental pollution from leaks is a major problem, as is the risk to employees' health and safety.

Seelampur is India's principal center for recycling and recycling of electronic garbage, and the nation's capital city of Delhi is home to Seelampur. Every day, children and adults alike devote eight to ten hours to the process of disassembling electronic garbage in the hope of recovering valuable resources like copper, gold, and other precious metals. Open burning and acid leaching are two of the recycling processes that are utilized by companies that deal with electronic trash. The implementation of awareness programs, the improvement of recycling unit infrastructure, and the reform of policies that are already in place are all possible answers to this problem. India's top ten generating States of electronic is shown in Table 18.1.

Table 18.1 India's top ten generating States of electronic trash.

States	E-waste (in MT)
MH	20270.59 (Twenty Thousand Two Hundred Seventy)
TN	13486.24 (Thirteen thousand Four Hundred Eighty Six)
AP	12780.33 (Twelve Thousand Seven Hundred Eighty)
UP	10381.11 (Ten Thousand Three Hundred Eighty One)
WB	10059.36 (Ten Thousand Fifty Nine)
DL	9729.15 (Nine Thousand Seven Hundred Twenty Nine)
KA	9118.74 (Nine Thousand One Hundred Eighteen)
GJ	8994.33 (Eight Thousand Nine Hundred Ninety-Four)
MP	7800.62 (Seven Thousand Eight Hundred)
PB	6958.46 (Six Thousand Nine Hundred Fifty Eight)

Source (Rajya Sabha 2001)

E-Waste: The Problem and the Solutions 381

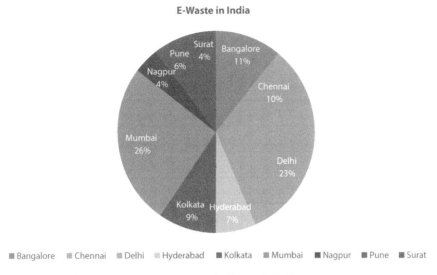

Figure 18.2 Electronic waste management challenges in India.

The informal economy in India is responsible for the disposal of the vast majority of electronic waste. Electronic waste management challenges in India is shown in Figure 18.2.

In addition, a sizeable portion of discarded electronic equipment is collected through unofficial recycling and reuse channels like repair shops, used goods dealers, and online shopping portal merchants for the purpose of reusing the equipment and cannibalizing it for its individual parts and components [16]. Top ten E-waste generating cities is shown in Table 18.2.

In India, the unorganized sector is responsible for the majority of the recycling of electronic trash. Numerous low-income households are able to make ends meet by rescuing usable items from landfills. Paper, plastic, clothes, and metal that have been discarded in urban homes belonging to middle-class families are typically recycled by being sold to small, unchecked purchasers called "kabadiwalas." These buyers then classify the trash and use it as a raw material for industrial processors.

A similar pattern can be observed in the way that India(In) handles its electronic garbage. An unregulated business in the recycling of electronic trash employs a large number of families living in non-rural areas to collect, categorize, mend, renew, and dismantle obsolete electrical and e-gadgets. The situation is not common in industrialized

Table 18.2 Top ten E-waste generating cities.

Cities	E-waste (in tons)
AHM	3286.5 (Three Thousand Two Hundred Eighty Six)
BLR	4648.4 (Four Thousand Six Hundred Forty Eight)
MAA	4132.2 (Four Thousand One Hundred Thirty Two)
DL	9730.3 (Nine Thousand Seven Hundred Thirty)
HYD	2833.5 (Two Thousand Eight Hundred Thirty Three)
KOL	4025.3 (Four Thousand Twenty Five)
MUM	11017.1 (Eleven Thousand Seventeen)
NG	1768.9 (One Thousand Seven Hundred Sixty Eight)
The Pune	2584.2 (Two thousand Five hundred Eighty Four)
The Surat	1836.5 (One thousand Eight Hundred Thirty Six)

Source (Rajya Sabha 2001)

nations and it is unheard of for individuals in India to voluntarily donate obsolete electrical and e-gadgets to official e-waste reprocessing facilities. In addition, there is no such thing as the concept of customers being required to pay for the disposal of the electronic rubbish that they make.

These major issues are brought on by the e-waste recycling industry's excessive reliance on the unorganized sector, as listed below in Figure 18.3:

- To start, it is ineffective to try and penalize people financially for breaking the laws for managing and processing e-waste.
- Second, because fewer well-paid people who perform this task lack the necessary skills, the general public is less aware of MRP and health and safety expenses connected with e-waste recycling.
- Third, in spite of the enormous growth in the amount of e-waste produced every year, little is being spent by large-scale industries on recovery and repairing infrastructure.

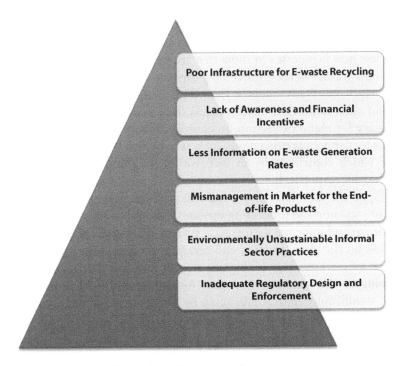

Figure 18.3 E-waste recycling industry's unorganized sector.

18.3 Inadequate Infrastructure for Refurbishing E-Waste

India(In) has a relatively limited number of buildings that are capable of handling large amounts of electronic trash. At the very few facilities that have been recognized by the government, only around one fifth of the entire amount of electronic trash that is produced each year in the country gets recycled. The co-funded grant program that is offered by the Indian government covers 25% to 50% of the project expenses for the purpose of improving the capacity of e-waste firms and constructing disposal facilities for electronic waste. On the other hand, practically very little of this plan has been implemented. Because the majority of informal e-waste collectors in India have poorly established supply networks, which causes them to operate significantly below their capacity, there is a scarcity of legally certified e-waste recycling facilities. This is due to the fact that there are not enough of these facilities.

In spite of this, the formal recycling sector in India is only capable of managing e-waste through the use of human sorting and mechanical disassembly of the devices. The management of industrial e-waste with appropriate ecological controls is currently inadequate, although this is a requirement for the big recovery of precious and the base metals. There are a few new businesses in India that are working to recover metals from electronic trash, but their processing capacity is limited. The vast majority of the electronic trash that is processed in an organized manner is brought in from other countries, which possess the extensive infrastructure that is necessary for metal extraction. In contrast, the informal industry obtains metals by dangerous procedures such as acid filtering and open-air burn, all of which contribute to an increase in environmental pollution and health risks, respectively.

Despite the fact that the e-waste rules provide a number of technologies for recycling and processing various types and components of e-waste, the formal and informal sectors have primarily focused on metal recovery and less on the glass, plastics, and ceramics that comprise the significant portion of e-waste. This is the case despite the fact that metal recovery accounts for the majority of e-waste. Recycling e-trash made of plastic is made significantly more challenging by the presence of flame retardants and other persistent organic pollutants [17][18].

18.3.1 Financial Incentives and Lack of Awareness

Because there is a lack of public knowledge of the dangers connected with e-waste in India, the country's rate of recycling is very low. As a result of this lack of understanding, India's recycling rate is relatively low. The majority of consumers are either oblivious to or have a limited awareness of the hazards connected with the components of e-waste, as well as the implications of improper disposal. They are not aware that the handling of electronic garbage in India is handled by either state or municipal governments.

Customers who voluntarily choose to dispose of their electronic garbage may only drop it off at a limited number of specialized collection warehouses or official recycling facilities in various cities. When purchasing any new e-gadgets from small-scale retail establishments, the majority of individuals and urban homeowners used to sell their old electronics in exchange for a discount or to obtain a discount in exchange for selling their old electronics. Because consumers lack market information on pricing for e-waste and the various components of e-waste, there are few financial incentives for consumers to dispose of their electronic trash in a responsible manner.

18.3.2 Limited Available Statistics on Rates of Creation of Electronic Trash

It is common knowledge that there are no electronic waste inventories and it is the exclusive responsibility of the various state pollution control boards to compile e-waste inventories for their respective states (SPCBs). Information on the sale of electronic products is a crucial component in the process of measuring the total amount of e-waste. Due to the fact that it is commonly aggregated at the Indian level, it is difficult to compile inventories at the state level. In addition to being manufactured on home soil, electronic garbage is frequently brought in illegally from countries that have a developed industrial sector [19][20].

Fewer people are familiar with the kind and extent of the electronic waste that is carried into the country. These systems are essential for the effective generation of garbage and the acquisition of knowledge regarding its composition, as well as for the collection, transportation, and processing of trash.

18.3.3 Market Mismanagement for End-of-Life Products

The capacity of private actors to establish e-waste management systems in a formal industry is constrained by the difficulty to consistently obtain sufficient amounts of e-waste to generate economies of scale. In India, for example, adopting efficient recycling methods for e-waste management may necessitate substantial up-front capital costs, which private firms cannot justify in the lack of confidence about where to get sufficient amounts of e-waste. These marketplaces are also hampered by information obstacles.

To begin, the recycling of electronic waste is still a relatively new business, therefore, there may be a barrier to entry in the market owing to a lack of information regarding effective recycling procedures. Second, the operation of markets is hindered by a lack of knowledge, which is largely caused by the limited access that customers have to trustworthy information on the disposal of electronic waste. However, public policy, which will have a greater influence, will make it feasible to create improved markets for e-waste [21][22].

18.3.4 Environmentally Irresponsible Practices in the Informal Sector

Although the official recycling and dismantling sector is growing, it still processes a small percentage of the world's rubbish. Most of these institutional facilities are operating at or below their allowed levels because they lack sufficient waste sources. Insufficient public education on e-waste and the costs involved with returning end-of-life equipment to formal collection facilities have contributed to the reluctance of residential and commercial customers to return their trash to the official sector. Most importantly, the informal sector makes it enticing for customers to return their rubbish by providing easy house pickup and financial incentives, while the official sector has not yet invested in dependable collection and processing equipment.

Despite its lack of oversight, the e-waste sector has provided employment for millions of people, many of whom are members of the most marginalized communities. On the other hand, the industry's waste management practices pose significant threats to public and worker health as well as the environment. Finding a moral solution to this potential moral deadlock is crucial to the long-term viability of any e-waste management in India [23].

18.3.5 Poorly Designed and Enforced Regulations

Because it lacks corresponding collection targets, which served as incentives for producers to assume authority, the mandatory take-back system had limited influence on e-waste management operations. Several changes were suggested that improved regulatory clarity by outlining more stringent collection goals. The regulatory framework places an excessive burden on agencies that are already understaffed. The EPR plan submitted by manufacturers must be examined, approved, and enforced by the appropriate authorities.

The regulations need to specify strict rules and processes for dismantlers, collectors, recyclers, and bulk consumers in addition to mandating that authorities strictly enforce compliance with the set standards. A lack of transparency and disinclination to make information on compliance and regulatory actions public must benefit the regulatory agencies. This persistent difficulty in enforcing environmental regulations in India extends even to recent efforts to limited e-waste production. It poses the greatest threat to future e-waste management public policy in India.

18.4 Enhancing India's E-Waste Management

There are a number of opportunities for India to better manage its electronic waste. In any case, there are essentially 5 main components that may be linked together to improve e-waste management in India.

18.4.1 Providing Price Information on the Market for E-Waste

There is a robust market for discarded electronics among and within both legal and illegal commercial enterprises. However, city dwellers are not often exposed to or made aware of the costs associated with e-waste and its components. An important market signal for customers who sell their e-waste to local sellers is weekly price changes, therefore this is essential information that should be communicated to the public.

Glass, metals, plastics, ceramics, batteries, and bulk e-waste are only some of the materials that should be accounted for on the price list. Much like commodity price lists or foreign exchange rates, this data must be made public by local newspapers and major cities on dedicated websites. To ensure that non-formal sector collectors may acquire and sell electronic waste to private processors or government-approved recycling and dismantling facilities at fair market rates, the pricing list must account for current market circumstances for e-waste component demand.

18.4.2 Promoting the Recycling of Formal E-Waste

The Indian(In) government has instituted a point-based reimbursement system known as E-waste Recycling Credits to incentivize commercial firms to recycle their e-waste at government-approved facilities (ERCs). The E-Trash Regulation has already established a system for categorizing and coding electronic waste. These subgroups need separate ERC reward tiers. Depending on the kind of e-waste provided, businesses must obtain the appropriate ERCs, which may be used to lower the prices of energy utilities. Also, an illegal e-waste business will have a strong incentive to formalize their activities and establish supply chain relationships with authorized recycling facilities if such a scheme were in place.

It's possible that the ERCs will be tested out as pilot projects for anywhere from three to five years so that their efficacy can be measured and tweaks can be made before they're widely implemented. The commercial and public sectors account for about 70% of urban areas' e-waste. Some major corporations and government organizations may try out the ERCs in Mumbai(Mum), Delhi(DL), or Bangalore(BLR).

The Indian(In) government may be able to boost formal electronic-waste recycling capacity by co-sponsoring infrastructure upgrades and processing systems at existing government-approved recycling plants. Governments may be eligible for co-funding incentives to build new recycling facilities if they form public-private partnerships with major e-waste firms. Governments at the state level may also institute incentive programs aimed at getting informal recycling centers for electronic debris to upgrade their facilities to meet current standards for environmental protection and worker safety. Many states have access to national financing programs for urban development that might be used to link the vast informal sector network of decentralized collection and small recycling firms with industrial recycling facilities on a massive scale.

18.4.3 Players in the Informal Sector are Trained and Upgraded

The majority of a community's informal e-waste recycling workforce requires upskilling, particularly in the areas of handling and dismantling dangerous devices. It must ensure the health and safety of both workers and the environment and provide a link between the supply chain and regulated processors. The Indian(In) government's National Skill Development Mission(NSDM) is working toward this goal. E-waste collectors, handlers, and dismantlers can benefit from the combined expertise of the E-Sector Skill Council, the Green Jobs Sector Skill Council(GJSSC), and regulatory bodies like the Central and State Pollution Control Boards(PCB) in the development of novel short courses and training programs [24][25].

Educating both parties and governments at the federal and state levels should roll out a unified, nationwide initiative to better equip workers in the informal economy to mitigate the risks associate with e-waste and eliminate the role of the informal economy in e-waste disposal and locations of government-sanctioned e-waste collecting facilities.

18.4.4 Utilizing Technology for Recycling that is Both Finished and Cutting-Edge

In addition to the country's present manual procedures, advanced recycling equipment is needed to maximize the recycling efficiency of India's massive e-waste production. Recovery of plastic from e-waste is only one example of how advanced and extensive India's plastics processing sector is.

The government of India should promote cooperative partnerships between local and international firms to facilitate the construction of large-scale industrial e-waste recovery facilities. Various sources of public and private financing may be pooled together to support these efforts.

18.4.5 Creating New Technology & Methods for Processing New Types of E-Waste

The content of e-waste is always changing as a result of the release of new electronic devices. Major investments in R&D for innovative conversion processes and technologies are required to future-proof India's e-waste legislation and management. For instance, smartphone use has skyrocketed in India over the last five years, yet the Li-ion batteries that power these devices are not yet included in the country's e-waste recycling legislation.

To produce the next wave of electronics, manufacturers will use a slew of cutting-edge battery and material technologies. The government of India should thus promote and fund efforts to develop innovative, future-proof methods of recycling and upcycling e-waste.

18.5 Effects of E-Waste Recycling in Developing India Like Nations

Many common recyclables, such as plastic (PVC), glass (GLS), and metals, are found in e-waste but cannot be reused because of improper disposal techniques and methods. The components of electronic trash, if improperly separated and disposed of, pose a significant threat to human health. Dismantling, then wet chemical processing and incineration are all methods used to get rid of trash, but they expose employees to harmful chemicals via direct skin contact and inhaled smoke. Employees sometimes lack the knowledge and training required to carry out their responsibilities properly and safety equipment like gloved and facemasks are rarely used. Furthermore, the individual physically removing the dangerous metals is subjected to potentially lethal compounds being absorbed into their body. There is a wide range of health concerns, from those in the brain to those in the kidneys and liver. Air, soil, and water pollution are all caused by the recycling of electronic waste trash. The burning of wires and cables to extract metal results in the production of brominated (Br) and chlorinated dioxins, as well as other carcinogens, which pollute the air and cause cancer in humans and animals. Dangerous substances that have no purpose in the recycling process are thrown away. When these harmful chemicals seep into the groundwater supply, they degrade the quality of the water to the point that it can't be used for drinking or farming. The dangerous and unusable soil that results from burying e-waste in landfills is caused by the presence of lead(Pb), mercury(Hg), cadmium(Cd), arsenic, and PCBs.

Rising concentrations of PCBs, dioxins and furans, plasticizers, bisphenol-A (BPA), polycyclic aromatic hydrocarbons (PAH), and heavy metals have been found in the topsoil of India's four major metropolises, including New Delhi, Kolkata, Mumbai, and Chennai, where e-waste is processed by the informal sector. Findings suggest that sites frequented for metal recovery are frequented by these persistent hazardous substances. Studies by the same team show that due to their semi-volatile nature, persistent organic contaminants are released into the environment when recycled.

18.6 Opportunity for Managing E-Waste in India

The Ministry of Environment, Forests, and Climate Change (MEFCC) issued the E-Waste (Management) Rules in 2016 to reduce e-waste production and increase recycling. Producers are obligated to collect 30% to 70% (during a seven-year period) of the e-waste they generate according these restrictions, which the government instituted through EPR.

A transparent recycling system that includes the informal sector is necessary for successful management of the consequences on environment and human health. Many efforts have been made to bring the existing informal sector into the new framework. Organizations like GIZ have to develop substitute business models to aid in the transition of informal sector groups to official authorization. In order to maximize profits from printed circuit boards, these b/s models advocate for a city-wide collection system that gives access to both a manual disassembly facility and the most advanced technological capabilities. When compared to the traditional wet chemical leaching method, exporting to integrated admixture and refineries results in safer operations and more money/unit of collected e-waste.

Recycling electronic trash might reintroduce precious metals like (Au) gold, (Ag)silver, and (Cu)copper into the economy. There is tremendous economic opportunity for individuals and corporations alike in the efficient regaining of valuable materials from e-waste. Indian authorities updated the E-Waste Management Rules, 2016 in March 2018 to make it easier to implement environmentally responsible strategies for dealing with e-waste. The amended rules are in force as of October 1, 2017 and reflect an update to the original rules' intentions regarding the collecting of EPRs. Objectives will be monitored and updated by CPCB to ensure efficient and better e-waste management [26][27][28].

18.6.1 What Assistance Can Governments, City Management, and Citizens Provide?

The report of ASSOCHAM, 2017 suggests that, in order to achieve the aim of reducing e-waste to an absolute minimum, the government should collaborate with the industry to adopt formal/standard operating procedures and a phased plan. The government may also recommend international best practices for e-waste recycling. As one of the world's leading producers of electronic goods, South Korea produced 8,000,000 tons of e-waste in 2015, and according to the research, just 21% of that was recycled.

The government should support new business owners by providing them with funding and technical guidance in order to mitigate the negative consequences of e-waste on land or water and the environment. The recycling and disposal of electronic trash is an area where startups can be encouraged by providing incentives. In the informal economy, a system of collectors has developed over time. However, significant financial resources are needed in the case of the organized industry. As a result, if the unorganized and organized sectors can work together, the unorganized sector's collected resources may be delivered to the organized sector for processing that would be good for the environment. The government may play a critical role between the two sectors for the effective processing of e-waste in this kind of situation. The government must act aggressively to recycle and dispose of electronic waste in a safe way in order to protect the environment, people, and other living things.

The EPR concept is rapidly being used for the management of e-waste since it has been shown to be reasonably efficient and fruitful in the countries of the European Union. Economic, regulatory, and voluntary/informational components may all be integrated into instruments for implementing EPR. For e-waste management to be successful, all relevant parties, including users, retailers, state government, municipality, NGOs, CSOs, SHGs, and local collection services like extracarbon.com, must play their part. Electronic trash disposal still rests with the original manufacturers (EPR) [29][30][31].

Design for Environment (DfE) is an innovative strategy for lowering pollution levels that is now receiving a lot of attention in many parts of the world. The DfE concept is an approach to minimize goods' environmental impact prior to their commercial launch. The weak implementation of India's tough rules is often cited as the reason for their ineffectiveness.

The public has a crucial role in e-waste management. We throw away a variety of small electronics with our regular garbage, and some people even go so far as to burn the resulting piles of rubbish in public. Breathing this in releases a wide variety of harmful compounds, such as dioxins and furans. This is a terrible practice that has to end right this second. Some of the most progressive RWAs provide clearly labelled bins for the collection of electronic garbage. All other neighborhoods should start doing the same thing. Student and Women SHGs may participate in this activity in their local RWAs.

18.7   Management of Electronic Waste

The "polluter pays concept" was established by the Environmental (Protection) Act, 1986 to hold those responsible for pollution accountable for the costs associated with environmental harm.

Principle 16 of the Rio Declaration on Environment and Development (RDED) makes reference to it in the context of international environmental law. Polluter pays are often called "extended producer responsibility" (EPR). The federal and state governments have the jurisdiction to pass legislation to safeguard citizens and ecosystems from toxic waste thanks to the Environment (Protection) Act of 1986. Any violation of this Act or its stated standards carries criminal penalties. This kind of punishment may be levied if conventions and legislation around e-waste are breached.

The CPCB India, which is finalizing some regulations, published an official set of guidelines for the proper and ecologically friendly handling and disposal of e-garbage. The Ministry of Environment and Forests (MEF) is now revising rules established by electronic equipment manufacturers with the help of private organizations. According to CPCB's revised 2007 guidelines, e-trash is classified as hazardous waste and must be handled in accordance with schedules 1, 2, and 3 of the Hazardous Waste (Management and Handling) Rules,2003 and schedule 1 of the Municipal Solid Waste Management Rules (MSWM), 2000. Each manufacturer of a computer, music system, mobile phone, or other electronic device will be "personally" responsible for the proper and environmentally friendly disposal of their product once it reaches the status of e-waste. The Department of Information Technology (DIT) and Ministry of Communication and Information Technology (MCIT), has also published and disseminated a detailed technical document titled "Environmental Management for the Information Technology Industry in India." The DIT has also launched pilot projects at Indian Telephone Industries to demonstrate the feasibility of copper(Cu) recovery from old PCBs.

Companies like Apple, Dell, and HP, among others, have started reprocessing initiatives to inform customers of the need to properly dispose of electronic equipment. Nokia India just announced their new "recycling initiative" for the country. The program encouraged consumers of all mobile phone brands to drop off their old devices and accessories at any of the 1300 green recycling bins set up at the campaign's priority dealers and care facilities. Nokia plans to provide a service to manage electronic waste as well.

In addition, the Department of Environment (DoE) under the Delhi (DL) government has made the decision to include ragpickers into the city's overall rubbish management as garbage collectors after receiving training, clothes, and ID cards. The department also aims to include the local ragpickers in this initiative, since these eco-clubs will be engaging with them. Over 1,600 of the Capital Region's public and private schools have established eco-clubs.

References

[1] Dahl R. Who pays for e-junk? Environ Health Perspect. 2002;110:A196–9. [PMC free article] [PubMed] [Google Scholar]

[2] CPCB. Guidelines for environmentally sound management of e-waste (As approved vide MoEF letter No. 23-23/2007-HSMD) Delhi: Ministry of Environment and Forests, Central Pollution Control Board, March 2008. Available from: http://www.cpcb.nic.in [last accessed on 2008 Mar 12]

[3] Baud I, Grafakos S, Hordjik M, Post J. Quality of life and alliances in solid waste management. Cities. 2001;18:3–12. [Google Scholar]

[4] Pandve HT. E-waste management in India: An emerging environmental and health issue. Indian J Occup Environ Med. 2007;11:116. [PMC free article] [PubMed] [Google Scholar]

[5] Puckett J, Byster L, Westervelt S, Gutierrez R, Davis S, Hussain A, et al. Exporting Harm: The high-tech Trashing of Asia. Seattle: Basal Action Network; Available from: http://www.ban.org [last accessed on 2002] [Google Scholar]

[6] World market for domestic electrical appliances. US: Euromonitor; Euromonitor. Available from: http://www.nautilus.org [last accessed on 2004 Feb] [Google Scholar]

[7] Kumar, A., Dubey, A.K., Ramírez, I.S., Muñoz del Río, A., Márquez, F.P.G. (2022). A Review and Analysis of Forecasting of Photovoltaic Power Generation Using Machine Learning. In: Xu, J., Altiparmak, F., Hassan, M.H.A., García Márquez, F.P., Hajiyev, A. (eds) Proceedings of the Sixteenth International Conference on Management Science and Engineering Management – Volume 1. ICMSEM 2022. Lecture Notes on Data Engineering and Communications Technologies, vol 144. Springer, Cham. https://doi.org/10.1007/978-3-031-10388-9_36

[8] Dubey, A.K., Kumar, A., Ramirez, I.S., Marquez, F.P.G. (2022). A Review of Intelligent Systems for the Prediction of Wind Energy Using Machine

Learning. In: Xu, J., Altiparmak, F., Hassan, M.H.A., García Márquez, F.P., Hajiyev, A. (eds) Proceedings of the Sixteenth International Conference on Management Science and Engineering Management – Volume 1. ICMSEM 2022. Lecture Notes on Data Engineering and Communications Technologies, vol 144. Springer, Cham. https://doi.org/10.1007/978-3-031-10388-9_35

[9] Rathore, P.S., Chatterjee, J.M., Kumar, A. et al. Energy-efficient cluster head selection through relay approach for WSN. J Supercomput 77, 7649–7675 (2021). https://doi.org/10.1007/s11227-020-03593-4

[10] A. Dubey, S. Narang, A. Srivastav, A. Kumar, V. Díaz, Woodhead Publishing, Science Direct, Artificial Intelligence for Renewable Energy Systems. Paperback ISBN: 9780323903967

[11] A. Dubey, S. Narang, A. Srivastav, A. Kumar, V. Díaz, Woodhead Publishing, Science Direct, a Visualization Techniques for Climate Change with Machine Learning and Artificial Intelligence. ISBN: 9780323997140

[12] Agarwal R, Ranjan R, Sarkar P. New Delhi: Toxics Link; 2003. Scrapping the hi-tech myth: Computer waste in India. [Google Scholar]

[13] ELCINA-DSIR. E-waste menace needs urgent technological and market interventions. Global SMT and Packaging India. Available from: http://www.global-smtindia.in/indexphp?option=com_contentandtask=viewandid=26548anditemid=7 [last accessed on 2009 Feb]

[14] Mehra HC. Tribune. 2004. PC waste leaves toxic taste. [Google Scholar]

[15] Widmer R, Oswald HK, Sinha DK, Schnellmann M, Heinz B. Global perspectives on e-waste. Environ Impact Assess Rev. 2004;25:436–58. [Google Scholar]

[16] Jang YC, Townsend TG. Leaching of lead from computer printed wire boards and cathode ray tubes by municipal solid waste landfill leachates. Environ Sci Technol. 2003;37:4778–4. [PubMed] [Google Scholar]

[17] Bathurst PA, McMichael AJ, Wigg NR, Vimpani GV, Robertson EF, Roberts RJ, et al. Environmental exposure to lead and children's intelligence at the age of seven years: The Port Pirie Cohort Study. N Engl J Med. 1992;327:1279–84. [PubMed] [Google Scholar]

[18] Brigden K, Labunska I, Santillo D, Allsopp M. Recycling of electronic wastes in China and India: workplace and environmental contamination. Greenpeace. Available from: http://www.greenpeace.org/india/press/reports/recycling-of-electronic-wastes [last accessed on 2005]

[19] Qiu B, Peng L, Xu X, Lin X, Hong J, Huo X. In: Proceedings of the International Conference on Electronic Waste and Extended Producer Responsibility, April 21-22, 2004. Beijing, China: Greenpeace and Chinese Society for Environmental Sciences; 2004. Medical investigation of e- waste demanufacturing industry in Guiyu town; pp. 79–83. [Google Scholar]

[20] Huo X, Peng L, Xu X, Zhang L, Qiu B, Qi Z, et al. Elevated Blood Lead levels of children in Guiyu, An Electronic Waste Recycling Town in China. Environ Health Perspect. 2007;115:1113–7. [PMC free article] [PubMed] [Google Scholar]

[21] Janet KY, Xing CG, Xu Y, Liang Y, Chen LX, Wu SC, *et al.* Body loadings and health risk assesments of polychlorinated dobenzo-p-dioxines and dibezofurans at an intensive electronic waste recycling site in China. Environ Sci Technol. 2007;41:7668–74. [PubMed] [Google Scholar]

[22] Wang T, Fu JJ, Wang Y, Liao C, Tao Y, Jiang G. Use of scalp hair as indicator of human exposure to heavy metals in an electronic waste recycling area. Environ Pollut. 2009;157:2445–51. [PubMed] [Google Scholar]

[23] Leung AO, Lutisemburg WJ, Wong AS, Wong MH. Spatial distribution of polybrominated diphenyl ethers and polychlorinated dibenzo-p-dioxins and bibenzofurans in soil and combined residue at Guiyu: An electronic-waste recycling site in southeast China. Environ Sci Technol. 2007;41:2730–7. [PubMed] [Google Scholar]

[24] Hicks C, Dietmar R, Eugster M. The recycling and disposal of electronic waste in China – legislative and market response. Environ Impact Assess Rev. 2005;25:459–71.

[25] Silicon India News Bureau. Only for Rs.5 e-waste workers risk lives (news) Silicon India. Available from http://www.siliconindia.com/ shownews/ Indias_ewaste_ hazard _only_for_Rs5_ workers_risk_ lives_-nid-63623-cid-1.html

[26] Roman LS, Puckett J. In: Proc. International Symposium on Electronics and the Environment, IEEE, May 6–9, 2002, San Francisco CA, USA. 2002. E-scrap exportation: Challenges and considerations; pp. 79–84. [Google Scholar]

[27] Williams E. In: Proc. Third Workshop on Materials Cycles and Waste Management in Asia. Tsukuba, Japan: National Institute of Environmental Sciences; 2005. International activities on E-waste and guidelines for future work. [Google Scholar]

[28] Yanez L, Ortiz D, Calderon J, Batres L, Carrizales L, Mejia J, *et al.* Overview of human health and chemical mixtures: Problems facing developing countries. Environ Health Perspect. 2002;110:901–9. [PMC free article] [PubMed] [Google Scholar]

[29] Kishore J. A Dictionary of Public Health. 2nd ed. New Delhi: Century Publications; 2007. p. 680.

[30] Kishore J. National Health Programs of India: National Policies and Legislations related to health. 8th ed. New Delhi: Century Publications; 2010. pp. 735–6. [Google Scholar]

[31] Environmental management for Information Technology industry in India. New Delhi: Department of Information Technology, Government of India; 2003. DIT; pp. 122–4.

19

Contribution of E-Waste Management in Green Computing

Shweta Sharma[1,2]* and Vishal Dutt[1,2]

[1]RK Patni College, Ajmer, India
[2]Chandigarh University, Ajmer, India

Abstract

An essential component of green computing is the efficient handling of e-waste. Computers, TVs, and cell phones are just a few examples of the electronic equipment that is referred to as "e-waste," or electronic garbage. Given that many of these gadgets include toxic elements, improper e-waste disposal can have a significant impact on the environment and human health. Green computing, on the other hand, aims to reduce the harmful effects of computers on the environment by increasing energy efficiency, lowering waste, and utilising non-toxic materials. E-waste management minimises the environmental impact of discarded electronic equipment, which is essential for green computing. By recycling valuable materials, proper e-waste management guarantees that hazardous items are disposed of securely and that natural resources are conserved. E-waste treatment can also aid in lowering greenhouse gas emissions linked to the production of new electronic gadgets. E-waste management has the potential to boost local economies and advance sustainable development by generating jobs, notably in the recycling and repair of electronic gadgets. In conclusion, e-waste management is crucial to the development of green computing. We can lessen technology's negative environmental effects and save resources by properly disposing of and recycling electronic equipment. For the rising issue of e-waste to be solved sustainably, governments and businesses must collaborate to develop effective e-waste management regulations and initiatives.

Keywords: Green computing, e-waste management, recycling, benefits, harmful effects of e-waste, e-waste, 2030 agenda

---

*Corresponding author: sharmashweta671@gmail.com

Abhishek Kumar, Pramod Singh Rathore, Ashutosh Kumar Dubey, Arun Lal Srivastav, T. Ananth Kumar and Vishal Dutt (eds.) Sustainable Management of Electronic Waste, (397–412) © 2024 Scrivener Publishing LLC

19.1 Introduction

Electronic garbage, often known as "e-waste," is the term used to describe abandoned electronic gadgets, including computers, TVs, and mobile phones. The quantity of electronic trash produced has increased dramatically as a result of people upgrading their gadgets more frequently and the rapid advancement of technology. Due to the fact that many of these gadgets include dangerous elements like lead, mercury, and cadmium, incorrect e-waste disposal can have a significant impact on the environment and human health [1][2].

The activity of designing, producing, utilising, and disposing of computers, servers, and related subsystems like monitors and printers in an ecologically responsible way is known as "green computing." By increasing energy efficiency, lowering waste, and employing non-toxic materials, it aims to reduce the detrimental effects of computers on the environment.

19.2 Concept of Green Computing

The term "green computing" refers to the ecologically responsible use of computers and related resources, which includes the use of renewable energy sources, the deployment of energy-efficient systems, and the appropriate disposal of electronic waste. Green computing aims to maximise technology's positive social and environmental effects while minimising its negative environmental effects. This involves cutting back on carbon emissions, protecting the environment, and producing less electronic trash [3][4][5]. This may be done by using data centres, hardware, and software that

Figure 19.1 Concept of green computing.

are energy-efficient, as well as renewable energy sources and eco-friendly corporate practises. Concept of green computing is shown in Figure 19.1.

19.2.1 E-Waste Management and Recycling

E-waste management minimises the environmental impact of discarded electronic equipment, which is essential for green computing. Hazardous garbage is properly disposed of and valuable items are repurposed with proper e-waste management. This conserves natural resources while simultaneously protecting the environment.

The management of e-waste can also aid in lowering greenhouse gas emissions linked to the production of new electronic equipment. Recycling materials reduces the need to mine new raw materials, which in turn lowers the energy needed to create new technology. As a result, less carbon dioxide and other greenhouse gases are emitted into the atmosphere. Additionally, managing e-waste has the potential to provide employment, notably in the recycling and refurbishing of electronic equipment. This might support sustainable development by boosting local economies. E-waste and recycling E-waste are shown in Figure 19.2 and Figure 19.3.

E-waste management and recycling is the term used to describe the ethical disposal and recycling of electronic garbage, which includes outdated

Figure 19.2 E-waste.

Figure 19.3 Recycling e-waste.

or malfunctioning consumer goods including computers, TVs, cell phones, and other electronics. Because many electronic gadgets include dangerous compounds like lead, mercury, and cadmium that can affect the environment and people's health if improperly disposed of, managing and recycling e-waste is crucial.

There are several ways to manage and recycle e-waste, including:

- Collection and Drop-off Initiatives: Many local governments and electronics stores provide collection and drop-off initiatives for e-waste, allowing citizens to bring their outdated electronics for recycling.
- Mail-back Initiatives: A few businesses provide mail-back initiatives where customers may bring in their used electronics for recycling.
- Professional E-waste Recycling Businesses: These businesses specialise in the ethical disposal of electronic trash.
- Reuse and Refurbishment: Some e-waste may be repaired or rebuilt, utilising its component parts to create new technology.

To prevent harm to the environment and human health, it's crucial to make sure that the e-waste is recycled by a reputable firm that has received certification in accordance with international standards [6][7].

19.3 A History of Green Computing

Green computing has its roots in the early days of computers, when environmental impact and energy usage were not big issues. The environmental effect of computing, however, started to be acknowledged as a serious concern as technology evolved and the number of computers and other electronic devices rose.

The technique of creating, utilising, and disposing of computers and other electronic equipment in an ecologically responsible way has been referred to as "green computing" since the 1990s. This entails employing non-toxic materials, consuming less energy, and correctly disposing of e-waste.

The Restriction of Hazardous Substances (RoHS) directive, which restricted the use of some hazardous elements in electronic equipment, was enacted by the European Union in 2007. Similar legislation was enacted in countries such as China and the United States.

The environmental effect of cloud computing and the Internet of Things (IoT) has come to light in recent years and attempts are now being made to lessen these technologies' energy usage. Additionally, data centres and other computer infrastructure are increasingly being powered by renewable energy sources.

In general, the history of green computing has witnessed a rise in attempts to lessen the environmental effect of technology through energy efficiency and the adoption of environmentally friendly products and practices, as well as a rising awareness of that impact [8][9][10].

19.4 Benefits of Recycling in Green Computing

Recycling in green computing can have several benefits, including:

- Recycling electronic trash can help conserve natural resources by lowering the need for fresh materials to make new electronic products.
- Recycling electronic trash can lower energy use by lowering the need to mine, refine, and create new materials.

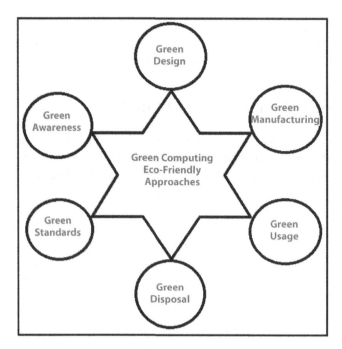

Figure 19.4 Green computing approaches.

- By not having to mine, prepare, and ship new materials, recycling electronic waste can help cut down on greenhouse gas emissions.
- Recycling electronic garbage can cut down on the quantity of e-waste that ends up in landfills, decreasing the impact of electronic waste on the atmosphere.
- The collection, transportation, and processing of electronic trash all result from recycling, meaning there is opportunity to create jobs.
- Recycling e-waste can save money because it can cut down on the need for new items and save money by reusing parts from old gadgets. Green computing approaches is shown in Figure 19.4.

19.5 E-Waste Management Steps

To guarantee that electronic trash is handled and disposed of properly and safely, e-waste management entails a number of stages. The six phases listed below are what e-waste management normally entails:

Collection: Electronic garbage has to be gathered and sent to a processing facility. Programs for collection and drop-off, mail-back, or expert e-waste management firms can all do this.

Sorting: After being brought into the plant, the electronic trash is divided into several groups, such as metals, plastics, and glass. This makes it possible to treat and recycle the materials independently.

Disassembly: To distinguish between valuable materials, such as metals and plastics, and non-value materials, such as glass and ceramics, electronic gadgets are deconstructed.

Recycling: The priceless components are afterwards recycled and repurposed to make new goods. Glass is crushed and used as aggregate, while metals are melted down and shaped into new shapes.

Disposal: Non-valuable goods are disposed of in an environmentally responsible way, like through a landfill or incinerated, like glass and ceramics.

Compliance: The e-waste management facility must adhere to all applicable environmental laws and standards, including those governing the handling of sensitive data and the disposal of hazardous materials. Steps of e-waste management is shown in Figure 19.5.

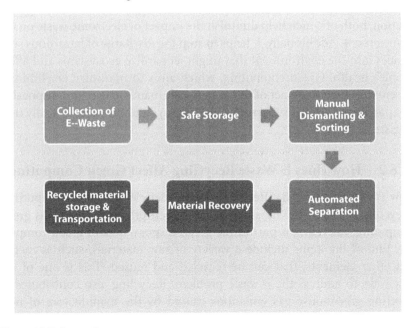

Figure 19.5 Steps of e-waste management.

19.6 E-Waste Recycling: An Approach Towards Green Computing

19.6.1 How is Green Computing Achieved?

From manufacturing to using the devices, everyone engaged can accomplish the fundamental goals of green computing.

Minimizing Energy Consumption: This goal strives to use IT equipment in an ecologically responsible manner while lowering its energy consumption (including that of its peripherals). For instance, while not in use, a computer should be put in hibernation or sleep mode.

Utilizing Green Energy: IT manufacturing businesses should concentrate on creating PCs, printers, servers, and other products that are energy-efficient. Furthermore, they should use green energy to produce gadgets and other subsystems with the least amount of waste possible.

Considering that it controls what happens to a discarded digital item, minimizing the requirements for equipment disposal is the least crucial of all. Outdated equipment that contains undesirable components should be used for recycling. Additionally, old computers that have been fixed or restored but are still functional ought to be put to other uses.

E-waste recycling involves taking salvageable parts out of old devices and using them to make new products. This approach reduces the need for extra raw materials and the amount of energy consumed during production, both of which help diminish the impact of electronic waste on the environment. Additionally, it helps to stop the discharge of hazardous substances into the environment that might jeopardize ecosystems and affect people's health. Green computing, which aims to minimize or eliminate the environmental impact of the design, use, manufacture, and disposal of computers and related subsystems like displays and printers, typically uses e-waste recycling.

19.6.2 How Does E-Waste Recycling Affect Green Computing?

How Do Companies That Recycle E-Waste Help With Green Computing? Recycling is one of the most significant green practices that helps green computing. Recycling is part of the green disposal area of green computing. End-of-life items include a variety of raw materials, such as metals and other elements, that can be recycled and reused. This is one of the best ways to address the e-waste problem. Recycling also contributes to lowering greenhouse gas emissions caused by the manufacture of new items.

Furthering the possibility for device reuse can also help with resource conservation, which is the best approach to lowering pollution levels. In the same way that functioning parts from old computers may be recycled throughout the refurbishment or repair process, green computing is a principle that encourages us to reuse our used computers, laptops, and other technology and it is the foundation of the complete recycling, refurbishing, and repair concept.

19.6.3 How Does Veracity World Differ?

Veracity World, one of the most well-known green companies in the UAE, focuses on treating e-waste in an eco-friendly manner. The pillars on which the entire company is founded are reduce, reuse, and recycle. Green computing is the company's primary business.

Veracity World provides very efficient recycling choices for old gadgets. For any equipment that is still functional, the company also employs a trained repair and maintenance team. The team recycles and fixes broken computers, laptops, and other technology in an effort to reduce carbon emissions. They also provide refurbishing alternatives for equipment that may be fixed and put back into use. Each of the company's activities is carried out with consideration for the environment. Energy efficiency, reliability, and sustainability provide wholehearted support for the production of the Veracity World IT operations division. You may thus experience the greatest green computing with the best solutions provided by the dedicated

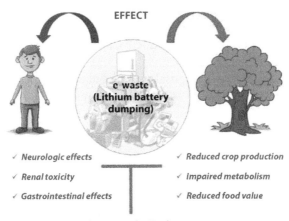

Figure 19.6 E-waste impact on the environment and human beings.

team at Veracity World. E-waste impact on the environment and human beings is shown in Figure 19.6.

19.7 Harmful Effects of E-Waste

When not in use or needed, electronics must be handled cautiously since they contain hazardous elements.

19.7.1 Effects on Air Quality

Disassembling, shredding, or melting electronic waste incorrectly results in the release of dust elements or chemicals, which damages the environment's air quality and the health of those who breathe it in.

Burning e-waste emits microscopic particles that might travel hundreds of kilometres and offer a multitude of risks to both human and animal health, increasing the likelihood of acquiring chronic diseases and cancer.

Although contamination from informal e-waste recycling can travel thousands of kilometres from recycling centers, the people who work with this material are the most vulnerable.

19.7.2 Effects on Humans

Lithium, barium, polybrominated flame retardants, lead, mercury, and cadmium are some of the toxic substances that can be found in electronic waste and are bad for people's health.

Humans exposed to these toxins run the risk of experiencing damage to their skeletal, liver, kidney, heart, and brain systems. It also significantly affects the human brain and reproductive systems, which can lead to disease and birth defects.

Since incorrect e-waste disposal puts global ecology in grave danger, it is critical to spread awareness of the threat it presents.

It's crucial to recycle properly so that electronics may be recycled, mended, resold, or utilised for anything else in order to prevent the negative consequences of e-waste.

19.7.3 Concerning Global Data on E-Waste

Every year, between 20 and 50 million metric tonnes of e-waste are disposed of globally.

Many precious metals, including gold and silver, are found in mobile phones and other electrical equipment.

Numerous "e-waste" items are really completely working pieces of useable electronic gear or parts that may be easily sold for reuse or recycled for material recovery.

Currently, just 12.5% of e-waste gets recycled. E-waste causes data theft, which makes the security issue worse.

19.8 E-Waste and the Sustainable Development Goals of the 2030 Agenda

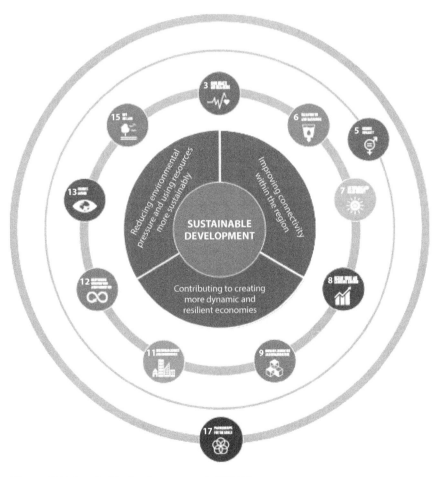

Figure 19.7 Sustainable development goals (SDGs).

The Sustainable Development Goals (SDGs) were adopted by the UN General Assembly in 2015, setting the foundation for the 2030 Agenda for Sustainable Development. 17 goals and 169 targets were established to be accomplished in the following 13 years in order to eradicate poverty, safeguard the environment, and ensure the prosperity of all people. Sustainable development goals (SDGs) is shown in Figure 19.7.

Every goal has an environmental component and some of them have e-waste as a direct connection. The global increase in e-waste makes it impossible to meet the 2030 Agenda for Sustainable Development. The UN system has to design an effective plan and coordinate its efforts in order to support the efforts of the states to manage their e-waste sustainably and to stop its growth and reduce the prevalence of illnesses and fatalities brought on by dangerous chemicals, as well as pollution and soil, water, and air contamination, considerably by 2030.

The objective of SDG Target 8.3 is to encourage policies that are focused on development and that support work-related activities, the development of decent jobs, entrepreneurship, innovation, and creativity, as well as the formalisation and growth of micro, small, and medium-sized businesses, including through access to financial services. It also ensures that all workers, including migrant workers, especially women migrant workers, and those with unstable jobs, have access to safe and secure working conditions.

SDG Target 12.5 says that by 2030, waste must be greatly reduced through repair, recycling, and reusing in order to reduce the bad things cities do to the environment per person by 2030, with a focus on air quality, managing municipal and other waste, and climate change.

19.9 Significant International E-Waste Agreements

19.9.1 The International Convention to Prevent Pollution from Ships (MARPOL)

MARPOL sets rules for how ships can pollute the air and how they can transport hazardous liquids in bulk, dangerous products, sewage, and trash.

The Basel Convention on the Control of Transboundary Movements of Hazardous Waste and the Disposal of Hazardous Waste (1989) was established with the goal of protecting the environment and general public from the harmful effects of the production, handling, transboundary transit, and disposal of hazardous and other wastes. Some of the most important parts of the Basel Convention are its rules on environmental management, mobility across borders, reducing waste, and getting rid of waste in ways that don't hurt people or the atmosphere.

19.9.2 The Ozone Depletion Protocol of the Montreal Protocol (1989)

An international agreement known as the Montreal Protocol restricts the production and use of substances that weaken the ozone layer (ODS). Some refrigerators and air conditioners still use ODS, CFCs, and HCFCs as refrigerants.

19.9.3 The Durban Declaration of 2008

In addition to international institutions, the statement advocated for an African regional e-waste platform or conference. According to the declaration, countries shall evaluate their present e-waste management regulations, improve their adherence to existing regulations, and revise existing regulations as appropriate.

19.9.4 India's E-Waste Production

However, the ability to deconstruct e-waste from 7.82 lakh metric tonnes did not improve in 2017-18. About 95% of India's e-waste is recycled in secret by scrap dealers who also throw it away in the wrong way by burning it or dissolving it in acids, according to information given to the court by the Ministry of Environment in 2018.

19.9.5 2016 E-Waste Management Regulations

The Ministry of Environment, Forestry, and Climate Change (MoEFCC) issued the E-Waste Management Rules, 2016, which replace the E-Waste (Management and Handling) Rules, 2011. The regulation included more than 21 items (Schedule I). There were also mercury-containing lamps, compact fluorescent lights (CFLs), and other similar devices. For the first time under the rules, producers and targets were both subject to Extended Producer Responsibility (EPR). E-waste exchange and collection are now the producer's duty. To oversee the collection of e-waste and its environmentally appropriate disposal, several producers can each establish their own Producer Responsibility Organization (PRO). There is also a punishment provision in place for violating the regulations. Urban local bodies (municipal committees, councils, and corporations) are in charge of gathering data and giving sufficient space to existing and prospective industrial units for the recycling and destruction of e-waste.

In the future: Managing e-waste is a major issue for the governments of many developing countries, including India. This is rapidly expanding and evolving into a significant public health issue. Combining the informal and formal sectors is essential to appropriately collect, process, and dispose of electronic waste in a way that avoids burning it in public and placing it in conventional landfills. The right people in developing countries need to figure out how to treat and get rid of e-waste in a safe and ethical way. The promotion of environmentally friendly e-waste management programs has to include more awareness-raising, capacity-building, and educational marketing. More effort has to be put into enhancing current practises, lessen the illicit e-waste trade, and consider collecting tactics and management solutions. It will be easier to deal with the various e-waste streams and will aid in prevention if there are less hazardous materials in e-products.

19.10 Conclusion

To sum up, e-waste management is an essential component of green computing. By properly disposing of and recycling electronic equipment, we can reduce technology's negative environmental consequences and conserve resources. Governments and businesses must work together to establish effective e-waste management rules and initiatives if the growing problem of e-waste is to be resolved sustainably. For a sustainable future, it is our duty to think about ecologically responsible practices. Innovation, inventiveness, and adaptability in the field of green technology are urgently required. The risk involved with handling this e-waste grows when green technology is used for final disposal. E-waste management practices encompass a variety of techniques for disposing of obsolete technology that have varying effects on the environment and, thus, public health. By using "green" technology, we help the Earth as a whole as well as ourselves by saving money, reducing waste, and fostering a more eco-friendly and clean environment. Although environmental preservation has come a long way since the advent of green computing, it is fair to say that this is a good thing. Reduced environmental damage caused by electronic waste is the aim of green computing. The basic goal of e-waste management is to keep society and the environment decent places to live. When attempting to mix green computing with e-waste, we run into challenges and problems. However, technological advancements have made it easier to collaborate with organizations that support e-waste management and green computing.

References

1. Kumar, A., Dubey, A.K., Ramírez, I.S., Muñoz del Río, A., Márquez, F.P.G. (2022). A Review and Analysis of Forecasting of Photovoltaic Power Generation Using Machine Learning. In: Xu, J., Altiparmak, F., Hassan, M.H.A., García Márquez, F.P., Hajiyev, A. (eds) Proceedings of the Sixteenth International Conference on Management Science and Engineering Management – Volume 1. ICMSEM 2022. Lecture Notes on Data Engineering and Communications Technologies, vol 144. Springer, Cham. https://doi.org/10.1007/978-3-031-10388-9_36
2. Dubey, A.K., Kumar, A., Ramirez, I.S., Marquez, F.P.G. (2022). A Review of Intelligent Systems for the Prediction of Wind Energy Using Machine Learning. In: Xu, J., Altiparmak, F., Hassan, M.H.A., García Márquez, F.P., Hajiyev, A. (eds) Proceedings of the Sixteenth International Conference on Management Science and Engineering Management – Volume 1. ICMSEM 2022. Lecture Notes on Data Engineering and Communications Technologies, vol 144. Springer, Cham. https://doi.org/10.1007/978-3-031-10388-9_35
3. Rathore, P.S., Chatterjee, J.M., Kumar, A. *et al.* Energy-efficient cluster head selection through relay approach for WSN. J Supercomput 77, 7649–7675 (2021). https://doi.org/10.1007/s11227-020-03593-4
4. A. Dubey, S. Narang, A. Srivastav, A. Kumar, V. Díaz, Woodhead Publishing, Science Direct, Artificial Intelligence for Renewable Energy Systems. Paperback ISBN: 9780323903967
5. A. Dubey, S. Narang, A. Srivastav, A. Kumar, V. Díaz, Woodhead Publishing, Science Direct, a Visualization Techniques for Climate Change with Machine Learning and Artificial Intelligence. ISBN: 9780323997140
6. A.-D. Pham, N.-T. Ngo, T.T.H. Truong, N.-T. Huynh, N.-S. Truong, Predicting energy consumption in multiple buildings using machine learning for improving energy efficiency and sustainability, J. Clean. Prod. 260 (2020) 121082.
7. C.-F. Chen, G.Z.D. Rubens, X. Xu, J. Li, Coronavirus comes home? Energy use, home energy management, and the social-psychological factors of covid-19, Energy Res. Soc. Sci. 68 (2020), https://doi.org/10.1016/j.erss.2020.101688 101688.
8. M. Berry, M. Gibson, A. Nelson, I. Richardson, How smart is smart? Smart homes and sustainability, Steering Sustain. Urbanizing World (2016) 239–251.

9. Y. Himeur, A. Alsalemi, A. Al-Kababji, F. Bensaali, A. Amira, Data fusion strategies for energy efficiency in buildings: Overview, challenges and novel orientations, Inf. Fusion 64 (2020) 99–120.
10. A. Alsalemi, C. Sardianos, F. Bensaali, I. Varlamis, A. Amira, G. Dimitrakopoulos, The role of micro-moments: A survey of habitual behavior change and recommender systems for energy saving, IEEE Syst. J. 13 (3) (2019) 3376–3387, https://doi.org/10.1109/jsyst.2019.2899832.

Index

2030 agenda 15

Artificial Intelligence 5, 7-8, 19-20, 25-29, 63-65, 72, 76, 87-90, 92, 102, 104-106, 108, 120, 122-123, 127, 135, 137, 169, 171-172, 174-176, 179-183, 185-186, 189-190, 225, 246-247, 261-262, 279, 302, 314, 392
Autoencoder 33

Basel Action Network 273, 311, 373, 404-405
Bioelectrochemical systems 356-357, 366
Biohydrometallurgy 13, 22, 329-330, 334-335, 353, 359, 361-362, 364, 387, 391, 399
Bioleaching 13, 22, 210, 329, 335-341, 352-356, 360-367, 385-387, 393-399, 401

Cathode ray tubes 229, 251, 287-288, 305, 324
Cloud 52, 61, 139, 144, 147-149, 151, 153, 165, 167, 189-190, 200
Constrained Application Protocol 149
Convolutional Neural Networks 33
CRT 202, 251, 288, 326, 403-404
Cutting-edge battery 403

Decision Tree 29-30, 80
Deep Belief Networks 33

Deep Learning 5, 7, 18, 27, 29, 32, 54, 66-68, 123-125, 127, 129, 135-136
Digital goods 21, 248-249
Digital Rubbish 286
Digital twin 181
Dispose

Electrical and electronic equipment 194-195, 201, 207, 210, 214, 218, 231, 237, 244, 249, 252, 255, 263, 268-269, 273, 283-284, 290, 292, 294, 300-301, 310, 315-317, 319, 321, 324, 364-365, 367, 370, 372, 375-376, 390, 394
Electronic garbage 21, 194-196, 198, 218, 222, 232, 248-249, 259, 264-267, 272-274, 277, 286, 322, 324, 376, 403-409, 412
Electronic trash 14, 220, 241, 272, 405
Electronic waste 3, 9-11, 14-15, 18, 21, 92, 198, 202, 207, 209-210, 213, 217-226, 234, 236-237, 245-246, 248-249, 252, 256-259, 261, 263-267, 271-278, 282-291, 293-295, 300-302, 304, 306-307, 314-317, 322-323, 327, 329, 363-366, 370, 374, 392-396, 398-401, 404, 406-407, 409, 411
Elicited expert 171

413

Energy 3, 5-8, 19-20, 25-33, 41-48, 50-53, 56-70, 72-77, 86-90, 92-94, 100, 102-109, 112-113, 115-120, 137, 139-156, 162-169, 171-181, 183-191, 198, 207, 210-212, 224-225, 242, 246, 249, 254-257, 261-262, 266, 278-280, 296, 298, 301-302, 314-316, 318, 330, 334-335, 338-340, 354-355, 359-362, 376, 380, 382-384, 386, 390, 392, 394-395, 399
Energy consumers 139
Energy efficiency 31, 92, 108, 118, 139, 154, 173-174, 176, 242
Energy payback time 256
Environment 8-9, 11-12, 14, 20-21, 26, 28, 42, 52, 94, 100, 104, 106, 114, 123, 146, 173-174, 179, 189-190, 193, 195, 197-198, 202, 207-208, 210-211, 213-215, 217-223, 226, 228, 231, 234-236, 239-240, 246-250, 255, 257, 260-261, 263, 267-270, 272-273, 275, 277, 279, 282-287, 290-291, 293, 295, 301, 304-307, 311-314, 318-319, 322-323, 327, 335, 355, 372, 374-375, 377-380, 383, 390-391, 393-394, 397, 399-401, 403-404, 406
Environmental management 209, 211, 224, 236, 245, 247, 314, 316-317, 366, 369, 393, 395
Environmental Protection Agency 194, 197
EPA 21, 93, 194, 264, 398
E-production 11, 264-266
E-trash 229, 237, 239, 264, 412
E-waste 6, 8-18, 20-22, 92-93, 193-204, 206-210, 212-226, 228-253, 257-262, 264-279, 282-302, 304-319, 321-327, 329-330, 332, 336-338, 340-341, 352-357, 359-363, 366-387, 390-401, 403-412
E-waste 8-12, 14-15, 18, 22, 193, 200-201, 204, 207, 209, 213, 218, 221, 224, 231-235, 238-240, 242, 244, 250, 257-258, 260-262, 269, 273, 276, 278, 282-283, 285, 291, 293, 298-300, 302, 305-312, 314-318, 324, 327, 332, 341, 354-355, 360, 368-370, 372, 374-375, 377, 379-382, 395, 397, 403, 405, 407-408, 410
E-waste classification 14
E-waste disposal 22, 197, 225, 235, 267, 274, 291, 295, 304, 311, 327, 369
E-waste management 8-9, 11, 207, 231, 234-235, 243, 296, 314, 400
E-waste recycling 10, 14, 210, 236, 242, 279, 288, 324, 397
Extended Producer Responsibility 201-202, 210, 219, 224, 237, 289-290, 293, 315-316

Fuzzy Logic 27, 29, 31, 41, 65, 94, 108, 175

Genetic Algorithm 31, 82, 89
Global Packet Radio Service 143
Green computing 15
Grid Management 45, 52, 63
GSM 142-143, 157, 159, 164, 232

Harmful effects of e-waste 15
Health hazards 9, 22, 310, 314-315, 369, 392, 407
Human health 12, 42, 92, 193, 195, 197, 208, 213, 217, 219, 222-223, 230, 235-236, 249, 263, 267, 282-283, 304-305, 311-312, 317, 322, 371-372, 374-375, 379, 390, 392-393, 399, 403-404, 406

Information and communications technology 228
Ingenious Forecasting of Wind Speeds 176
Internet of Energy 139, 162
Internet of Things 118-119, 136, 139-140, 155, 163, 165-167, 169, 181, 225
ISO 11, 264, 270

K-Nearest Neighbour 32, 88

Life-Cycle-Analysis 282, 284
Light Dependent Resistors 152
LSTM 33, 47, 50, 68, 128-130, 132, 134, 136

Machine Learning 5, 7, 18-20, 27, 29-30, 46, 54, 64-69, 87-90, 94, 102-109, 112, 118-120, 122-125, 127, 130, 134-137, 152, 168-169, 176, 180, 190, 225, 246, 262, 279, 301-302, 314-315, 317, 392, 394, 396
Massachusetts Institute of Technology 46, 232
Maximum Power Point 143, 148
Message Queue Telemetry Transfer 149
Metal extraction 13, 315-316, 329, 334, 337, 363, 384, 386-387, 395-396, 400, 412
Monitoring systems 7, 20, 123, 139-140, 142, 145, 151, 156, 162

Naïve Bayes 30-31

Owl Search Algorithm 154

Platinum group metals 359, 366, 383, 399
Principal Component Analysis 32, 87

Printed circuit boards 10, 202, 229-230, 236, 254, 262, 288, 305-306, 330, 364-366, 374, 393, 395-397, 400-401
Producer Responsibility Organization 201, 293

Random Forest 30, 51, 69, 103
Rare earth elements 13, 21, 282, 337, 359, 367, 379
RDED 403
Recurrent Neural Networks 7, 33, 122, 126-127, 181
Recycle 11, 21, 195, 222-223, 233, 248-249, 252, 256, 266-267, 289, 309, 316, 324, 374, 403, 405, 407
Renewable Energy 3, 5-8, 25-26, 28, 41-43, 45-47, 52-53, 60-66, 69, 72-73, 75-77, 90, 120, 137, 139-141, 163-164, 166, 169, 171-177, 179-180, 183-184, 187-190, 212, 225, 246, 255, 262, 279, 298, 302, 314, 392
Renewable energy sources 26, 28, 45-47, 61-63, 75, 163-164, 166, 171-172, 176-177
RoHS 203, 223, 247, 264, 282, 298, 374

Seagull Optimization Algorithm 154
Sensors 20, 57, 59, 67, 122-123, 126-129, 131-132, 135-136, 141, 143, 147-151, 153-154, 167
Sustainable development 10, 15, 20, 171, 193, 264, 316, 326
Sustainable development goals 15, 171, 193

Technology readiness level 22, 329-330

Technology Take-Back
   Coalition universal mobile
   telecommunications system
   232

Unmanned aerial vehicles 150, 167
Urban mining 193, 209, 221, 278,
   317-318, 363, 393, 400

Waste, Waste Electrical and Electronic
   Equipment 201
Wavelet Neural Network 30, 49

Weather forecasting 7, 43, 45, 122-123,
   126, 147, 176-177
WEEE 11, 21, 201, 203, 209-210,
   214-215, 218, 229, 231-233, 237,
   239, 246, 265, 269, 282-285, 288,
   300, 302, 309, 316, 319, 362,
   364-366, 370, 375-376, 382-383,
   387-388, 390-391, 398-399, 405-406
Wind Energy Conversion System 153,
   168

X-Ray Diffraction 336

Also of Interest

Check out these other related titles from Scrivener Publishing

QUANTUM COMPUTING IN CYBERSECURITY, Edited by Romil Rawat, Rajesh Kumar Chakrawarti, Sanjaya Kumar Sarangi, Jaideep Patel, and Vivek Bhardwaj, ISBN: 9781394166336. This cutting-edge new volume provides a comprehensive exploration of emerging technologies and trends in quantum computing and how it is used in cybersecurity, covering everything from artificial intelligence to how quantum computing can be used to secure networks and prevent cyber crime.

ROBOTIC PROCESS AUTOMATION, Edited by Romil Rawat, Rajesh Kumar Chakrawarti, Sanjaya Kumar Sarangi, Rahul Choudhary, Anand Singh Gadwal, and Vivek Bhardwaj. ISBN: 9781394166183. Presenting the latest technologies and practices in this ever-changing field, this groundbreaking new volume covers the theoretical challenges and practical solutions for using robotics across a variety of industries, encompassing many disciplines, including mathematics, computer science, electrical engineering, information technology, mechatronics, electronics, bioengineering, and command and software engineering.

AUTONOMOUS VEHICLES VOLUME 1: Using Machine Intelligence, Edited by Romil Rawat, A. Mary Sowjanya, Syed Imran Patel, Varshali Jaiswal, Imran Khan, and Allam Balaram. ISBN: 9781119871958. Addressing the current challenges, approaches and applications relating to autonomous vehicles, this groundbreaking new volume presents the research and techniques in this growing area, using Internet of Things, Machine Learning, Deep Learning, and Artificial Intelligence.

AUTONOMOUS VEHICLES VOLUME 2: Smart Vehicles for Communication, Edited by Romil Rawat, Purvee Bhardwaj, Upinder Kaur, Shrikant Telang, Mukesh Chouhan, and K. Sakthidasan Sankaran, ISBN: 9781394152254. The companion to *Autonomous Vehicles Volume 1: Using Machine Intelligence,* this second volume in the two-volume set covers intelligent techniques utilized for designing, controlling and managing vehicular systems based on advanced algorithms of computing like machine learning, artificial Intelligence, data analytics, and Internet of Things with prediction approaches to avoid accidental damages, security threats, and theft.

CONVERGENCE OF CLOUD WITH AI FOR BIG DATA ANALYTICS: Foundations and Innovation, Edited by Danda B. Rawat, Lalit K Awasthi, Valentina Emilia Balas, Mohit Kumar and Jitendra Kumar Samriya, ISBN: 9781119904885. This book covers the foundations and applications of cloud computing, AI, and Big Data and analyses their convergence for improved development and services.

SWARM INTELLIGENCE: An Approach from Natural to Artificial, By Kuldeep Singh Kaswan, Jagjit Singh Dhatterwal and Avadhesh Kumar, ISBN: 9781119865063. This important authored book presents valuable new insights by exploring the boundaries shared by cognitive science, social psychology, artificial life, artificial intelligence, and evolutionary computation by applying these insights to solving complex engineering problems.

FACTORIES OF THE FUTURE: Technological Advances in the Manufacturing Industry, Edited by Chandan Deep Singh and Harleen Kaur, ISBN: 9781119864943. The book provides insight into various technologies adopted and to be adopted in the future by industries and measures the impact of these technologies on manufacturing performance and their sustainability.

AI AND IOT-BASED INTELLIGENT AUTOMATION IN ROBOTICS, Edited by Ashutosh Kumar Dubey, Abhishek Kumar, S. Rakesh Kumar, N. Gayathri, Prasenjit Das, ISBN: 9781119711209. The 24 chapters in this book provide a deep overview of robotics and the application of AI and IoT in robotics across several industries such as healthcare, defense. education, etc.

SMART GRIDS AND INTERNET OF THINGS, Edited by Sanjeevikumar Padmanaban, Jens Bo Holm-Nielsen, Rajesh Kumar Dhanaraj, Malathy Sathyamoorthy, and Balamurugan Balusamy, ISBN: 9781119812449. Written and edited by a team of international professionals, this groundbreaking new volume covers the latest technologies in automation, tracking, energy distribution and consumption of Internet of Things (IoT) devices with smart grids.

Printed and bound by CPI Group (UK) Ltd, Croydon, CR0 4YY
26/02/2024